VOLUME FIVE HUNDRED AND TWENTY EIGHT

METHODS IN ENZYMOLOGY

Hydrogen Peroxide and Cell Signaling, Part C

METHODS IN ENZYMOLOGY

VOLUME FIVE HUNDRED AND TWENTY EIGHT

METHODS IN ENZYMOLOGY

Hydrogen Peroxide and Cell Signaling, Part C

Edited by

ENRIQUE CADENAS and LESTER PACKER
Pharmacology & Pharmaceutical Sciences
School of Pharmacy
University of Southern California
Los Angeles, CA, USA

AMSTERDAM • BOSTON • HEIDELBERG • LONDON
NEW YORK • OXFORD • PARIS • SAN DIEGO
SAN FRANCISCO • SINGAPORE • SYDNEY • TOKYO
Academic Press is an imprint of Elsevier

Academic Press is an imprint of Elsevier
225 Wyman Street, Waltham, MA 02451, USA
525 B Street, Suite 1800, San Diego, CA 92101-4495, USA
Radarweg 29, PO Box 211, 1000 AE Amsterdam, The Netherlands
The Boulevard, Langford Lane, Kidlington, Oxford, OX5 1GB, UK
32 Jamestown Road, London NW1 7BY, UK

First edition 2013

ISBN: 978-0-12-405881-1
ISSN: 0076-6879

Printed and bound in United States of America
13 14 15 16 11 10 9 8 7 6 5 4 3 2 1

CONTENTS

Contributors *xi*
Preface *xv*
Volumes in Series *xvii*

Section I
H_2O_2 Regulation of Cell Signaling

1. The Biological Chemistry of Hydrogen Peroxide **3**
Christine C. Winterbourn

1. Introduction 4
2. Chemical Properties 4
3. Antioxidant Defenses Against H_2O_2 11
4. Kinetics and Identification of Biological Targets for H_2O_2 13
5. Transmission of Redox Signals Initiated by H_2O_2 15
6. Diffusion Distances and Compartmentalization 17
7. Biological Detection of H_2O_2 19
8. Conclusion 20

References 21

2. Reactive Oxygen Species in the Activation of MAP Kinases **27**
Yong Son, Sangduck Kim, Hun-Taeg Chung, and Hyun-Ock Pae

1. Introduction 28
2. Reactive Oxygen Species 29
3. Mitogen-Activated Protein Kinases 33
4. Roles of ROS in MAPK Activation 39
5. Summary 44

Acknowledgment 45
References 45

3. Hydrogen Peroxide Signaling Mediator in the Activation of p38 MAPK in Vascular Endothelial Cells **49**
Rosa Bretón-Romero and Santiago Lamas

1. Introduction 50
2. Materials and Methods 51

References 58

4. *In Vivo* Imaging of Nitric Oxide and Hydrogen Peroxide in Cardiac Myocytes **61**

Juliano L. Sartoretto, Hermann Kalwa, Natalia Romero, and Thomas Michel

1. Introduction 62
2. Isolation and Culture of Adult Mouse Ventricular Cardiac Myocytes 63
3. Live Cell Imaging of Cardiac Myocytes 65
4. Imaging Intracellular NO with Cu_2(FL2E) Dye 66
5. Production and *In Vivo* Expression of Lentivirus Expressing the HyPer2 H_2O_2 Biosensor 68
6. Imaging Intracellular H_2O_2 in Cardiac Myocytes and Endothelial Cells Expressing HyPer2 76

Acknowledgments 77
References 77

5. Methods for Studying Oxidative Regulation of Protein Kinase C **79**

Rayudu Gopalakrishna, Thomas H. McNeill, Albert A. Elhiani, and Usha Gundimeda

1. Introduction 80
2. Materials 83
3. Direct Oxidative Modification of PKC Isoenzymes by H_2O_2 85
4. Indirect Cellular Regulation of PKC Isoenzymes by Sublethal Levels of H_2O_2 88
5. H_2O_2-Induced Signaling in GTPP-Induced Preconditioning for Cerebral Ischemia 90
6. Summary 95

Acknowledgments 95
References 95

6. p66Shc, Mitochondria, and the Generation of Reactive Oxygen Species **99**

Mirella Trinei, Enrica Migliaccio, Paolo Bernardi, Francesco Paolucci, Piergiuseppe Pelicci, and Marco Giorgio

1. Introduction 100
2. The *P66* Gene and Protein 100
3. The Mitochondrial Function of p66Shc 101
4. Preparation of Recombinant p66Shc Protein 102
5. Mitochondrial Swelling Assay 104
6. Mitochondrial ROS Formation by p66Shc 105
7. Conclusions: Role of p66Shc ROS 107

Acknowledgments 108
References 108

7. Detecting Disulfide-Bound Complexes and the Oxidative Regulation of Cyclic Nucleotide-Dependent Protein Kinases by H_2O_2 **111**

Joseph R. Burgoyne and Philip Eaton

1. Introduction 112
2. Experimental Considerations and Procedures 121
3. Summary 124

Acknowledgments 125
References 125

8. Redox Regulation of Protein Tyrosine Phosphatases: Methods for Kinetic Analysis of Covalent Enzyme Inactivation **129**

Zachary D. Parsons and Kent S. Gates

1. Introduction 130
2. Rate Expressions Describing Covalent Enzyme Inactivation 133
3. Ensuring That the Enzyme Activity Assay Accurately Reflects the Amount of Active Enzyme 134
4. Assays for Time-Dependent Inactivation of PTPs 139
5. Analysis of the Kinetic Data 142
6. Obtaining an Inactivation Rate Constant from the Data 150
7. Summary 152

References 152

Section II
H_2O_2 in the Redox Regulation of Transcription and Cell-Surface Receptors

9. Activation of Nrf2 by H_2O_2: *De Novo* Synthesis Versus Nuclear Translocation **157**

Gonçalo Covas, H. Susana Marinho, Luísa Cyrne, and Fernando Antunes

1. Introduction 158
2. Experimental Conditions and Considerations 159
3. Pilot Experiments 163
4. Experimental H_2O_2 Exposure 164
5. Data Handling and Analysis 166
6. Summary 170

Acknowledgments 170
References 170

10. H_2O_2 in the Induction of NF-κB-Dependent Selective Gene Expression **173**

Luísa Cyrne, Virgínia Oliveira-Marques, H. Susana Marinho, and Fernando Antunes

1. Introduction 174
2. Experimental Components and Considerations 175
3. Pilot Experiments 179
4. Steady-State Titration Experiments 181
5. NF-κB Family Protein Levels 182
6. NF-κB-Dependent Gene Expression 184
7. Summary 187

Acknowledgments 187
References 187

11. Detection of H_2O_2-Mediated Phosphorylation of Kinase-Inactive PDGFRα **189**

Hetian Lei and Andrius Kazlauskas

1. Construction of Kinase-Dead PDGFRα 190
2. Characterization of the Kinase-Inactive Receptor 191
3. Detection of H_2O_2-Mediated Phosphorylation of Kinase-Inactive PDGFRα 192
4. Implication 194

Acknowledgment 194
References 194

Section III
H_2O_2 and Regulation of Cellular Processes

12. Genetic Modifier Screens to Identify Components of a Redox-Regulated Cell Adhesion and Migration Pathway **197**

Thomas Ryan Hurd, Michelle Gail Leblanc, Leonard Nathaniel Jones, Matthew DeGennaro, and Ruth Lehmann

1. Introduction 198
2. Mutations in a *D. melanogaster* Gene Encoding a Peroxiredoxin Cause Germ Cell Adhesion and Migration Defects 199
3. Dominant Modifier Screens 200
4. Conducting a Dominant Modifier Screen to Identify Missing Components of a Redox-Regulated Germ Cell Migration Pathway 201
5. Limitations to Dominant Modifiers Screens 211

6. Concluding Remarks 212
Acknowledgments 213
References 213

13. Investigating the Role of Reactive Oxygen Species in Regulating Autophagy **217**

Spencer B. Gibson

1. Introduction 218
2. Regulation of Autophagy 218
3. ROS and Autophagy 220
4. Mechanisms for ROS Regulation of Autophagy 221
5. Methods for the Detection of Autophagy 225
6. Consideration When Using Oxidative Stress and Detecting ROS Under Autophagy Conditions 229
7. Conclusions 231
Acknowledgment 231
References 232

14. H_2O_2: A Chemoattractant? **237**

Balázs Enyedi and Philipp Niethammer

1. Introduction 238
2. The Zebrafish Tail Fin Wounding Assay 242
3. Measuring H_2O_2 Signals in Zebrafish 246
4. Imaging H_2O_2 Production by Wide-Field Microscopy 248
5. Imaging H_2O_2 Production by Confocal Microscopy 249
Acknowledgments 253
References 253

15. Measuring Mitochondrial Uncoupling Protein-2 Level and Activity in Insulinoma Cells **257**

Jonathan Barlow, Verena Hirschberg, Martin D. Brand, and Charles Affourtit

1. Introduction 258
2. Tissue Culture 258
3. UCP2 Protein Detection 259
4. UCP2 Protein Knockdown 260
5. UCP2 Activity 261
Acknowledgments 267
References 267

16. Effects of H_2O_2 on Insulin Signaling the Glucose Transport System in Mammalian Skeletal Muscle **269**

Erik J. Henriksen

1. Introduction 270
2. *In Vitro* Exposure to H_2O_2 271
3. Effects of H_2O_2 on the Glucose Transport System in Isolated Skeletal Muscle 274
4. Summary 275

References 276

17. Monitoring of Hydrogen Peroxide and Other Reactive Oxygen and Nitrogen Species Generated by Skeletal Muscle **279**

Malcolm J. Jackson

1. Introduction 280
2. Monitoring Extracellular ROS Using Microdialysis Techniques 284
3. Assessment of Intracellular ROS Activities 290
4. Concluding Remarks 296

Acknowledgments 297

References 297

Author Index *301*

Subject Index *323*

CONTRIBUTORS

Charles Affourtit
School of Biomedical and Biological Sciences, Plymouth University, Drake Circus, Plymouth, United Kingdom

Fernando Antunes
Departamento de Química e Bioquímica and Centro de Química e Bioquímica, Faculdade de Ciências, Universidade de Lisboa, Lisboa, Portugal

Jonathan Barlow
School of Biomedical and Biological Sciences, Plymouth University, Drake Circus, Plymouth, United Kingdom

Paolo Bernardi
Department of Biomedical Sciences, University of Padova, Padova, Italy

Martin D. Brand
Buck Institute for Research on Aging, Novato, California, USA

Rosa Bretón-Romero
Centro de Biología Molecular "Severo Ochoa" (CSIC-UAM), Madrid, Spain

Joseph R. Burgoyne
King's College London, Cardiovascular Division, The Rayne Institute, London, United Kingdom

Hun-Taeg Chung
Department of Biological Science, University of Ulsan, Ulsan, Republic of Korea

Gonçalo Covas*
Departamento de Química e Bioquímica and Centro de Química e Bioquímica, Faculdade de Ciências, Universidade de Lisboa, Lisboa, Portugal

Luísa Cyrne
Departamento de Química e Bioquímica and Centro de Química e Bioquímica, Faculdade de Ciências, Universidade de Lisboa, Lisboa, Portugal

Matthew DeGennaro
Laboratory of Neurogenetics and Behavior, The Rockefeller University, New York, USA

Philip Eaton
King's College London, Cardiovascular Division, The Rayne Institute, London, United Kingdom

Albert A. Elhiani
Department of Cell and Neurobiology, Keck school of Medicine, University of Southern California, Los Angeles, California, USA

*Present address: Bacterial Cell Surface and Pathogenesis, ITQB, Av. da República (EAN), Oeiras, Portugal.

Balázs Enyedi
Cell Biology Program, Memorial Sloan-Kettering Cancer Center, New York, USA

Kent S. Gates
Department of Chemistry, and Department of Biochemistry, University of Missouri, Columbia, Missouri, USA

Spencer B. Gibson
Manitoba Institute of Cell Biology, and Department of Biochemistry and Medical Genetics, Faculty of Medicine, University of Manitoba, Winnipeg, Manitoba, Canada

Marco Giorgio
Department of experimental Oncology, European Institute of Oncology, Milan, Italy

Rayudu Gopalakrishna
Department of Cell and Neurobiology, Keck school of Medicine, University of Southern California, Los Angeles, California, USA

Usha Gundimeda
Department of Cell and Neurobiology, Keck school of Medicine, University of Southern California, Los Angeles, California, USA

Erik J. Henriksen
Department of Physiology, Muscle Metabolism Laboratory, University of Arizona College of Medicine, Tucson, Arizona, USA

Verena Hirschberg
School of Biomedical and Biological Sciences, Plymouth University, Drake Circus, Plymouth, United Kingdom

Thomas Ryan Hurd
Department of Cell Biology, HHMI and Kimmel Center for Biology and Medicine of the Skirball Institute, New York University School of Medicine, New York, USA

Malcolm J. Jackson
MRC-Arthritis Research UK Centre for Integrated Research into Musculoskeletal Ageing, Institute of Ageing and Chronic Disease, University of Liverpool, Liverpool, United Kingdom

Leonard Nathaniel Jones
Department of Cell Biology, HHMI and Kimmel Center for Biology and Medicine of the Skirball Institute, New York University School of Medicine, New York, USA

Hermann Kalwa
Cardiovascular Division, Department of Medicine, Brigham and Women's Hospital, Harvard Medical School, Boston, Massachusetts, USA

Andrius Kazlauskas
The Schepens Eye Research Institute, Massachusetts Eye and Ear Infirmary, Department of Ophthalmology, Harvard Medical School, Boston, Massachusetts, USA

Sangduck Kim
Department of Ophthalmology, Wonkwang University School of Medicine, Iksan, Republic of Korea

Santiago Lamas
Centro de Biología Molecular “Severo Ochoa” (CSIC-UAM), Madrid, Spain

Michelle Gail Leblanc
Department of Cell Biology, HHMI and Kimmel Center for Biology and Medicine of the Skirball Institute, New York University School of Medicine, New York, USA

Ruth Lehmann
Department of Cell Biology, HHMI and Kimmel Center for Biology and Medicine of the Skirball Institute, New York University School of Medicine, New York, USA

Hetian Lei
The Schepens Eye Research Institute, Massachusetts Eye and Ear Infirmary, Department of Ophthalmology, Harvard Medical School, Boston, Massachusetts, USA

H. Susana Marinho
Departamento de Química e Bioquímica and Centro de Química e Bioquímica, Faculdade de Ciências, Universidade de Lisboa, Lisboa, Portugal

Thomas H. McNeill
Department of Cell and Neurobiology, Keck school of Medicine, University of Southern California, Los Angeles, California, USA

Thomas Michel
Cardiovascular Division, Department of Medicine, Brigham and Women’s Hospital, Harvard Medical School, Boston, Massachusetts, USA

Enrica Migliaccio
Department of experimental Oncology, European Institute of Oncology, Milan, Italy

Philipp Niethammer
Cell Biology Program, Memorial Sloan-Kettering Cancer Center, New York, USA

Virgínia Oliveira-Marques
Thelial Technologies S.A., Cantanhede, Portugal

Hyun-Ock Pae
Department of Microbiology and Immunology, Wonkwang University School of Medicine, Iksan, Republic of Korea

Francesco Paolucci
Department of Chemistry “G. Ciamician”, University of Bologna, Bologna, Italy

Zachary D. Parsons
Department of Chemistry, University of Missouri, Columbia, Missouri, USA

Piergiuseppe Pelicci
Department of experimental Oncology, European Institute of Oncology, Milan, Italy

Natalia Romero
Cardiovascular Division, Department of Medicine, Brigham and Women’s Hospital, Harvard Medical School, Boston, Massachusetts, USA

Juliano L. Sartoretto
Cardiovascular Division, Department of Medicine, Brigham and Women's Hospital, Harvard Medical School, Boston, Massachusetts, USA

Yong Son
Department of Anesthesiology and Pain Medicine, Wonkwang University School of Medicine, Iksan, Republic of Korea

Mirella Trinei
Department of experimental Oncology, European Institute of Oncology, Milan, Italy

Christine C. Winterbourn
Department of Pathology, Centre for Free Radical Research, University of Otago Christchurch, Christchurch, New Zealand

PREFACE

The identification of hydrogen peroxide in regulation of cell signaling and gene expression was a significant breakthrough in oxygen biology. Hydrogen peroxide is probably the most important redox signaling molecule that, among others, can activate NFκB, Nrf2, and other universal transcription factors and is involved in the regulation of insulin- and MAPK signaling. These pleiotropic effects of hydrogen peroxide are largely accounted for by changes in the thiol/disulfide status of the cell, an important determinant of the cell's redox status. Moreover disruption of redox signaling and control recognizes the occurrence of compartmentalized cell redox circuits.

Hydrogen peroxide signaling has been of central importance in cell research for some time and some previous volumes of *Methods in Enzymology* have covered in part some aspects of the physiological roles of hydrogen peroxide. However, there have been new developments and techniques that warrant these three volumes of *Methods in Enzymology*, which were designed to be the premier place for a compendium of hydrogen peroxide detection and delivery methods, microdomain imaging, and determinants of hydrogen peroxide steady-state levels; in addition, the role of hydrogen peroxide in cellular processes entailing redox regulation of cell signaling and transcription was covered by experts in mammalian and plant biochemistry and physiology.

In bringing this volume to fruition credit must be given to the experts in various aspects of hydrogen peroxide signaling research, whose thorough and innovative work is the basis of these three *Methods in Enzymology* volumes. Special thanks to the Advisory Board Members—Christopher J. Chang, So Goo Rhee, and Balyanaraman Kalyanaram—who provided guidance in the selection of topics and contributors. We hope that these volumes would be of help to both new and established investigators in this field.

Enrique Cadenas
Lester Packer
May 2013

METHODS IN ENZYMOLOGY

VOLUME I. Preparation and Assay of Enzymes
Edited by SIDNEY P. COLOWICK AND NATHAN O. KAPLAN

VOLUME II. Preparation and Assay of Enzymes
Edited by SIDNEY P. COLOWICK AND NATHAN O. KAPLAN

VOLUME III. Preparation and Assay of Substrates
Edited by SIDNEY P. COLOWICK AND NATHAN O. KAPLAN

VOLUME IV. Special Techniques for the Enzymologist
Edited by SIDNEY P. COLOWICK AND NATHAN O. KAPLAN

VOLUME V. Preparation and Assay of Enzymes
Edited by SIDNEY P. COLOWICK AND NATHAN O. KAPLAN

VOLUME VI. Preparation and Assay of Enzymes (*Continued*)
Preparation and Assay of Substrates
Special Techniques
Edited by SIDNEY P. COLOWICK AND NATHAN O. KAPLAN

VOLUME VII. Cumulative Subject Index
Edited by SIDNEY P. COLOWICK AND NATHAN O. KAPLAN

VOLUME VIII. Complex Carbohydrates
Edited by ELIZABETH F. NEUFELD AND VICTOR GINSBURG

VOLUME IX. Carbohydrate Metabolism
Edited by WILLIS A. WOOD

VOLUME X. Oxidation and Phosphorylation
Edited by RONALD W. ESTABROOK AND MAYNARD E. PULLMAN

VOLUME XI. Enzyme Structure
Edited by C. H. W. HIRS

VOLUME XII. Nucleic Acids (Parts A and B)
Edited by LAWRENCE GROSSMAN AND KIVIE MOLDAVE

VOLUME XIII. Citric Acid Cycle
Edited by J. M. LOWENSTEIN

VOLUME XIV. Lipids
Edited by J. M. LOWENSTEIN

VOLUME XXXII. Biomembranes (Part B)
Edited by SIDNEY FLEISCHER AND LESTER PACKER

VOLUME XXXIII. Cumulative Subject Index Volumes I-XXX
Edited by MARTHA G. DENNIS AND EDWARD A. DENNIS

VOLUME XXXIV. Affinity Techniques (Enzyme Purification: Part B)
Edited by WILLIAM B. JAKOBY AND MEIR WILCHEK

VOLUME XXXV. Lipids (Part B)
Edited by JOHN M. LOWENSTEIN

VOLUME XXXVI. Hormone Action (Part A: Steroid Hormones)
Edited by BERT W. O'MALLEY AND JOEL G. HARDMAN

VOLUME XXXVII. Hormone Action (Part B: Peptide Hormones)
Edited by BERT W. O'MALLEY AND JOEL G. HARDMAN

VOLUME XXXVIII. Hormone Action (Part C: Cyclic Nucleotides)
Edited by JOEL G. HARDMAN AND BERT W. O'MALLEY

VOLUME XXXIX. Hormone Action (Part D: Isolated Cells, Tissues, and Organ Systems)
Edited by JOEL G. HARDMAN AND BERT W. O'MALLEY

VOLUME XL. Hormone Action (Part E: Nuclear Structure and Function)
Edited by BERT W. O'MALLEY AND JOEL G. HARDMAN

VOLUME XLI. Carbohydrate Metabolism (Part B)
Edited by W. A. WOOD

VOLUME XLII. Carbohydrate Metabolism (Part C)
Edited by W. A. WOOD

VOLUME XLIII. Antibiotics
Edited by JOHN H. HASH

VOLUME XLIV. Immobilized Enzymes
Edited by KLAUS MOSBACH

VOLUME XLV. Proteolytic Enzymes (Part B)
Edited by LASZLO LORAND

VOLUME XLVI. Affinity Labeling
Edited by WILLIAM B. JAKOBY AND MEIR WILCHEK

VOLUME XLVII. Enzyme Structure (Part E)
Edited by C. H. W. HIRS AND SERGE N. TIMASHEFF

Volume 79. Interferons (Part B)
Edited by Sidney Pestka

Volume 80. Proteolytic Enzymes (Part C)
Edited by Laszlo Lorand

Volume 81. Biomembranes (Part H: Visual Pigments and Purple Membranes, I)
Edited by Lester Packer

Volume 82. Structural and Contractile Proteins (Part A: Extracellular Matrix)
Edited by Leon W. Cunningham and Dixie W. Frederiksen

Volume 83. Complex Carbohydrates (Part D)
Edited by Victor Ginsburg

Volume 84. Immunochemical Techniques (Part D: Selected Immunoassays)
Edited by John J. Langone and Helen Van Vunakis

Volume 85. Structural and Contractile Proteins (Part B: The Contractile Apparatus and the Cytoskeleton)
Edited by Dixie W. Frederiksen and Leon W. Cunningham

Volume 86. Prostaglandins and Arachidonate Metabolites
Edited by William E. M. Lands and William L. Smith

Volume 87. Enzyme Kinetics and Mechanism (Part C: Intermediates, Stereo-chemistry, and Rate Studies)
Edited by Daniel L. Purich

Volume 88. Biomembranes (Part I: Visual Pigments and Purple Membranes, II)
Edited by Lester Packer

Volume 89. Carbohydrate Metabolism (Part D)
Edited by Willis A. Wood

Volume 90. Carbohydrate Metabolism (Part E)
Edited by Willis A. Wood

Volume 91. Enzyme Structure (Part I)
Edited by C. H. W. Hirs and Serge N. Timasheff

Volume 92. Immunochemical Techniques (Part E: Monoclonal Antibodies and General Immunoassay Methods)
Edited by John J. Langone and Helen Van Vunakis

VOLUME 93. Immunochemical Techniques (Part F: Conventional Antibodies, Fc Receptors, and Cytotoxicity)
Edited by JOHN J. LANGONE AND HELEN VAN VUNAKIS

VOLUME 94. Polyamines
Edited by HERBERT TABOR AND CELIA WHITE TABOR

VOLUME 95. Cumulative Subject Index Volumes 61–74, 76–80
Edited by EDWARD A. DENNIS AND MARTHA G. DENNIS

VOLUME 96. Biomembranes [Part J: Membrane Biogenesis: Assembly and Targeting (General Methods; Eukaryotes)]
Edited by SIDNEY FLEISCHER AND BECCA FLEISCHER

VOLUME 97. Biomembranes [Part K: Membrane Biogenesis: Assembly and Targeting (Prokaryotes, Mitochondria, and Chloroplasts)]
Edited by SIDNEY FLEISCHER AND BECCA FLEISCHER

VOLUME 98. Biomembranes (Part L: Membrane Biogenesis: Processing and Recycling)
Edited by SIDNEY FLEISCHER AND BECCA FLEISCHER

VOLUME 99. Hormone Action (Part F: Protein Kinases)
Edited by JACKIE D. CORBIN AND JOEL G. HARDMAN

VOLUME 100. Recombinant DNA (Part B)
Edited by RAY WU, LAWRENCE GROSSMAN, AND KIVIE MOLDAVE

VOLUME 101. Recombinant DNA (Part C)
Edited by RAY WU, LAWRENCE GROSSMAN, AND KIVIE MOLDAVE

VOLUME 102. Hormone Action (Part G: Calmodulin and Calcium-Binding Proteins)
Edited by ANTHONY R. MEANS AND BERT W. O'MALLEY

VOLUME 103. Hormone Action (Part H: Neuroendocrine Peptides)
Edited by P. MICHAEL CONN

VOLUME 104. Enzyme Purification and Related Techniques (Part C)
Edited by WILLIAM B. JAKOBY

VOLUME 105. Oxygen Radicals in Biological Systems
Edited by LESTER PACKER

VOLUME 106. Posttranslational Modifications (Part A)
Edited by FINN WOLD AND KIVIE MOLDAVE

VOLUME 107. Posttranslational Modifications (Part B)
Edited by FINN WOLD AND KIVIE MOLDAVE

VOLUME 108. Immunochemical Techniques (Part G: Separation and Characterization of Lymphoid Cells)
Edited by GIOVANNI DI SABATO, JOHN J. LANGONE, AND HELEN VAN VUNAKIS

VOLUME 109. Hormone Action (Part I: Peptide Hormones)
Edited by LUTZ BIRNBAUMER AND BERT W. O'MALLEY

VOLUME 110. Steroids and Isoprenoids (Part A)
Edited by JOHN H. LAW AND HANS C. RILLING

VOLUME 111. Steroids and Isoprenoids (Part B)
Edited by JOHN H. LAW AND HANS C. RILLING

VOLUME 112. Drug and Enzyme Targeting (Part A)
Edited by KENNETH J. WIDDER AND RALPH GREEN

VOLUME 113. Glutamate, Glutamine, Glutathione, and Related Compounds
Edited by ALTON MEISTER

VOLUME 114. Diffraction Methods for Biological Macromolecules (Part A)
Edited by HAROLD W. WYCKOFF, C. H. W. HIRS, AND SERGE N. TIMASHEFF

VOLUME 115. Diffraction Methods for Biological Macromolecules (Part B)
Edited by HAROLD W. WYCKOFF, C. H. W. HIRS, AND SERGE N. TIMASHEFF

VOLUME 116. Immunochemical Techniques
(Part H: Effectors and Mediators of Lymphoid Cell Functions)
Edited by GIOVANNI DI SABATO, JOHN J. LANGONE, AND HELEN VAN VUNAKIS

VOLUME 117. Enzyme Structure (Part J)
Edited by C. H. W. HIRS AND SERGE N. TIMASHEFF

VOLUME 118. Plant Molecular Biology
Edited by ARTHUR WEISSBACH AND HERBERT WEISSBACH

VOLUME 119. Interferons (Part C)
Edited by SIDNEY PESTKA

VOLUME 120. Cumulative Subject Index Volumes 81–94, 96–101

VOLUME 121. Immunochemical Techniques (Part I: Hybridoma Technology and Monoclonal Antibodies)
Edited by JOHN J. LANGONE AND HELEN VAN VUNAKIS

VOLUME 122. Vitamins and Coenzymes (Part G)
Edited by FRANK CHYTIL AND DONALD B. MCCORMICK

VOLUME 123. Vitamins and Coenzymes (Part H)
Edited by FRANK CHYTIL AND DONALD B. MCCORMICK

VOLUME 124. Hormone Action (Part J: Neuroendocrine Peptides)
Edited by P. MICHAEL CONN

VOLUME 125. Biomembranes (Part M: Transport in Bacteria, Mitochondria, and Chloroplasts: General Approaches and Transport Systems)
Edited by SIDNEY FLEISCHER AND BECCA FLEISCHER

VOLUME 126. Biomembranes (Part N: Transport in Bacteria, Mitochondria, and Chloroplasts: Protonmotive Force)
Edited by SIDNEY FLEISCHER AND BECCA FLEISCHER

VOLUME 127. Biomembranes (Part O: Protons and Water: Structure and Translocation)
Edited by LESTER PACKER

VOLUME 128. Plasma Lipoproteins (Part A: Preparation, Structure, and Molecular Biology)
Edited by JERE P. SEGREST AND JOHN J. ALBERS

VOLUME 129. Plasma Lipoproteins (Part B: Characterization, Cell Biology, and Metabolism)
Edited by JOHN J. ALBERS AND JERE P. SEGREST

VOLUME 130. Enzyme Structure (Part K)
Edited by C. H. W. HIRS AND SERGE N. TIMASHEFF

VOLUME 131. Enzyme Structure (Part L)
Edited by C. H. W. HIRS AND SERGE N. TIMASHEFF

VOLUME 132. Immunochemical Techniques (Part J: Phagocytosis and Cell-Mediated Cytotoxicity)
Edited by GIOVANNI DI SABATO AND JOHANNES EVERSE

VOLUME 133. Bioluminescence and Chemiluminescence (Part B)
Edited by MARLENE DELUCA AND WILLIAM D. MCELROY

VOLUME 134. Structural and Contractile Proteins (Part C: The Contractile Apparatus and the Cytoskeleton)
Edited by RICHARD B. VALLEE

VOLUME 135. Immobilized Enzymes and Cells (Part B)
Edited by KLAUS MOSBACH

VOLUME 136. Immobilized Enzymes and Cells (Part C)
Edited by KLAUS MOSBACH

VOLUME 137. Immobilized Enzymes and Cells (Part D)
Edited by KLAUS MOSBACH

VOLUME 138. Complex Carbohydrates (Part E)
Edited by VICTOR GINSBURG

VOLUME 139. Cellular Regulators (Part A: Calcium- and Calmodulin-Binding Proteins)
Edited by ANTHONY R. MEANS AND P. MICHAEL CONN

VOLUME 140. Cumulative Subject Index Volumes 102–119, 121–134

VOLUME 141. Cellular Regulators (Part B: Calcium and Lipids)
Edited by P. MICHAEL CONN AND ANTHONY R. MEANS

VOLUME 142. Metabolism of Aromatic Amino Acids and Amines
Edited by SEYMOUR KAUFMAN

VOLUME 143. Sulfur and Sulfur Amino Acids
Edited by WILLIAM B. JAKOBY AND OWEN GRIFFITH

VOLUME 144. Structural and Contractile Proteins (Part D: Extracellular Matrix)
Edited by LEON W. CUNNINGHAM

VOLUME 145. Structural and Contractile Proteins (Part E: Extracellular Matrix)
Edited by LEON W. CUNNINGHAM

VOLUME 146. Peptide Growth Factors (Part A)
Edited by DAVID BARNES AND DAVID A. SIRBASKU

VOLUME 147. Peptide Growth Factors (Part B)
Edited by DAVID BARNES AND DAVID A. SIRBASKU

VOLUME 148. Plant Cell Membranes
Edited by LESTER PACKER AND ROLAND DOUCE

VOLUME 149. Drug and Enzyme Targeting (Part B)
Edited by RALPH GREEN AND KENNETH J. WIDDER

VOLUME 150. Immunochemical Techniques (Part K: *In Vitro* Models of B and T Cell Functions and Lymphoid Cell Receptors)
Edited by GIOVANNI DI SABATO

VOLUME 151. Molecular Genetics of Mammalian Cells
Edited by MICHAEL M. GOTTESMAN

VOLUME 152. Guide to Molecular Cloning Techniques
Edited by SHELBY L. BERGER AND ALAN R. KIMMEL

VOLUME 153. Recombinant DNA (Part D)
Edited by RAY WU AND LAWRENCE GROSSMAN

VOLUME 154. Recombinant DNA (Part E)
Edited by RAY WU AND LAWRENCE GROSSMAN

VOLUME 155. Recombinant DNA (Part F)
Edited by RAY WU

VOLUME 156. Biomembranes (Part P: ATP-Driven Pumps and Related Transport: The Na, K-Pump)
Edited by SIDNEY FLEISCHER AND BECCA FLEISCHER

VOLUME 157. Biomembranes (Part Q: ATP-Driven Pumps and Related Transport: Calcium, Proton, and Potassium Pumps)
Edited by SIDNEY FLEISCHER AND BECCA FLEISCHER

VOLUME 158. Metalloproteins (Part A)
Edited by JAMES F. RIORDAN AND BERT L. VALLEE

VOLUME 159. Initiation and Termination of Cyclic Nucleotide Action
Edited by JACKIE D. CORBIN AND ROGER A. JOHNSON

VOLUME 160. Biomass (Part A: Cellulose and Hemicellulose)
Edited by WILLIS A. WOOD AND SCOTT T. KELLOGG

VOLUME 161. Biomass (Part B: Lignin, Pectin, and Chitin)
Edited by WILLIS A. WOOD AND SCOTT T. KELLOGG

VOLUME 162. Immunochemical Techniques (Part L: Chemotaxis and Inflammation)
Edited by GIOVANNI DI SABATO

VOLUME 163. Immunochemical Techniques (Part M: Chemotaxis and Inflammation)
Edited by GIOVANNI DI SABATO

VOLUME 164. Ribosomes
Edited by HARRY F. NOLLER, JR., AND KIVIE MOLDAVE

VOLUME 165. Microbial Toxins: Tools for Enzymology
Edited by SIDNEY HARSHMAN

Volume 166. Branched-Chain Amino Acids
Edited by Robert Harris and John R. Sokatch

Volume 167. Cyanobacteria
Edited by Lester Packer and Alexander N. Glazer

Volume 168. Hormone Action (Part K: Neuroendocrine Peptides)
Edited by P. Michael Conn

Volume 169. Platelets: Receptors, Adhesion, Secretion (Part A)
Edited by Jacek Hawiger

Volume 170. Nucleosomes
Edited by Paul M. Wassarman and Roger D. Kornberg

Volume 171. Biomembranes (Part R: Transport Theory: Cells and Model Membranes)
Edited by Sidney Fleischer and Becca Fleischer

Volume 172. Biomembranes (Part S: Transport: Membrane Isolation and Characterization)
Edited by Sidney Fleischer and Becca Fleischer

Volume 173. Biomembranes [Part T: Cellular and Subcellular Transport: Eukaryotic (Nonepithelial) Cells]
Edited by Sidney Fleischer and Becca Fleischer

Volume 174. Biomembranes [Part U: Cellular and Subcellular Transport: Eukaryotic (Nonepithelial) Cells]
Edited by Sidney Fleischer and Becca Fleischer

Volume 175. Cumulative Subject Index Volumes 135–139, 141–167

Volume 176. Nuclear Magnetic Resonance (Part A: Spectral Techniques and Dynamics)
Edited by Norman J. Oppenheimer and Thomas L. James

Volume 177. Nuclear Magnetic Resonance (Part B: Structure and Mechanism)
Edited by Norman J. Oppenheimer and Thomas L. James

Volume 178. Antibodies, Antigens, and Molecular Mimicry
Edited by John J. Langone

Volume 179. Complex Carbohydrates (Part F)
Edited by Victor Ginsburg

VOLUME 180. RNA Processing (Part A: General Methods)
Edited by JAMES E. DAHLBERG AND JOHN N. ABELSON

VOLUME 181. RNA Processing (Part B: Specific Methods)
Edited by JAMES E. DAHLBERG AND JOHN N. ABELSON

VOLUME 182. Guide to Protein Purification
Edited by MURRAY P. DEUTSCHER

VOLUME 183. Molecular Evolution: Computer Analysis of Protein and Nucleic Acid Sequences
Edited by RUSSELL F. DOOLITTLE

VOLUME 184. Avidin-Biotin Technology
Edited by MEIR WILCHEK AND EDWARD A. BAYER

VOLUME 185. Gene Expression Technology
Edited by DAVID V. GOEDDEL

VOLUME 186. Oxygen Radicals in Biological Systems (Part B: Oxygen Radicals and Antioxidants)
Edited by LESTER PACKER AND ALEXANDER N. GLAZER

VOLUME 187. Arachidonate Related Lipid Mediators
Edited by ROBERT C. MURPHY AND FRANK A. FITZPATRICK

VOLUME 188. Hydrocarbons and Methylotrophy
Edited by MARY E. LIDSTROM

VOLUME 189. Retinoids (Part A: Molecular and Metabolic Aspects)
Edited by LESTER PACKER

VOLUME 190. Retinoids (Part B: Cell Differentiation and Clinical Applications)
Edited by LESTER PACKER

VOLUME 191. Biomembranes (Part V: Cellular and Subcellular Transport: Epithelial Cells)
Edited by SIDNEY FLEISCHER AND BECCA FLEISCHER

VOLUME 192. Biomembranes (Part W: Cellular and Subcellular Transport: Epithelial Cells)
Edited by SIDNEY FLEISCHER AND BECCA FLEISCHER

VOLUME 193. Mass Spectrometry
Edited by JAMES A. MCCLOSKEY

Volume 194. Guide to Yeast Genetics and Molecular Biology
Edited by Christine Guthrie and Gerald R. Fink

Volume 195. Adenylyl Cyclase, G Proteins, and Guanylyl Cyclase
Edited by Roger A. Johnson and Jackie D. Corbin

Volume 196. Molecular Motors and the Cytoskeleton
Edited by Richard B. Vallee

Volume 197. Phospholipases
Edited by Edward A. Dennis

Volume 198. Peptide Growth Factors (Part C)
Edited by David Barnes, J. P. Mather, and Gordon H. Sato

Volume 199. Cumulative Subject Index Volumes 168–174, 176–194

Volume 200. Protein Phosphorylation (Part A: Protein Kinases: Assays, Purification, Antibodies, Functional Analysis, Cloning, and Expression)
Edited by Tony Hunter and Bartholomew M. Sefton

Volume 201. Protein Phosphorylation (Part B: Analysis of Protein Phosphorylation, Protein Kinase Inhibitors, and Protein Phosphatases)
Edited by Tony Hunter and Bartholomew M. Sefton

Volume 202. Molecular Design and Modeling: Concepts and Applications (Part A: Proteins, Peptides, and Enzymes)
Edited by John J. Langone

Volume 203. Molecular Design and Modeling: Concepts and Applications (Part B: Antibodies and Antigens, Nucleic Acids, Polysaccharides, and Drugs)
Edited by John J. Langone

Volume 204. Bacterial Genetic Systems
Edited by Jeffrey H. Miller

Volume 205. Metallobiochemistry (Part B: Metallothionein and Related Molecules)
Edited by James F. Riordan and Bert L. Vallee

Volume 206. Cytochrome P450
Edited by Michael R. Waterman and Eric F. Johnson

Volume 207. Ion Channels
Edited by Bernardo Rudy and Linda E. Iverson

VOLUME 208. Protein–DNA Interactions
Edited by ROBERT T. SAUER

VOLUME 209. Phospholipid Biosynthesis
Edited by EDWARD A. DENNIS AND DENNIS E. VANCE

VOLUME 210. Numerical Computer Methods
Edited by LUDWIG BRAND AND MICHAEL L. JOHNSON

VOLUME 211. DNA Structures (Part A: Synthesis and Physical Analysis of DNA)
Edited by DAVID M. J. LILLEY AND JAMES E. DAHLBERG

VOLUME 212. DNA Structures (Part B: Chemical and Electrophoretic Analysis of DNA)
Edited by DAVID M. J. LILLEY AND JAMES E. DAHLBERG

VOLUME 213. Carotenoids (Part A: Chemistry, Separation, Quantitation, and Antioxidation)
Edited by LESTER PACKER

VOLUME 214. Carotenoids (Part B: Metabolism, Genetics, and Biosynthesis)
Edited by LESTER PACKER

VOLUME 215. Platelets: Receptors, Adhesion, Secretion (Part B)
Edited by JACEK J. HAWIGER

VOLUME 216. Recombinant DNA (Part G)
Edited by RAY WU

VOLUME 217. Recombinant DNA (Part H)
Edited by RAY WU

VOLUME 218. Recombinant DNA (Part I)
Edited by RAY WU

VOLUME 219. Reconstitution of Intracellular Transport
Edited by JAMES E. ROTHMAN

VOLUME 220. Membrane Fusion Techniques (Part A)
Edited by NEJAT DÜZGÜNEŞ

VOLUME 221. Membrane Fusion Techniques (Part B)
Edited by NEJAT DÜZGÜNEŞ

VOLUME 222. Proteolytic Enzymes in Coagulation, Fibrinolysis, and Complement Activation (Part A: Mammalian Blood Coagulation

Factors and Inhibitors)
Edited by Laszlo Lorand and Kenneth G. Mann

Volume 223. Proteolytic Enzymes in Coagulation, Fibrinolysis, and Complement Activation (Part B: Complement Activation, Fibrinolysis, and Nonmammalian Blood Coagulation Factors)
Edited by Laszlo Lorand and Kenneth G. Mann

Volume 224. Molecular Evolution: Producing the Biochemical Data
Edited by Elizabeth Anne Zimmer, Thomas J. White, Rebecca L. Cann, and Allan C. Wilson

Volume 225. Guide to Techniques in Mouse Development
Edited by Paul M. Wassarman and Melvin L. DePamphilis

Volume 226. Metallobiochemistry (Part C: Spectroscopic and Physical Methods for Probing Metal Ion Environments in Metalloenzymes and Metalloproteins)
Edited by James F. Riordan and Bert L. Vallee

Volume 227. Metallobiochemistry (Part D: Physical and Spectroscopic Methods for Probing Metal Ion Environments in Metalloproteins)
Edited by James F. Riordan and Bert L. Vallee

Volume 228. Aqueous Two-Phase Systems
Edited by Harry Walter and Göte Johansson

Volume 229. Cumulative Subject Index Volumes 195–198, 200–227

Volume 230. Guide to Techniques in Glycobiology
Edited by William J. Lennarz and Gerald W. Hart

Volume 231. Hemoglobins (Part B: Biochemical and Analytical Methods)
Edited by Johannes Everse, Kim D. Vandegriff, and Robert M. Winslow

Volume 232. Hemoglobins (Part C: Biophysical Methods)
Edited by Johannes Everse, Kim D. Vandegriff, and Robert M. Winslow

Volume 233. Oxygen Radicals in Biological Systems (Part C)
Edited by Lester Packer

Volume 234. Oxygen Radicals in Biological Systems (Part D)
Edited by Lester Packer

Volume 235. Bacterial Pathogenesis (Part A: Identification and Regulation of Virulence Factors)
Edited by Virginia L. Clark and Patrik M. Bavoil

Volume 236. Bacterial Pathogenesis (Part B: Integration of Pathogenic Bacteria with Host Cells)
Edited by Virginia L. Clark and Patrik M. Bavoil

Volume 237. Heterotrimeric G Proteins
Edited by Ravi Iyengar

Volume 238. Heterotrimeric G-Protein Effectors
Edited by Ravi Iyengar

Volume 239. Nuclear Magnetic Resonance (Part C)
Edited by Thomas L. James and Norman J. Oppenheimer

Volume 240. Numerical Computer Methods (Part B)
Edited by Michael L. Johnson and Ludwig Brand

Volume 241. Retroviral Proteases
Edited by Lawrence C. Kuo and Jules A. Shafer

Volume 242. Neoglycoconjugates (Part A)
Edited by Y. C. Lee and Reiko T. Lee

Volume 243. Inorganic Microbial Sulfur Metabolism
Edited by Harry D. Peck, Jr., and Jean LeGall

Volume 244. Proteolytic Enzymes: Serine and Cysteine Peptidases
Edited by Alan J. Barrett

Volume 245. Extracellular Matrix Components
Edited by E. Ruoslahti and E. Engvall

Volume 246. Biochemical Spectroscopy
Edited by Kenneth Sauer

Volume 247. Neoglycoconjugates (Part B: Biomedical Applications)
Edited by Y. C. Lee and Reiko T. Lee

Volume 248. Proteolytic Enzymes: Aspartic and Metallo Peptidases
Edited by Alan J. Barrett

Volume 249. Enzyme Kinetics and Mechanism (Part D: Developments in Enzyme Dynamics)
Edited by Daniel L. Purich

Volume 250. Lipid Modifications of Proteins
Edited by Patrick J. Casey and Janice E. Buss

VOLUME 251. Biothiols (Part A: Monothiols and Dithiols, Protein Thiols, and Thiyl Radicals)
Edited by LESTER PACKER

VOLUME 252. Biothiols (Part B: Glutathione and Thioredoxin; Thiols in Signal Transduction and Gene Regulation)
Edited by LESTER PACKER

VOLUME 253. Adhesion of Microbial Pathogens
Edited by RON J. DOYLE AND ITZHAK OFEK

VOLUME 254. Oncogene Techniques
Edited by PETER K. VOGT AND INDER M. VERMA

VOLUME 255. Small GTPases and Their Regulators (Part A: Ras Family)
Edited by W. E. BALCH, CHANNING J. DER, AND ALAN HALL

VOLUME 256. Small GTPases and Their Regulators (Part B: Rho Family)
Edited by W. E. BALCH, CHANNING J. DER, AND ALAN HALL

VOLUME 257. Small GTPases and Their Regulators (Part C: Proteins Involved in Transport)
Edited by W. E. BALCH, CHANNING J. DER, AND ALAN HALL

VOLUME 258. Redox-Active Amino Acids in Biology
Edited by JUDITH P. KLINMAN

VOLUME 259. Energetics of Biological Macromolecules
Edited by MICHAEL L. JOHNSON AND GARY K. ACKERS

VOLUME 260. Mitochondrial Biogenesis and Genetics (Part A)
Edited by GIUSEPPE M. ATTARDI AND ANNE CHOMYN

VOLUME 261. Nuclear Magnetic Resonance and Nucleic Acids
Edited by THOMAS L. JAMES

VOLUME 262. DNA Replication
Edited by JUDITH L. CAMPBELL

VOLUME 263. Plasma Lipoproteins (Part C: Quantitation)
Edited by WILLIAM A. BRADLEY, SANDRA H. GIANTURCO, AND JERE P. SEGREST

VOLUME 264. Mitochondrial Biogenesis and Genetics (Part B)
Edited by GIUSEPPE M. ATTARDI AND ANNE CHOMYN

VOLUME 265. Cumulative Subject Index Volumes 228, 230–262

Volume 266. Computer Methods for Macromolecular Sequence Analysis
Edited by Russell F. Doolittle

Volume 267. Combinatorial Chemistry
Edited by John N. Abelson

Volume 268. Nitric Oxide (Part A: Sources and Detection of NO; NO Synthase)
Edited by Lester Packer

Volume 269. Nitric Oxide (Part B: Physiological and Pathological Processes)
Edited by Lester Packer

Volume 270. High Resolution Separation and Analysis of Biological Macromolecules (Part A: Fundamentals)
Edited by Barry L. Karger and William S. Hancock

Volume 271. High Resolution Separation and Analysis of Biological Macromolecules (Part B: Applications)
Edited by Barry L. Karger and William S. Hancock

Volume 272. Cytochrome P450 (Part B)
Edited by Eric F. Johnson and Michael R. Waterman

Volume 273. RNA Polymerase and Associated Factors (Part A)
Edited by Sankar Adhya

Volume 274. RNA Polymerase and Associated Factors (Part B)
Edited by Sankar Adhya

Volume 275. Viral Polymerases and Related Proteins
Edited by Lawrence C. Kuo, David B. Olsen, and Steven S. Carroll

Volume 276. Macromolecular Crystallography (Part A)
Edited by Charles W. Carter, Jr., and Robert M. Sweet

Volume 277. Macromolecular Crystallography (Part B)
Edited by Charles W. Carter, Jr., and Robert M. Sweet

Volume 278. Fluorescence Spectroscopy
Edited by Ludwig Brand and Michael L. Johnson

Volume 279. Vitamins and Coenzymes (Part I)
Edited by Donald B. McCormick, John W. Suttie, and Conrad Wagner

Volume 280. Vitamins and Coenzymes (Part J)
Edited by Donald B. McCormick, John W. Suttie, and Conrad Wagner

VOLUME 281. Vitamins and Coenzymes (Part K)
Edited by DONALD B. MCCORMICK, JOHN W. SUTTIE, AND CONRAD WAGNER

VOLUME 282. Vitamins and Coenzymes (Part L)
Edited by DONALD B. MCCORMICK, JOHN W. SUTTIE, AND CONRAD WAGNER

VOLUME 283. Cell Cycle Control
Edited by WILLIAM G. DUNPHY

VOLUME 284. Lipases (Part A: Biotechnology)
Edited by BYRON RUBIN AND EDWARD A. DENNIS

VOLUME 285. Cumulative Subject Index Volumes 263, 264, 266–284, 286–289

VOLUME 286. Lipases (Part B: Enzyme Characterization and Utilization)
Edited by BYRON RUBIN AND EDWARD A. DENNIS

VOLUME 287. Chemokines
Edited by RICHARD HORUK

VOLUME 288. Chemokine Receptors
Edited by RICHARD HORUK

VOLUME 289. Solid Phase Peptide Synthesis
Edited by GREGG B. FIELDS

VOLUME 290. Molecular Chaperones
Edited by GEORGE H. LORIMER AND THOMAS BALDWIN

VOLUME 291. Caged Compounds
Edited by GERARD MARRIOTT

VOLUME 292. ABC Transporters: Biochemical, Cellular, and Molecular Aspects
Edited by SURESH V. AMBUDKAR AND MICHAEL M. GOTTESMAN

VOLUME 293. Ion Channels (Part B)
Edited by P. MICHAEL CONN

VOLUME 294. Ion Channels (Part C)
Edited by P. MICHAEL CONN

VOLUME 295. Energetics of Biological Macromolecules (Part B)
Edited by GARY K. ACKERS AND MICHAEL L. JOHNSON

VOLUME 296. Neurotransmitter Transporters
Edited by SUSAN G. AMARA

VOLUME 297. Photosynthesis: Molecular Biology of Energy Capture
Edited by LEE MCINTOSH

VOLUME 298. Molecular Motors and the Cytoskeleton (Part B)
Edited by RICHARD B. VALLEE

VOLUME 299. Oxidants and Antioxidants (Part A)
Edited by LESTER PACKER

VOLUME 300. Oxidants and Antioxidants (Part B)
Edited by LESTER PACKER

VOLUME 301. Nitric Oxide: Biological and Antioxidant Activities (Part C)
Edited by LESTER PACKER

VOLUME 302. Green Fluorescent Protein
Edited by P. MICHAEL CONN

VOLUME 303. cDNA Preparation and Display
Edited by SHERMAN M. WEISSMAN

VOLUME 304. Chromatin
Edited by PAUL M. WASSARMAN AND ALAN P. WOLFFE

VOLUME 305. Bioluminescence and Chemiluminescence (Part C)
Edited by THOMAS O. BALDWIN AND MIRIAM M. ZIEGLER

VOLUME 306. Expression of Recombinant Genes in Eukaryotic Systems
Edited by JOSEPH C. GLORIOSO AND MARTIN C. SCHMIDT

VOLUME 307. Confocal Microscopy
Edited by P. MICHAEL CONN

VOLUME 308. Enzyme Kinetics and Mechanism (Part E: Energetics of Enzyme Catalysis)
Edited by DANIEL L. PURICH AND VERN L. SCHRAMM

VOLUME 309. Amyloid, Prions, and Other Protein Aggregates
Edited by RONALD WETZEL

VOLUME 310. Biofilms
Edited by RON J. DOYLE

VOLUME 311. Sphingolipid Metabolism and Cell Signaling (Part A)
Edited by ALFRED H. MERRILL, JR., AND YUSUF A. HANNUN

VOLUME 312. Sphingolipid Metabolism and Cell Signaling (Part B)
Edited by ALFRED H. MERRILL, JR., AND YUSUF A. HANNUN

VOLUME 313. Antisense Technology
(Part A: General Methods, Methods of Delivery, and RNA Studies)
Edited by M. IAN PHILLIPS

VOLUME 314. Antisense Technology (Part B: Applications)
Edited by M. IAN PHILLIPS

VOLUME 315. Vertebrate Phototransduction and the Visual Cycle
(Part A)
Edited by KRZYSZTOF PALCZEWSKI

VOLUME 316. Vertebrate Phototransduction and the Visual Cycle (Part B)
Edited by KRZYSZTOF PALCZEWSKI

VOLUME 317. RNA–Ligand Interactions (Part A: Structural Biology
Methods)
Edited by DANIEL W. CELANDER AND JOHN N. ABELSON

VOLUME 318. RNA–Ligand Interactions (Part B: Molecular Biology
Methods)
Edited by DANIEL W. CELANDER AND JOHN N. ABELSON

VOLUME 319. Singlet Oxygen, UV-A, and Ozone
Edited by LESTER PACKER AND HELMUT SIES

VOLUME 320. Cumulative Subject Index Volumes 290–319

VOLUME 321. Numerical Computer Methods (Part C)
Edited by MICHAEL L. JOHNSON AND LUDWIG BRAND

VOLUME 322. Apoptosis
Edited by JOHN C. REED

VOLUME 323. Energetics of Biological Macromolecules (Part C)
Edited by MICHAEL L. JOHNSON AND GARY K. ACKERS

VOLUME 324. Branched-Chain Amino Acids (Part B)
Edited by ROBERT A. HARRIS AND JOHN R. SOKATCH

VOLUME 325. Regulators and Effectors of Small GTPases
(Part D: Rho Family)
Edited by W. E. BALCH, CHANNING J. DER, AND ALAN HALL

VOLUME 326. Applications of Chimeric Genes and Hybrid Proteins
(Part A: Gene Expression and Protein Purification)
Edited by JEREMY THORNER, SCOTT D. EMR, AND JOHN N. ABELSON

VOLUME 327. Applications of Chimeric Genes and Hybrid Proteins (Part B: Cell Biology and Physiology)
Edited by JEREMY THORNER, SCOTT D. EMR, AND JOHN N. ABELSON

VOLUME 328. Applications of Chimeric Genes and Hybrid Proteins (Part C: Protein–Protein Interactions and Genomics)
Edited by JEREMY THORNER, SCOTT D. EMR, AND JOHN N. ABELSON

VOLUME 329. Regulators and Effectors of Small GTPases (Part E: GTPases Involved in Vesicular Traffic)
Edited by W. E. BALCH, CHANNING J. DER, AND ALAN HALL

VOLUME 330. Hyperthermophilic Enzymes (Part A)
Edited by MICHAEL W. W. ADAMS AND ROBERT M. KELLY

VOLUME 331. Hyperthermophilic Enzymes (Part B)
Edited by MICHAEL W. W. ADAMS AND ROBERT M. KELLY

VOLUME 332. Regulators and Effectors of Small GTPases (Part F: Ras Family I)
Edited by W. E. BALCH, CHANNING J. DER, AND ALAN HALL

VOLUME 333. Regulators and Effectors of Small GTPases (Part G: Ras Family II)
Edited by W. E. BALCH, CHANNING J. DER, AND ALAN HALL

VOLUME 334. Hyperthermophilic Enzymes (Part C)
Edited by MICHAEL W. W. ADAMS AND ROBERT M. KELLY

VOLUME 335. Flavonoids and Other Polyphenols
Edited by LESTER PACKER

VOLUME 336. Microbial Growth in Biofilms (Part A: Developmental and Molecular Biological Aspects)
Edited by RON J. DOYLE

VOLUME 337. Microbial Growth in Biofilms (Part B: Special Environments and Physicochemical Aspects)
Edited by RON J. DOYLE

VOLUME 338. Nuclear Magnetic Resonance of Biological Macromolecules (Part A)
Edited by THOMAS L. JAMES, VOLKER DÖTSCH, AND ULI SCHMITZ

VOLUME 339. Nuclear Magnetic Resonance of Biological Macromolecules (Part B)
Edited by THOMAS L. JAMES, VOLKER DÖTSCH, AND ULI SCHMITZ

VOLUME 340. Drug–Nucleic Acid Interactions
Edited by JONATHAN B. CHAIRES AND MICHAEL J. WARING

VOLUME 341. Ribonucleases (Part A)
Edited by ALLEN W. NICHOLSON

VOLUME 342. Ribonucleases (Part B)
Edited by ALLEN W. NICHOLSON

VOLUME 343. G Protein Pathways (Part A: Receptors)
Edited by RAVI IYENGAR AND JOHN D. HILDEBRANDT

VOLUME 344. G Protein Pathways (Part B: G Proteins and Their Regulators)
Edited by RAVI IYENGAR AND JOHN D. HILDEBRANDT

VOLUME 345. G Protein Pathways (Part C: Effector Mechanisms)
Edited by RAVI IYENGAR AND JOHN D. HILDEBRANDT

VOLUME 346. Gene Therapy Methods
Edited by M. IAN PHILLIPS

VOLUME 347. Protein Sensors and Reactive Oxygen Species (Part A: Selenoproteins and Thioredoxin)
Edited by HELMUT SIES AND LESTER PACKER

VOLUME 348. Protein Sensors and Reactive Oxygen Species (Part B: Thiol Enzymes and Proteins)
Edited by HELMUT SIES AND LESTER PACKER

VOLUME 349. Superoxide Dismutase
Edited by LESTER PACKER

VOLUME 350. Guide to Yeast Genetics and Molecular and Cell Biology (Part B)
Edited by CHRISTINE GUTHRIE AND GERALD R. FINK

VOLUME 351. Guide to Yeast Genetics and Molecular and Cell Biology (Part C)
Edited by CHRISTINE GUTHRIE AND GERALD R. FINK

VOLUME 352. Redox Cell Biology and Genetics (Part A)
Edited by CHANDAN K. SEN AND LESTER PACKER

VOLUME 353. Redox Cell Biology and Genetics (Part B)
Edited by CHANDAN K. SEN AND LESTER PACKER

VOLUME 354. Enzyme Kinetics and Mechanisms (Part F: Detection and Characterization of Enzyme Reaction Intermediates)
Edited by DANIEL L. PURICH

VOLUME 355. Cumulative Subject Index Volumes 321–354

VOLUME 356. Laser Capture Microscopy and Microdissection
Edited by P. MICHAEL CONN

VOLUME 357. Cytochrome P450, Part C
Edited by ERIC F. JOHNSON AND MICHAEL R. WATERMAN

VOLUME 358. Bacterial Pathogenesis (Part C: Identification, Regulation, and Function of Virulence Factors)
Edited by VIRGINIA L. CLARK AND PATRIK M. BAVOIL

VOLUME 359. Nitric Oxide (Part D)
Edited by ENRIQUE CADENAS AND LESTER PACKER

VOLUME 360. Biophotonics (Part A)
Edited by GERARD MARRIOTT AND IAN PARKER

VOLUME 361. Biophotonics (Part B)
Edited by GERARD MARRIOTT AND IAN PARKER

VOLUME 362. Recognition of Carbohydrates in Biological Systems (Part A)
Edited by YUAN C. LEE AND REIKO T. LEE

VOLUME 363. Recognition of Carbohydrates in Biological Systems (Part B)
Edited by YUAN C. LEE AND REIKO T. LEE

VOLUME 364. Nuclear Receptors
Edited by DAVID W. RUSSELL AND DAVID J. MANGELSDORF

VOLUME 365. Differentiation of Embryonic Stem Cells
Edited by PAUL M. WASSAUMAN AND GORDON M. KELLER

VOLUME 366. Protein Phosphatases
Edited by SUSANNE KLUMPP AND JOSEF KRIEGLSTEIN

VOLUME 367. Liposomes (Part A)
Edited by NEJAT DÜZGÜNEŞ

VOLUME 368. Macromolecular Crystallography (Part C)
Edited by CHARLES W. CARTER, JR., AND ROBERT M. SWEET

VOLUME 369. Combinational Chemistry (Part B)
Edited by GUILLERMO A. MORALES AND BARRY A. BUNIN

VOLUME 370. RNA Polymerases and Associated Factors (Part C)
Edited by SANKAR L. ADHYA AND SUSAN GARGES

VOLUME 371. RNA Polymerases and Associated Factors (Part D)
Edited by SANKAR L. ADHYA AND SUSAN GARGES

VOLUME 372. Liposomes (Part B)
Edited by NEJAT DÜZGÜNEŞ

VOLUME 373. Liposomes (Part C)
Edited by NEJAT DÜZGÜNEŞ

VOLUME 374. Macromolecular Crystallography (Part D)
Edited by CHARLES W. CARTER, JR., AND ROBERT W. SWEET

VOLUME 375. Chromatin and Chromatin Remodeling Enzymes (Part A)
Edited by C. DAVID ALLIS AND CARL WU

VOLUME 376. Chromatin and Chromatin Remodeling Enzymes (Part B)
Edited by C. DAVID ALLIS AND CARL WU

VOLUME 377. Chromatin and Chromatin Remodeling Enzymes (Part C)
Edited by C. DAVID ALLIS AND CARL WU

VOLUME 378. Quinones and Quinone Enzymes (Part A)
Edited by HELMUT SIES AND LESTER PACKER

VOLUME 379. Energetics of Biological Macromolecules (Part D)
Edited by JO M. HOLT, MICHAEL L. JOHNSON, AND GARY K. ACKERS

VOLUME 380. Energetics of Biological Macromolecules (Part E)
Edited by JO M. HOLT, MICHAEL L. JOHNSON, AND GARY K. ACKERS

VOLUME 381. Oxygen Sensing
Edited by CHANDAN K. SEN AND GREGG L. SEMENZA

VOLUME 382. Quinones and Quinone Enzymes (Part B)
Edited by HELMUT SIES AND LESTER PACKER

VOLUME 383. Numerical Computer Methods (Part D)
Edited by LUDWIG BRAND AND MICHAEL L. JOHNSON

VOLUME 384. Numerical Computer Methods (Part E)
Edited by LUDWIG BRAND AND MICHAEL L. JOHNSON

VOLUME 385. Imaging in Biological Research (Part A)
Edited by P. MICHAEL CONN

VOLUME 386. Imaging in Biological Research (Part B)
Edited by P. MICHAEL CONN

VOLUME 387. Liposomes (Part D)
Edited by NEJAT DÜZGÜNEŞ

VOLUME 388. Protein Engineering
Edited by DAN E. ROBERTSON AND JOSEPH P. NOEL

VOLUME 389. Regulators of G-Protein Signaling (Part A)
Edited by DAVID P. SIDEROVSKI

VOLUME 390. Regulators of G-Protein Signaling (Part B)
Edited by DAVID P. SIDEROVSKI

VOLUME 391. Liposomes (Part E)
Edited by NEJAT DÜZGÜNEŞ

VOLUME 392. RNA Interference
Edited by ENGELKE ROSSI

VOLUME 393. Circadian Rhythms
Edited by MICHAEL W. YOUNG

VOLUME 394. Nuclear Magnetic Resonance of Biological Macromolecules (Part C)
Edited by THOMAS L. JAMES

VOLUME 395. Producing the Biochemical Data (Part B)
Edited by ELIZABETH A. ZIMMER AND ERIC H. ROALSON

VOLUME 396. Nitric Oxide (Part E)
Edited by LESTER PACKER AND ENRIQUE CADENAS

VOLUME 397. Environmental Microbiology
Edited by JARED R. LEADBETTER

VOLUME 398. Ubiquitin and Protein Degradation (Part A)
Edited by RAYMOND J. DESHAIES

VOLUME 399. Ubiquitin and Protein Degradation (Part B)
Edited by RAYMOND J. DESHAIES

VOLUME 400. Phase II Conjugation Enzymes and Transport Systems
Edited by HELMUT SIES AND LESTER PACKER

VOLUME 401. Glutathione Transferases and Gamma Glutamyl Transpeptidases
Edited by HELMUT SIES AND LESTER PACKER

VOLUME 402. Biological Mass Spectrometry
Edited by A. L. BURLINGAME

VOLUME 403. GTPases Regulating Membrane Targeting and Fusion
Edited by WILLIAM E. BALCH, CHANNING J. DER, AND ALAN HALL

VOLUME 404. GTPases Regulating Membrane Dynamics
Edited by WILLIAM E. BALCH, CHANNING J. DER, AND ALAN HALL

VOLUME 405. Mass Spectrometry: Modified Proteins and Glycoconjugates
Edited by A. L. BURLINGAME

VOLUME 406. Regulators and Effectors of Small GTPases: Rho Family
Edited by WILLIAM E. BALCH, CHANNING J. DER, AND ALAN HALL

VOLUME 407. Regulators and Effectors of Small GTPases: Ras Family
Edited by WILLIAM E. BALCH, CHANNING J. DER, AND ALAN HALL

VOLUME 408. DNA Repair (Part A)
Edited by JUDITH L. CAMPBELL AND PAUL MODRICH

VOLUME 409. DNA Repair (Part B)
Edited by JUDITH L. CAMPBELL AND PAUL MODRICH

VOLUME 410. DNA Microarrays (Part A: Array Platforms and Web-Bench Protocols)
Edited by ALAN KIMMEL AND BRIAN OLIVER

VOLUME 411. DNA Microarrays (Part B: Databases and Statistics)
Edited by ALAN KIMMEL AND BRIAN OLIVER

VOLUME 412. Amyloid, Prions, and Other Protein Aggregates (Part B)
Edited by INDU KHETERPAL AND RONALD WETZEL

VOLUME 413. Amyloid, Prions, and Other Protein Aggregates (Part C)
Edited by INDU KHETERPAL AND RONALD WETZEL

VOLUME 414. Measuring Biological Responses with Automated Microscopy
Edited by JAMES INGLESE

VOLUME 415. Glycobiology
Edited by MINORU FUKUDA

VOLUME 416. Glycomics
Edited by MINORU FUKUDA

VOLUME 417. Functional Glycomics
Edited by MINORU FUKUDA

VOLUME 418. Embryonic Stem Cells
Edited by IRINA KLIMANSKAYA AND ROBERT LANZA

VOLUME 419. Adult Stem Cells
Edited by IRINA KLIMANSKAYA AND ROBERT LANZA

VOLUME 420. Stem Cell Tools and Other Experimental Protocols
Edited by IRINA KLIMANSKAYA AND ROBERT LANZA

VOLUME 421. Advanced Bacterial Genetics: Use of Transposons and Phage for Genomic Engineering
Edited by KELLY T. HUGHES

VOLUME 422. Two-Component Signaling Systems, Part A
Edited by MELVIN I. SIMON, BRIAN R. CRANE, AND ALEXANDRINE CRANE

VOLUME 423. Two-Component Signaling Systems, Part B
Edited by MELVIN I. SIMON, BRIAN R. CRANE, AND ALEXANDRINE CRANE

VOLUME 424. RNA Editing
Edited by JONATHA M. GOTT

VOLUME 425. RNA Modification
Edited by JONATHA M. GOTT

VOLUME 426. Integrins
Edited by DAVID CHERESH

VOLUME 427. MicroRNA Methods
Edited by JOHN J. ROSSI

VOLUME 428. Osmosensing and Osmosignaling
Edited by HELMUT SIES AND DIETER HAUSSINGER

VOLUME 429. Translation Initiation: Extract Systems and Molecular Genetics
Edited by JON LORSCH

VOLUME 430. Translation Initiation: Reconstituted Systems and Biophysical Methods
Edited by JON LORSCH

VOLUME 431. Translation Initiation: Cell Biology, High-Throughput and Chemical-Based Approaches
Edited by JON LORSCH

VOLUME 432. Lipidomics and Bioactive Lipids: Mass-Spectrometry–Based Lipid Analysis
Edited by H. ALEX BROWN

Volume 433. Lipidomics and Bioactive Lipids: Specialized Analytical Methods and Lipids in Disease
Edited by H. Alex Brown

Volume 434. Lipidomics and Bioactive Lipids: Lipids and Cell Signaling
Edited by H. Alex Brown

Volume 435. Oxygen Biology and Hypoxia
Edited by Helmut Sies and Bernhard Brüne

Volume 436. Globins and Other Nitric Oxide-Reactive Protiens (Part A)
Edited by Robert K. Poole

Volume 437. Globins and Other Nitric Oxide-Reactive Protiens (Part B)
Edited by Robert K. Poole

Volume 438. Small GTPases in Disease (Part A)
Edited by William E. Balch, Channing J. Der, and Alan Hall

Volume 439. Small GTPases in Disease (Part B)
Edited by William E. Balch, Channing J. Der, and Alan Hall

Volume 440. Nitric Oxide, Part F Oxidative and Nitrosative Stress in Redox Regulation of Cell Signaling
Edited by Enrique Cadenas and Lester Packer

Volume 441. Nitric Oxide, Part G Oxidative and Nitrosative Stress in Redox Regulation of Cell Signaling
Edited by Enrique Cadenas and Lester Packer

Volume 442. Programmed Cell Death, General Principles for Studying Cell Death (Part A)
Edited by Roya Khosravi-Far, Zahra Zakeri, Richard A. Lockshin, and Mauro Piacentini

Volume 443. Angiogenesis: *In Vitro* Systems
Edited by David A. Cheresh

Volume 444. Angiogenesis: *In Vivo* Systems (Part A)
Edited by David A. Cheresh

Volume 445. Angiogenesis: *In Vivo* Systems (Part B)
Edited by David A. Cheresh

Volume 446. Programmed Cell Death, The Biology and Therapeutic Implications of Cell Death (Part B)
Edited by Roya Khosravi-Far, Zahra Zakeri, Richard A. Lockshin, and Mauro Piacentini

VOLUME 447. RNA Turnover in Bacteria, Archaea and Organelles
Edited by LYNNE E. MAQUAT AND CECILIA M. ARRAIANO

VOLUME 448. RNA Turnover in Eukaryotes: Nucleases, Pathways and Analysis of mRNA Decay
Edited by LYNNE E. MAQUAT AND MEGERDITCH KILEDJIAN

VOLUME 449. RNA Turnover in Eukaryotes: Analysis of Specialized and Quality Control RNA Decay Pathways
Edited by LYNNE E. MAQUAT AND MEGERDITCH KILEDJIAN

VOLUME 450. Fluorescence Spectroscopy
Edited by LUDWIG BRAND AND MICHAEL L. JOHNSON

VOLUME 451. Autophagy: Lower Eukaryotes and Non-Mammalian Systems (Part A)
Edited by DANIEL J. KLIONSKY

VOLUME 452. Autophagy in Mammalian Systems (Part B)
Edited by DANIEL J. KLIONSKY

VOLUME 453. Autophagy in Disease and Clinical Applications (Part C)
Edited by DANIEL J. KLIONSKY

VOLUME 454. Computer Methods (Part A)
Edited by MICHAEL L. JOHNSON AND LUDWIG BRAND

VOLUME 455. Biothermodynamics (Part A)
Edited by MICHAEL L. JOHNSON, JO M. HOLT, AND GARY K. ACKERS (RETIRED)

VOLUME 456. Mitochondrial Function, Part A: Mitochondrial Electron Transport Complexes and Reactive Oxygen Species
Edited by WILLIAM S. ALLISON AND IMMO E. SCHEFFLER

VOLUME 457. Mitochondrial Function, Part B: Mitochondrial Protein Kinases, Protein Phosphatases and Mitochondrial Diseases
Edited by WILLIAM S. ALLISON AND ANNE N. MURPHY

VOLUME 458. Complex Enzymes in Microbial Natural Product Biosynthesis, Part A: Overview Articles and Peptides
Edited by DAVID A. HOPWOOD

VOLUME 459. Complex Enzymes in Microbial Natural Product Biosynthesis, Part B: Polyketides, Aminocoumarins and Carbohydrates
Edited by DAVID A. HOPWOOD

VOLUME 474. Thiol Redox Transitions in Cell Signaling, Part B Cellular Localization and Signaling
Edited by ENRIQUE CADENAS AND LESTER PACKER

VOLUME 475. Single Molecule Tools, Part B: Super-Resolution, Particle Tracking, Multiparameter, and Force Based Methods
Edited by NILS G. WALTER

VOLUME 476. Guide to Techniques in Mouse Development, Part A Mice, Embryos, and Cells, 2nd Edition
Edited by PAUL M. WASSARMAN AND PHILIPPE M. SORIANO

VOLUME 477. Guide to Techniques in Mouse Development, Part B Mouse Molecular Genetics, 2nd Edition
Edited by PAUL M. WASSARMAN AND PHILIPPE M. SORIANO

VOLUME 478. Glycomics
Edited by MINORU FUKUDA

VOLUME 479. Functional Glycomics
Edited by MINORU FUKUDA

VOLUME 480. Glycobiology
Edited by MINORU FUKUDA

VOLUME 481. Cryo-EM, Part A: Sample Preparation and Data Collection
Edited by GRANT J. JENSEN

VOLUME 482. Cryo-EM, Part B: 3-D Reconstruction
Edited by GRANT J. JENSEN

VOLUME 483. Cryo-EM, Part C: Analyses, Interpretation, and Case Studies
Edited by GRANT J. JENSEN

VOLUME 484. Constitutive Activity in Receptors and Other Proteins, Part A
Edited by P. MICHAEL CONN

VOLUME 485. Constitutive Activity in Receptors and Other Proteins, Part B
Edited by P. MICHAEL CONN

VOLUME 486. Research on Nitrification and Related Processes, Part A
Edited by MARTIN G. KLOTZ

VOLUME 487. Computer Methods, Part C
Edited by MICHAEL L. JOHNSON AND LUDWIG BRAND

VOLUME 488. Biothermodynamics, Part C
Edited by MICHAEL L. JOHNSON, JO M. HOLT, AND GARY K. ACKERS

VOLUME 503. Protein Engineering for Therapeutics, Part B
Edited by K. DANE WITTRUP AND GREGORY L. VERDINE

VOLUME 504. Imaging and Spectroscopic Analysis of Living Cells
Optical and Spectroscopic Techniques
Edited by P. MICHAEL CONN

VOLUME 505. Imaging and Spectroscopic Analysis of Living Cells
Live Cell Imaging of Cellular Elements and Functions
Edited by P. MICHAEL CONN

VOLUME 506. Imaging and Spectroscopic Analysis of Living Cells
Imaging Live Cells in Health and Disease
Edited by P. MICHAEL CONN

VOLUME 507. Gene Transfer Vectors for Clinical Application
Edited by THEODORE FRIEDMANN

VOLUME 508. Nanomedicine
Cancer, Diabetes, and Cardiovascular, Central Nervous System, Pulmonary and Inflammatory Diseases
Edited by NEJAT DÜZGÜNEŞ

VOLUME 509. Nanomedicine
Infectious Diseases, Immunotherapy, Diagnostics, Antifibrotics, Toxicology and Gene Medicine
Edited by NEJAT DÜZGÜNEŞ

VOLUME 510. Cellulases
Edited by HARRY J. GILBERT

VOLUME 511. RNA Helicases
Edited by ECKHARD JANKOWSKY

VOLUME 512. Nucleosomes, Histones & Chromatin, Part A
Edited by CARL WU AND C. DAVID ALLIS

VOLUME 513. Nucleosomes, Histones & Chromatin, Part B
Edited by CARL WU AND C. DAVID ALLIS

VOLUME 514. Ghrelin
Edited by MASAYASU KOJIMA AND KENJI KANGAWA

VOLUME 515. Natural Product Biosynthesis by Microorganisms and Plants, Part A
Edited by DAVID A. HOPWOOD

Volume 516. Natural Product Biosynthesis by Microorganisms and Plants, Part B
Edited by David A. Hopwood

Volume 517. Natural Product Biosynthesis by Microorganisms and Plants, Part C
Edited by David A. Hopwood

Volume 518. Fluorescence Fluctuation Spectroscopy (FFS), Part A
Edited by Sergey Y. Tetin

Volume 519. Fluorescence Fluctuation Spectroscopy (FFS), Part B
Edited by Sergey Y. Tetin

Volume 520. G Protein Couple Receptors
Structure
Edited by P. Michael Conn

Volume 521. G Protein Couple Receptors
Trafficking and Oligomerization
Edited by P. Michael Conn

Volume 522. G Protein Couple Receptors
Modeling, Activation, Interactions and Virtual Screening
Edited by P. Michael Conn

Volume 523. Methods in Protein Design
Edited by Amy E. Keating

Volume 524. Cilia, Part A
Edited by Wallace F. Marshall

Volume 525. Cilia, Part B
Edited by Wallace F. Marshall

Volume 526. Hydrogen Peroxide and Cell Signaling, Part A
Edited by Enrique Cadenas and Lester Packer

Volume 527. Hydrogen Peroxide and Cell Signaling, Part B
Edited by Enrique Cadenas and Lester Packer

Volume 528. Hydrogen Peroxide and Cell Signaling, Part C
Edited by Enrique Cadenas and Lester Packer

H_2O_2 Regulation of Cell Signaling

CHAPTER ONE

The Biological Chemistry of Hydrogen Peroxide

Christine C. Winterbourn[1]

Department of Pathology, Centre for Free Radical Research, University of Otago Christchurch, Christchurch, New Zealand

[1]Corresponding author: e-mail address: christine.winterbourn@otago.ac.nz

Contents

1. Introduction 4
2. Chemical Properties 4
   2.1 Two-electron oxidations 5
   2.2 Reactions with transition metals and one-electron oxidations 8
3. Antioxidant Defenses Against H_2O_2 11
4. Kinetics and Identification of Biological Targets for H_2O_2 13
5. Transmission of Redox Signals Initiated by H_2O_2 15
6. Diffusion Distances and Compartmentalization 17
   6.1 Diffusion 17
   6.2 Compartmentalization and membrane permeability 17
7. Biological Detection of H_2O_2 19
8. Conclusion 20
References 21

Abstract

Hydrogen peroxide is generated in numerous biological processes and is implicated as the main transmitter of redox signals. Although a strong oxidant, high activation energy barriers make it unreactive with most biological molecules. It reacts directly with thiols, but for low-molecular-weight thiols and cysteine residues in most proteins, the reaction is slow. The most favored reactions of hydrogen peroxide are with transition metal centers, selenoproteins, and selected thiol proteins. These include proteins such as catalase, glutathione peroxidases, and peroxiredoxins, which, as well as providing antioxidant defense, are increasingly being considered as targets for signal transmission. This overview describes the main biological reactions of hydrogen peroxide and takes a kinetic approach to identifying likely targets in the cell. It also considers diffusion of hydrogen peroxide and constraints to its acting at localized sites.

Methods in Enzymology, Volume 528
ISSN 0076-6879
http://dx.doi.org/10.1016/B978-0-12-405881-1.00001-X

1. INTRODUCTION

Hydrogen peroxide is a major biological reactive oxygen species, excess of which can cause damage to cells and tissues. It is a by-product of respiration, an end product of a number of metabolic reactions, particularly peroxisomal oxidation pathways, and a likely transmitter of cellular signals. There are many examples of exogenous H_2O_2 initiating redox signals or stress responses, and receptor-mediated redox signaling is widely regarded as involving endogenously generated H_2O_2. The mechanism of signal transmission is widely considered to involve oxidation of thiol proteins. However, in most instances, initial targets for H_2O_2 and the specific reactions involved in transmitting the signal are poorly understood. There are also uncertainties about how redox signaling pathways operate in cells that are rich in antioxidant defenses and which may be generating a substantial amount of H_2O_2 as a metabolic end product. Therefore, to understand the role of H_2O_2 in cell signaling, it is important to know where and when it is produced and with what it can react. This chapter gives an overview of the biological chemistry of H_2O_2 and considers how it is likely to react under the conditions of the cell.

2. CHEMICAL PROPERTIES

H_2O_2 is a strong two-electron oxidant, with a standard reduction potential (E'_o) of 1.32 V at pH 7.0. It is therefore more oxidizing than hypochlorous acid (OCl^-/Cl^-) or peroxynitrite ($ONOO^-/NO_2^-$), for which the equivalent values are 1.28 and 1.20 V, respectively. However, in contrast to the two reactive species, H_2O_2 reacts poorly or not at all with most biological molecules, including low-molecular-weight antioxidants. This is because a high activation energy barrier must be overcome to release its oxidizing power, or in other words, the reactions of H_2O_2 are kinetically rather than thermodynamically driven (Fig. 1.1). H_2O_2 is a weak one-electron oxidant ($E'_o = 0.32$V). However, the one-electron reduction product, the hydroxyl radical, is one of the strongest oxidants known ($E'_o = 2.31$V). Its reactions have a very low activation energy barrier and occur at close to diffusion-controlled rates.

H_2O_2 has a pK_a of 11.6 so is mostly uncharged at physiological pH. The anion (HOO—) is a strong nucleophile, but its reactions are limited by the high pK_a. Due to the polarizability of the O—O bond, H_2O_2 can also act as

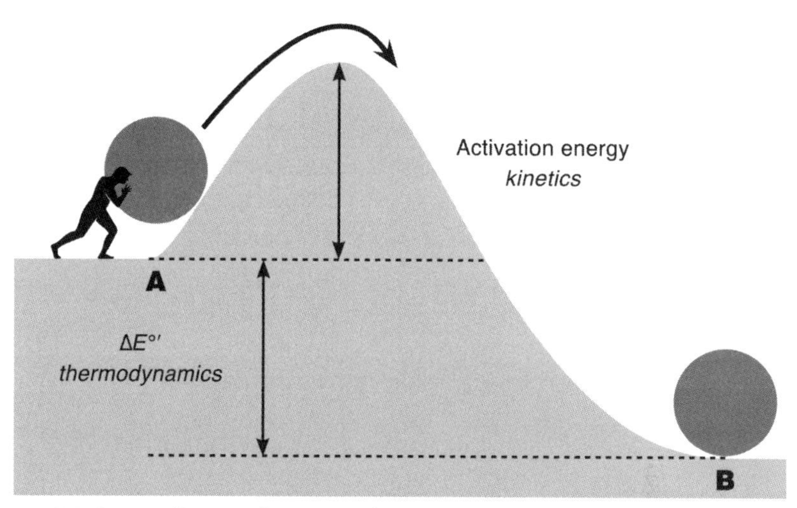

Figure 1.1 Energy diagram illustrating the kinetic constraint for a thermodynamically favorable reaction of H_2O_2. *Reproduced from Winterbourn (2012) with permission.* (For color version of this figure, the reader is referred to the online version of this chapter.)

an electrophile, and as discussed below, proteins that facilitate polarization enhance its reactivity by many orders of magnitude. The O—O bond is also relatively weak and is susceptible to homolysis when H_2O_2 is subjected to heating, radiolysis, photolysis, or redox metals. This gives rise to the hydroxyl radical, or in some cases, a higher oxidation state of the metal. These secondary products are responsible for many of the strong oxidizing (or disinfecting) actions of H_2O_2.

2.1. Two-electron oxidations

2.1.1 Thiols

Most biological molecules that lack a transition metal center do not react directly with H_2O_2. Cys residues in proteins and low-molecular-weight thiols are among its few direct targets, although with minor but important exceptions, they react relatively slowly (Winterbourn & Metodiewa, 1999). The reaction is exclusively with the thiolate anion, so at physiological pH, thiols with a low pK_a are more ionized and therefore more reactive. Reactivity is measured as the rate constant (k) for the reaction, and when rate constants are expressed in terms of thiolate rather than total thiol concentration, values for low-molecular-weight thiols are similar ($\sim 20\ M^{-1}\ s^{-1}$) (Winterbourn & Metodiewa, 1999). Thus, GSH which is less ionized than

Table 1.1 pK_a values and reaction rates with H_2O_2 for selected low-molecular-weight thiols and thiol proteins

Thiol	pK_a	Rate constant (M^{-1} s^{-1})
GSH[a]	8.8	0.89
Cysteine[a]	8.3	2.9
N-acetylcysteine[a]	9.5	0.16
Thioredoxin	6.5	1.05
GAPDH	8.2	~500
PTP1B	5.4	20
Cdc25B	6.1	160
Peroxiredoxins	5–6	$1–4 \times 10^7$

Rate constants are for pH 7.4–7.6, measured at 37 or 20–25 °C. Data are taken from Stone (2004) and Winterbourn and Hampton (2008), where references to the original papers reporting these values are given. Peroxiredoxin values are for the typical 2-Cys peroxiredoxins, human peroxiredoxins 2 and 3, yeast Tsa1 bacterial AhpC; human peroxiredoxin 5 (atypical 2-Cys form) has a value of 3×10^5.
[a]Microscopic pK_a values are given (see Nagy and Winterbourn, 2010 for explanation).

Cys reacts more slowly at pH 7.4 (Table 1.1). The same general relationship holds for many protein thiols, including low pK_a protein tyrosine phosphatases (PTPs), which react as predicted for a low pK_a thiol. A few thiol proteins are known to be more reactive, most notably the peroxiredoxins, which have rate constants up to five orders of magnitude higher. Clearly (as discussed in Section 2.1.4), low pK_a is not sufficient to confer this reactivity. Methionine residues also react with H_2O_2, but ~100 times more slowly than thiols.

The initial product from the thiolate is the sulfenic acid (Fig. 1.2). Sulfenic acids, which generally have a lower pK_a and are therefore more ionized than their corresponding thiol (Claiborne, Miller, Parsonage, & Ross, 1993), react rapidly with thiols to form the disulfide, and unless shielded from such reactions, as in serum albumin (Carballal et al., 2003), they are transient species. Sulfenic acids are also oxidized by H_2O_2 to give the sulfinic acid, but rates tend to be 1000-fold slower than for the equivalent thiol. In some proteins, including PTP1B (Salmeen et al., 2003), the sulfenic acid can react reversibly with a neighboring amide group to form a sulfenyl amide, which protects it from further oxidation. Sulfenic acids can be derivatized, for example, by dimedone, and can be detected using procedures based on these reagents (Nelson et al., 2010; Truong, Garcia, Seo, & Carroll, 2011).

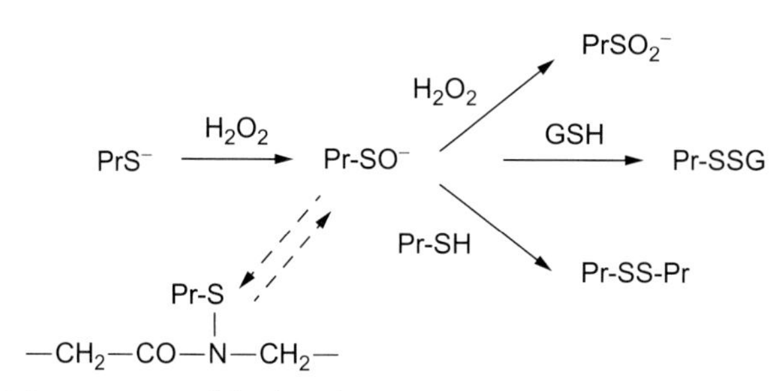

Figure 1.2 Major reactions of thiols with H_2O_2. The thiol (RS^-), sulfenic acid (RSO^-), and sulfinic acid (RSO_2^-) are represented as anions as it is these rather than the protonated forms that react with it. The dashed arrows show the possibility of the sulfenate reacting reversibly with an amide nitrogen to give the sulfenyl amide (as occurs in some proteins).

2.1.2 Keto acids

Another biologically relevant reaction of H_2O_2 is the oxidation of keto acids. Pyruvate is oxidized to acetate and carbon dioxide with a rate constant of 2.2 M^{-1} s^{-1} (Vasquez-Vivar, Denicola, Radi, & Augusto, 1997). Rapid removal of H_2O_2 from cell culture media containing added or secreted pyruvate has been observed (O'Donnell-Tormey, Nathan, Lanks, DeBoer, & de la Harpe, 1987), and at a typical intracellular concentration of 0.1–0.5 mM, pyruvate would be competitive with most thiols for H_2O_2.

2.1.3 Carbon dioxide

H_2O_2 reacts with carbon dioxide to form peroxymonocarbonate (reaction 1.1)

$$H_2O_2 + CO_2 \rightleftarrows HCO_4^- + H^+. \qquad [1.1]$$

As peroxymonocarbonate reacts approximately 200 times faster than H_2O_2 with thiols and methionine (Bakhmutova-Albert, Yao, Denevan, & Richardson, 2010; Trindade, Cerchiaro, & Augusto, 2006), this reaction is of interest as a potential mechanism for enhancing its reactivity. However, the reaction is an equilibrium ($K = 0.4$ M^{-1}) with only ~1% of the H_2O_2 present as peroxymonocarbonate in a physiological bicarbonate buffer, so enhancement is limited to a fewfold. Also, the forward step is slow ($k = 0.02$ M^{-1} s^{-1}) so once preexisting peroxymonocarbonate had reacted, it would take seconds to regenerate under these conditions. Reaction (1.1) can be accelerated by carbonic anhydrase (Bakhmutova-Albert et al., 2010), and this may enhance its physiological relevance.

2.1.4 *Explaining the high reactivity of thiol peroxidases*

A select group of thiol proteins, most notably the peroxiredoxins and glutathione peroxidases, and the bacterial H_2O_2-responsive transcriptional activator oxyR, react with H_2O_2 many orders of magnitude faster than can be explained by the presence of an ionized thiol (Ferrer-Sueta et al., 2011; Winterbourn & Hampton, 2008). It is becoming evident that this exceptional reactivity requires additional activation of the H_2O_2 in the transition state of the enzyme (Hall, Parsonage, Poole, & Karplus, 2010; Nagy et al., 2011). It is noteworthy that these proteins are highly reactive with other peroxides and peroxynitrite, all of which also contain an O—O bond, but not with other oxidants or reagents such as maleimides and iodo compounds (Peskin et al., 2007). Evidence from structural and kinetic studies of the peroxiredoxins supports the formation of a transition state with the H_2O_2 aligned so that hydrogen bonding polarizes the —O—O bond to facilitate electrophilic attack on the thiolate by one end of the molecule and release of the other as OH^- or H_2O. Some of the critical amino acid residues in the peroxiredoxins have been identified (Hall et al., 2010; Nagy et al., 2011) and it will be interesting to see if other thiol proteins contain similar motifs.

2.2. Reactions with transition metals and one-electron oxidations

Many of the biologically damaging effects of H_2O_2 are dependent on transition metals such as iron and copper, which cleave the O—O bond to generate hydroxyl radicals or activated metal complexes (Imlay, 2008). These species are more reactive and less discriminating than H_2O_2 itself and are prime initiators of free radical reactions. Physiologically, the transition metal centers may be low-molecular-weight chelates, heme peroxidases, or other redox-active metalloproteins such as iron/sulfur proteins. Metal-catalyzed or free radical reactions have received little attention as participants in cell signaling, as they are generally considered to lack the required specificity. However, these reactions are more facile than most two-electron oxidations and it is possible that reactions involving metal complexes or metalloproteins could contribute. Regardless of whether this is the case, it is likely that H_2O_2 can react with metals and form radicals under conditions where redox signaling is observed.

2.2.1 *Transition metals and Fenton chemistry*

The Fenton reaction has been much studied as a source of hydroxyl radicals and initiator of biological damage (Imlay, 2003; Winterbourn, 1995). In its simplest form, it can be written (where iron is complexed to ligand L) as

$$L - Fe^{2+} + H_2O_2 \rightarrow L - Fe^{3+} + OH^{\bullet} + OH^{-} \quad [1.2]$$

Fenton chemistry strictly involves iron and the name originates from observations by Fenton in the 1880s that a mixture of acidified iron and H_2O_2 generates a strong oxidant. The term Fenton-like is also used to describe comparable reactions of other transition metals such as copper. Rate constants for reaction (1.2) depend on the metal complex and are typically in the 5–20 × 10^3 M^{-1} s^{-1} range. Over the years, there has been considerable debate as to whether the product is the hydroxyl radical or a higher oxidation state of the metal (such as ferryl $[FeO]^{2+}$ or Cu(III) complexes), with product analyses and competition kinetic studies with a variety of iron chelates and detector systems leading to differing conclusions (Sutton & Winterbourn, 1989; Walling, 1975). The complex situation is best described by reactions (1.3–1.6) (Goldstein, Meyerstein, & Czapski, 1993; Winterbourn, 1995):

$$L - Fe^{2+} + H_2O_2 \rightarrow L - Fe^{2+}(H_2O_2) \quad [1.3]$$

$$L - Fe^{2+}(H_2O_2) \rightarrow (L - FeO)^{4+} + H_2O \quad [1.4]$$

$$L - Fe^{2+}(H_2O_2) \rightarrow L - Fe^{3+} + OH^{-} + OH^{\cdot} \quad [1.5]$$

$$L - Fe^{2+}(H_2O_2) \text{or } (L - FeO)^{4+} \text{ or } OH^{\bullet} + \text{substrate} \rightarrow \text{products.} \quad [1.6]$$

Rather than reaction (1.2), it is likely that the initial step is formation of a ferrous peroxide (Fe[IV]) complex (reaction 1.3). This can convert to a ferryl species (reaction 1.4), breakdown to give hydroxyl radicals (reaction 1.5) or directly oxidize a substrate. The rate of reaction (1.3) varies depending on the chelator which, in biological systems, could be a small molecule such as citrate or ATP or a macromolecule-binding site. Whether the initial complex undergoes reaction (1.4) or (1.5) or reacts directly with a substrate also depends on the chelator as well as the nature of the substrate. So, in different circumstances, one or all three of the reactive species may be the oxidant in reaction (1.6). These species are difficult, if not impossible, to distinguish kinetically. However, they are all strong oxidants, so regardless of which is involved, the products from a particular substrate should be similar and it is best to describe the reactive species as the Fenton oxidant. An important variation of the Fenton reaction is where the metal is bound to a biological molecule that can be oxidized, in which case site-localized reactions by the bound oxidant can occur. Examples of this are copper-catalyzed oxidation of DNA and protein oxidation by an iron/ascorbate system (Chevion, 1988; Stadtman, 1990).

Chelation or sequestration of transition metals to limit their redox activity (such as iron in transferrin or ferritin) is an important protective mechanism against damage by H_2O_2 (Imlay, 2003). Nevertheless, these metals are in a dynamic state in cells, and micromolar concentrations of low-molecular-weight iron have been measured. To maintain Fenton chemistry at these concentrations, the oxidized metal needs to be recycled. This was once thought to be an important function for superoxide, in what is commonly referred to as the superoxide-driven Fenton reaction or Haber–Weiss reaction. However, reductants such as ascorbate or GSH are much more prominent in this role. Higher iron concentrations are likely in lysosomes, where metalloproteins are degraded and the stability and reactivity of iron salts are enhanced by the low pH (Yu, Persson, Eaton, & Brunk, 2003). These organelles are favored sites for Fenton chemistry, and cytoprotective effects of chelators such as desferrioxamine may be largely attributed to complexing of lysosomal iron (Kurz, Gustafsson, & Brunk, 2006).

2.2.2 Interaction of H_2O_2 with heme peroxidases and other metalloproteins

H_2O_2 reacts rapidly with heme peroxidases, for example, myeloperoxidase, eosinophil peroxidase, and lactoperoxidase, with rate constants typically 10^7–10^8 M^{-1} s^{-1} (Davies, Hawkins, Pattison, & Rees, 2008). The "activated" heme–peroxide complexes are then able to oxidize a range of substrates, to generate free radicals or in some cases two-electron oxidants such as hypochlorous acid. These major peroxidases are largely restricted to specialist cells involved in host defense, so while important for dictating the reactivity of H_2O_2 in inflammatory conditions, they are unlikely to have a universal role as targets for H_2O_2 in cell signaling. However, other cellular heme proteins including cytochromes have surrogate peroxidase activity. Although they are less reactive than dedicated peroxidases (e.g., the rate constant for cyt *c* is 200 M^{-1} s^{-1}; Kagan et al., 2005), their prevalence makes them feasible intracellular targets and regulators of H_2O_2 action.

Other metalloproteins, for example, Cu/Zn superoxide dismutase, cytochrome oxidase, and Fe/S proteins react with H_2O_2 (Jang & Imlay, 2007; Zhang, Joseph, Gurney, Becker, & Kalyanaraman, 2002). Inactivation of the protein can result from the oxidized metal-modifying amino acid(s) at the binding site. It can also result in peroxidase activity with exogenous substrates that access the bound metal, as in the case of conversion of CO_2 to the carbonate radical by superoxide dismutase.

3. ANTIOXIDANT DEFENSES AGAINST H_2O_2

Cells are endowed with multiple defenses against H_2O_2. These include catalase, glutathione peroxidases, and the more recently recognized peroxiredoxins. Plants also use heme peroxidases such as ascorbate peroxidase (Foyer & Noctor, 2009). Catalase, which breaks down H_2O_2 to oxygen and water, is confined to peroxisomes in most cells (erythrocytes and neutrophils being exceptions). Its prime function is to remove H_2O_2 generated by peroxisomal oxidases but it can also remove any H_2O_2 that diffuses into these organelles (Schrader & Fahimi, 2006). Glutathione peroxidase and peroxiredoxin family members are more widely distributed in the different cell compartments (Rhee, Woo, Kil, & Bae, 2012; Ursini et al., 1995).

Glutathione peroxidases contain an active site selenocysteine (or in some cases cysteine), which reacts rapidly with peroxides. The senelenic acid (—SeOH) and/or selenyl amide are proposed intermediates (Fig. 1.3A), and the catalytic cycle is completed by GSH which is regenerated by glutathione reductase and NADPH (Flohe, Toppo, Cozza, & Ursini, 2011). The peroxiredoxins are highly abundant thiol proteins that can be classified as typical 2-Cys, atypical 2-Cys, or 1-Cys forms (Nelson et al., 2011; Rhee et al., 2012). As well as H_2O_2, they react readily with peroxynitrite and organic peroxides (Trujillo, Ferrer-Sueta, Thomson, Flohe, & Radi, 2007), initially to form the sulfenic acid (Hall, Karplus, & Poole, 2009). The typical 2-Cys peroxiredoxins (Prxs 1–4 in mammals) form decameric structures made up of tight dimers and form interchain disulfides by condensation of the sulfenic acid with the resolving cysteine on the opposing chain. Atypical 2-Cys peroxiredoxins form intramolecular disulfides. The disulfides are recycled by thioredoxin, which in turn is regenerated by thioredoxin reductase and NADPH (Fig. 1.3). Other proteins with thioredoxin domains, for example, cyclophilins (Lee et al., 2001) and protein disulfide isomerases (Tavender, Springate, & Bulleid, 2010) may also carry out this step. The 1-Cys peroxiredoxins are recycled by GSH. For Prx6, the only mammalian form, this involves the participation of glutathione *S*-transferase π (Fisher, 2011). A significant feature of the 2-Cys peroxiredoxins (particularly in eukaryotes) is that they can be inactivated by a second peroxide molecule reacting with the sulfenic acid to give the sulfinic acid. Oxidation appears to occur more readily than with other sulfenic acids and regulates both activity and oligomerization state (Rhee et al., 2012). It is noteworthy that both the glutathione peroxidases and

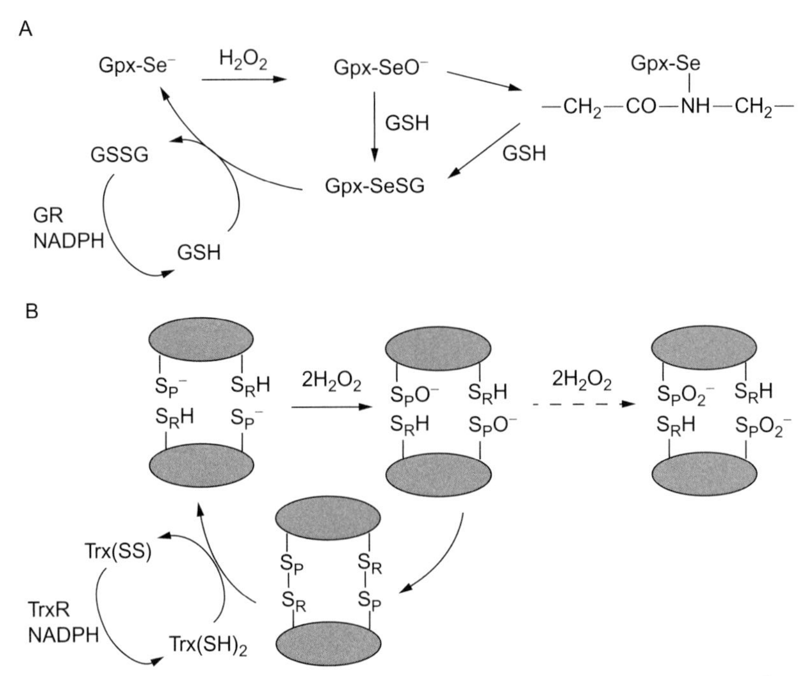

Figure 1.3 Redox reactions of (A) seleno-glutathione peroxidases and (B) typical 2-Cys peroxiredoxins. For (A), the initial reaction with peroxides is proposed to generate a selenenic acid intermediate that can be stabilized by condensation with the adjacent amide nitrogen (Flohe et al., 2011). Either can be recycled by glutathione reductase (GR) and NADPH. For (B), H_2O_2 (and other peroxides) reacts with the active site Cys (S_P) of the peroxiredoxin to form the sulfenic acid which condenses with the resolving Cys (S_R) to give a disulfide. Further oxidation (hyperoxidation) by another H_2O_2 gives the sulfinic acid (dashed arrow) which can be slowly recycled by sulfiredoxin. The disulfide is recycled by thioredoxin (Trx) which is coupled to thioredoxin reductase (TrxR) and NADPH. See Rhee et al. (2012) for further detail.

peroxiredoxins ultimately depend on NADPH to maintain their catalytic activity, making NADPH regeneration a key feature of effective antioxidant defense (Fig. 1.3).

All these proteins react extremely rapidly with H_2O_2. The rate constants for catalases, seleno-glutathione peroxidases, and many of the peroxiredoxins are in the 10^7–10^8 M^{-1} s^{-1} range (Cox, Peskin, Paton, Winterbourn, & Hampton, 2009; Peskin et al., 2007; Toppo, Flohe, Ursini, Vanin, & Maiorino, 2009). The glutathione peroxidases that use Cys and some of the peroxiredoxins (Baker & Poole, 2003; Hugo et al., 2009) react 100- to 1000-fold more slowly, but even these values are orders of magnitude faster than for typical low pK_a thiols. This raises the interesting questions of how H_2O_2 escapes these defenses and reacts with other targets, and why there is an apparent redundancy of defenses against H_2O_2. It is commonly stated that glutathione peroxidases eliminate low concentrations of H_2O_2, whereas catalase is more efficient at higher concentrations. However, this is erroneously argued on the basis of the K_m for H_2O_2 being higher for catalase, whereas in fact neither enzyme can be saturated with H_2O_2 (Flohe, 1982). Instead, the difference in efficiency of the two enzymes is seen at higher H_2O_2 concentrations where glutathione peroxidase becomes less effective due to recycling of GSH becoming rate limiting. One possible explanation for the apparent redundancy in H_2O_2-consuming enzymes is different compartmentalization; whereas this holds to some extent with catalase, it does not take into account the colocalization of multiple peroxiredoxins and glutathione peroxidases at different sites.

The reason why there are multiple enzymes for metabolizing H_2O_2 may lie in the evolving view that prooxidant and antioxidant activities are not strictly opposing processes, but rather there is a continuum of oxidative damage, oxidative stress, and antioxidant protection (Flohe, 2010; Foyer & Noctor, 2009; Rhee et al., 2012; Winterbourn & Hampton, 2008). Thus, antioxidants should be considered not simply as agents for removing an oxidant but more as regulators of its level in the cell and thereby of redox metabolism. In that capacity, antioxidants could control the extent to which an oxidant participates in a metabolic, signaling, or destructive pathway. Additionally, the "antioxidants" themselves (glutathione peroxidase, glutathione, peroxiredoxin, or thioredoxin) could regulate cell pathways by undergoing redox-dependent interactions with other proteins. There are examples of this mechanism, especially in yeast, where an oxidized peroxiredoxin activates a transcription pathway (Delaunay, Pflieger, Barrault, Vinh, & Toledano, 2002; Neumann, Cao, & Manevich, 2009; Veal et al., 2004). It is also possible, although not as yet described, that varying the proportions of the different cellular peroxiredoxins and glutathione peroxidases could differentially affect redox-regulated pathways without changing total H_2O_2 scavenging ability.

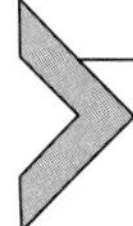

4. KINETICS AND IDENTIFICATION OF BIOLOGICAL TARGETS FOR H_2O_2

Although H_2O_2 undergoes relatively few reactions, it still oxidizes numerous biological molecules. However, these will not all be relevant targets in a complex biological context, as some will outcompete others. The key factors that define whether a particular substrate will be oxidized are how fast it reacts (defined by the rate constant) and its concentration. Thus, a constituent with a high rate constant may not be favored if its abundance is low. It is possible to predict likely reactions using the simple kinetic expression that describes the ratio (r) of an oxidant reacting with one substrate (A) in competition with another (B) (where k_a and k_b are the rate constants and [A] and [B] are the concentrations; or if there are multiple competitors for A, the denominator will be the sum of the k[substrate] terms):

$$r = k_a[A]/k_b[B].$$

This kinetic analysis can be used to predict likely targets for H_2O_2 in a cell (Winterbourn, 2008). Although rate constants and concentrations for all possible targets are not available, useful insight can be obtained by considering just a selection of reactions.

Modeling a simple homogeneous system containing a peroxiredoxin, a seleno-glutathione peroxidase, selected PTPs, and GSH at estimated cell concentrations gives a clear picture (Fig. 1.4). The peroxiredoxin and glutathione peroxidase (which is slightly more reactive but less abundant than peroxiredoxins) trap almost all the peroxide, an almost negligible proportion reacts directly with GSH, and the PTPs compete poorly even with GSH. Modeling a mitochondrial environment with more potential targets

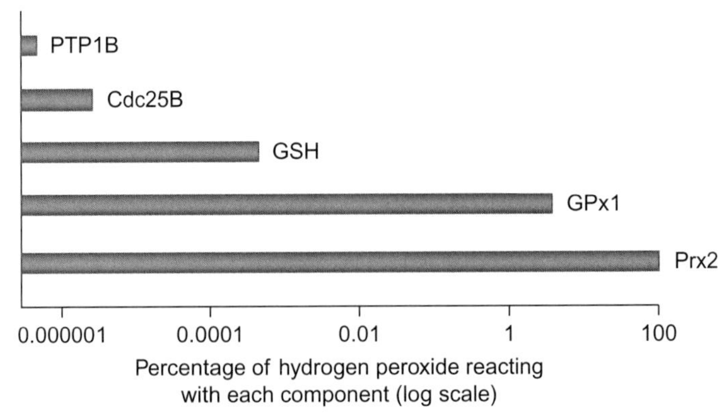

Figure 1.4 Kinetic heirarchy for cellular targets of H_2O_2. Simulation of competition between a peroxiredoxin (Prx2), glutathione peroxidase (GPx 1), GSH, and two protein tyrosine phosphatases (PTP1B and Cdc25B). Kinetic modeling was performed using rate constants and estimated cell concentrations. *Adapted from Winterbourn (2008) where more details of the modeling procedure and the parameters used are given.* (For color version of this figure, the reader is referred to the online version of this chapter.)

tells a similar story (Cox, Winterbourn, & Hampton, 2009). Thus, the low pK_a PTPs and related enzymes that are widely considered as peroxide targets in cell signaling would not be competitive kinetically. For other proteins to be direct targets, it would seem that they would need to have reactivity approaching that of a peroxiredoxin. At least in mammalian cells, no other thiol proteins with this reactivity have been identified. It should be noted, however, that while this example focuses on thiol proteins, in view of their involvement in redox signaling, some metal centers will react with H_2O_2 just as well as most thiols.

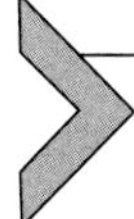

5. TRANSMISSION OF REDOX SIGNALS INITIATED BY H_2O_2

Although kinetic modeling implies that thiol proteins involved in signaling pathways are unlikely to be oxidized directly by H_2O_2 (I in Fig. 1.5), oxidative inactivation of proteins such as PTPs has been observed in cells treated with low doses of H_2O_2 and during cell signaling (for example Haque, Andersen, Salmeen, Barford, & Tonks, 2011; Rhee, Bae, Lee, & Kwon, 2000; Tonks, 2005; Truong & Carroll, 2012). A number of alternative mechanisms can be proposed. One possibility that is receiving increased attention (II in Fig. 1.5) is indirect oxidation via a highly reactive sensor protein such as a peroxiredoxin that transmits the oxidizing equivalents of H_2O_2 (Forman, Maiorino, & Ursini, 2010; Rhee et al., 2012; Winterbourn & Hampton, 2008). Several examples of this mechanism have been described (e.g., by D'Autreaux & Toledano, 2007; Neumann et al., 2009). This interchange would be facilitated by protein–protein interaction with the sensor, and could therefore provide the selectivity required for a signaling process. Activation of a signaling pathway could also occur by a related mechanism involving redox-sensitive protein binding, if this altered the activity or location of the signaling protein (III in Fig. 1.5).

Another possibility (IV in Fig. 1.5) is that the thiol compound that recycles the sensor (e.g., oxidized thioredoxin or glutathione) transmits the signal to another protein, either by thiol exchange or selective binding of the oxidized or reduced form. The latter mechanism is seen with apoptosis signaling kinase (ASK1) which is inactive when bound to reduced thioredoxin and released when the thioredoxin is oxidized (Saitoh et al., 1998). Although glutathione and thioredoxin were regarded for a long time as redox buffers that equilibrate with other thiol/disulfide couples, thiol exchange reactions are generally slow and it is now established that cellular redox couples are not

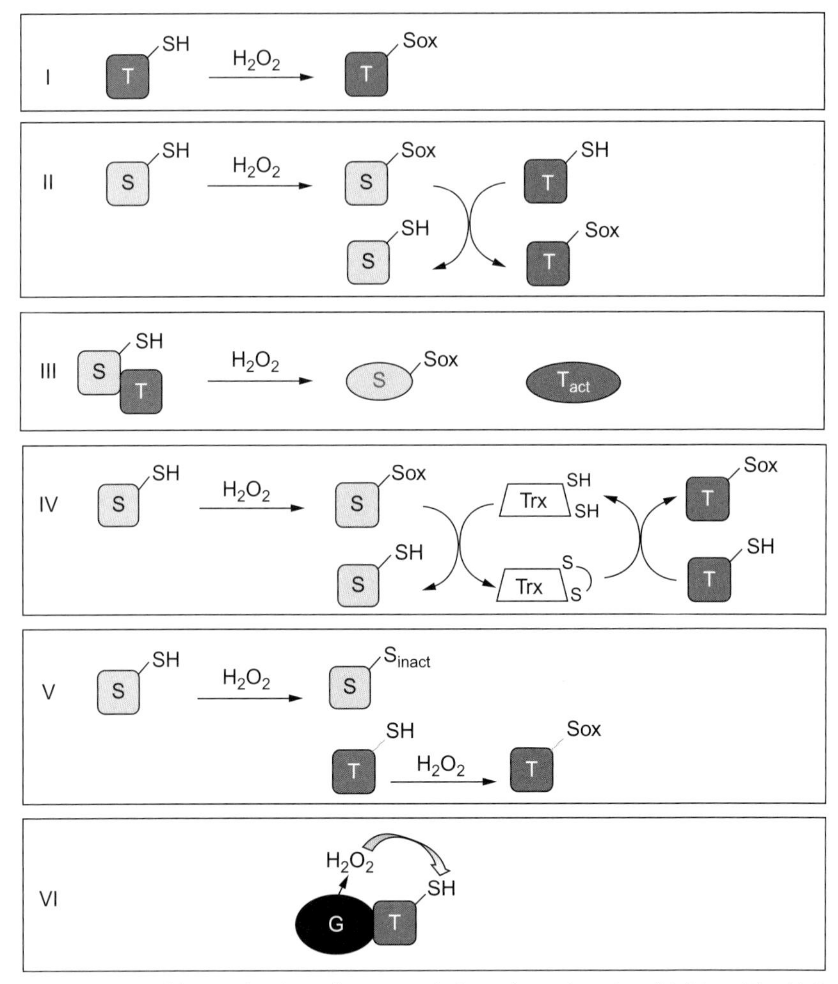

Figure 1.5 Possible mechanisms for transmission of a redox signal initiated by H_2O_2. I, direct oxidation of target protein (T); II, oxidation via a highly reactive sensor protein (S); III, activation of T by dissociation from oxidized S; IV, oxidation of T via a secondary product of S such as thioredoxin (Trx); V, inactivation of scavenging protein S to allow oxidation of T (floodgate model); VI, association of T with H_2O_2-generating protein to allow site-directed oxidation. (See Color Insert.)

in thermodynamic equilibrium (Kemp, Go, & Jones, 2008). Relevant exchange reactions are therefore likely to require facilitation, for example, by preferential binding of the protein to thioredoxin or catalysis by glutaredoxin (Gallogly, Starke, & Mieyal, 2009).

An important caveat to predictions made from the kinetic analyses described earlier is that they assume homogeneous conditions. But the cell, with its numerous membranous structures and organelles, is not homogeneous, and H_2O_2 production and targets are not necessarily evenly distributed. This might allow site-localized oxidation of less-favored targets in a particular region or organelle where reactive proteins such as peroxiredoxins are absent or have been inactivated (V in Fig. 1.5). Such a mechanism provides the basis for the floodgate model of redox regulation (Wood, Poole, & Karplus, 2003). The likelihood of site-localized action can be addressed by considering how far H_2O_2 could diffuse in the presence of reactive targets, and the ability of membranes to restrict this diffusion, as described in Section 6.

6. DIFFUSION DISTANCES AND COMPARTMENTALIZATION

6.1. Diffusion

It is possible to model how far H_2O_2 would diffuse in the presence of various targets and relate this to the dimensions of the cell (Winterbourn, 2008). For a homogeneous situation with no membrane barriers, the distance a species travels before it is consumed by reactive targets is inversely related to the concentrations of the targets and the rate constants of their reactions (see Footnote 1 for equation). On this basis, it can be calculated that a peroxiredoxin or glutathione peroxidase at a typical cell concentration would limit the range of H_2O_2 to between 50% and 10% of the cell diameter (Winterbourn, 2008). Other thiol proteins such as PTPs would not restrict diffusion to within the diameter of the cell. This implies that they would not undergo selective oxidation at the site of H_2O_2 generation, even if more reactive proteins such as peroxiredoxins were inactivated. Based on this analysis, site-localized oxidation would require an association between oxidant generator and target that facilitates transfer of the oxidant (VI in Fig. 1.5).

6.2. Compartmentalization and membrane permeability

One way of promoting a reaction between H_2O_2 and a target would be to confine them to the same compartment or organelle. This is seen in peroxisomes and in neutrophil phagosomes (Schrader & Fahimi, 2006;

[1] Diffusion distance is described by the relationship $\ln C/C_o = l\sqrt{\sum k[S]/D}$ (where k and [S] are as described for competition kinetics, l is the distance over which the oxidant drops from an initial concentration of C_o to C, and D is the diffusion coefficient (Lancaster, 1996). D is not known for cellular milieu, but for small reactive species is estimated to be not greatly less than for aqueous solution. For multiple substrates, the square root term in the denominator includes the sum of all the k[S] values, and the distance is correspondingly smaller.

Winterbourn, Hampton, Livesey, & Kettle, 2006). However, these are special cases where a very high concentration of a reactive substrate (catalase and myeloperoxidase, respectively) is able to consume most of the H_2O_2 before it can diffuse through the cell membrane. The ability of a membrane to confine a reaction between H_2O_2 and a less-reactive substrate is less likely. However, membranes do retard the diffusion of H_2O_2 (Antunes & Cadenas, 2000; Bienert, Schjoerring, & Jahn, 2006) and transport is facilitated by some of the aquaporin isoforms (Bienert et al., 2007). Therefore, aquaporins may be important for regulating H_2O_2 movement, and if they were absent, more confinement within a membrane-bound compartment might be possible. Although such a mechanism has not as yet been described, there is evidence for an aquaporin regulating the ability of H_2O_2 to act intracellularly. For example, induction of T-cell migration by H_2O_2 has been reported to depend on aquaporin-3-mediated entry into the cells (Hara-Chikuma et al., 2012).

6.2.1 NADPH oxidases

Diffusion and membrane permeability are particularly relevant for understanding the fate of H_2O_2 generated by the NADPH oxidases (NOXs). This family of membrane-associated multiprotein complexes is an important source of H_2O_2 and NOX activation is strongly implicated in redox signaling (Lambeth, 2004). NOXs catalyze the transfer of electrons from NADPH to oxygen to give superoxide. In most cases, superoxide is released and dismutates

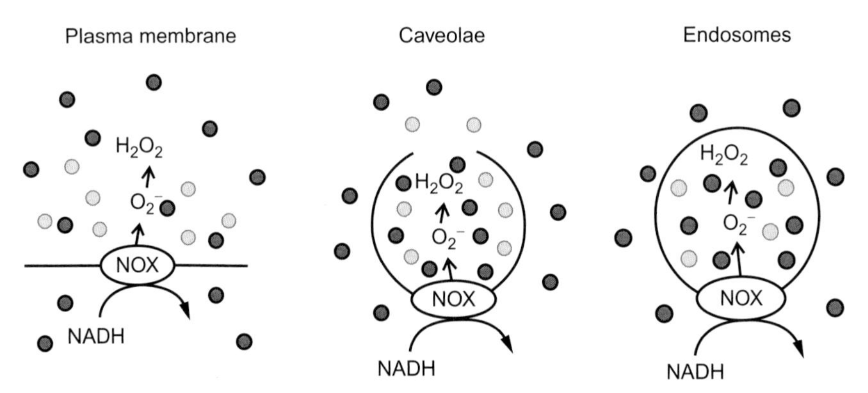

Figure 1.6 Schematic representation of release and diffusion of superoxide (light circles, pink fill) and H_2O_2 (black circles, blue fill) from NADPH oxidase activity at the cell surface, at caveolae, or in endosomes. The membrane is represented as a dark line. NADPH is oxidized on the cytoplasmic surface and superoxide released at the exosurface. Low membrane permeability restricts superoxide to the compartment where it is generated. The membrane retards but does not prevent H_2O_2 diffusion. (See Color Insert.)

to give H_2O_2 or (as with NOX4) it may be converted to H_2O_2 before leaving the active site (Leto, Morand, Hurt, & Ueyama, 2009). Electron flow is directional, with NADPH oxidized in the cytoplasm and superoxide (or H_2O_2) released on the external side of the membrane. This may be to the outside of the cell, possibly in a localized area such as caveolae or into an internal organelle such as an endosome (Fig. 1.6). An interesting unresolved conundrum is how these oxidants exert their effect. Is H_2O_2 in some way restricted to react with a target present in the organelle? Could superoxide, which has very low membrane permeability, undergo a localized reaction? Or does external H_2O_2 traverse the cell membrane, perhaps, via an aquaporin in the vicinity, and interact with an intracellular target? Can this occur locally, especially if there is nothing to prevent diffusion in other directions?

7. BIOLOGICAL DETECTION OF H_2O_2

A critical requirement for elucidating the details of how H_2O_2 is involved in redox signaling is to be able to detect it in biological systems. In fact, there are major limitations to detecting H_2O_2 specifically and quantitatively, and in many studies, evidence that H_2O_2 is responsible for observed redox changes is equivocal. Detection methods are covered in other chapters of this volume and problems have been highlighted in recent reviews (Kalyanaraman et al., 2012; Maghzal, Krause, Stocker, & Jaquet, 2012; Murphy et al., 2011; Wardman, 2007) so only brief points will be made here. Measuring extracellular H_2O_2 is less problematic than intracellular detection, and with appropriate controls, quantitative data can be obtained using a peroxidase with a detector such as Amplex red. Intracellular H_2O_2 is more of a challenge and every method has its limitations. Two general approaches are currently used: one using genetically encoded green fluorescent protein (GFP)-derivatives that change fluorescence when oxidized (Belousov et al., 2006; Meyer & Dick, 2010) and the other using small molecule fluorescent probes. The GFP probes (Hyper or RoGFP) have the advantage of containing highly reactive dithiols that react directly and selectively with peroxides, and they can be expressed at specific sites in the cell. However, caution is still required to take into account pH sensitivity and reversibility by reductants such as thioredoxin.

Most low-molecular-weight probes, including dihydrorhodamine and the much-used dichlorofluorescin (DCF) have major limitations. They can (with appropriate controls) give some indication of whether a cell shifts to a more oxidizing state. However, they have little value for identifying oxidants or mechanisms, and they are *not* good detectors for H_2O_2 (Wardman, 2007). Most importantly, H_2O_2 does not react directly with DCF and related compounds. The reaction has an absolutely requirement for a transition metal catalyst, which could be a low-molecular-weight chelate, a peroxidase or cytochrome, or some other poorly characterized metal center. Therefore, obtaining a signal depends on a catalyst being present, and any variation could arise from a change in availability of the catalyst (such as cytochrome *c* release from mitochondria (Burkitt & Wardman, 2001), iron release from lysosomes (Karlsson, Kurz, Brunk, Nilsson, & Frennesson, 2010), or inactivation of a metalloprotein) without any change in H_2O_2 generation. As these probes are initially oxidized to a radical intermediate, any change in radical scavenging capacity will also affect the signal. Additionally, they are oxidized by multiple species including peroxynitrite or hypochlorous acid or by exposure to light.

Boronate compounds are a more recently developed class of probes that have the advantages of reacting directly with H_2O_2 to give a fluorescent product and of not forming radicals (Miller, Albers, Pralle, Isacoff, & Chang, 2005). One limitation is that the currently available probes react slowly and would compete poorly for H_2O_2. However, high fluorescence intensity to some extent overcomes this inefficient trapping. Also, because these probes cause little perturbation to the system by removing H_2O_2, they can provide useful information on changes in steady-state concentration. However, a major caution is that the boronates are several orders of magnitude more reactive with peroxynitrite than H_2O_2 (Zielonka et al., 2012).

In summary, DCF and related probes are not appropriate for the specific detection of intracellular H_2O_2. Genetically encoded GFP derivatives and boronate compounds, if used with appropriate precautions, are able to detect intracellular H_2O_2 and provide useful information, particularly for comparative purposes. However, none of these methods measure absolute amounts of H_2O_2 formed within cells, and further advances are needed before this can be achieved.

8. CONCLUSION

H_2O_2 would be expected to react with selected biological targets, with transition metal centers, selenoproteins, and a small selection of thiol proteins being most favored. It is usual to attribute biological damage by H_2O_2 to (usually metal-dependent) one-electron reactions and free radical production, and redox signaling to oxidation of thiol proteins in regulatory pathways. However, although thiol protein oxidation is seen in cells that have been treated with H_2O_2 or when redox-regulated pathways are activated, molecular mechanisms of signaling by H_2O_2 are not well characterized. No thiol proteins in signaling pathways have yet been shown to react fast enough with H_2O_2 to be competitive with other much more reactive cell constituents. A mechanism of facilitated oxidation therefore seems necessary. Of the possibilities considered in Fig. 1.5, oxidation via a sensor protein or direct association between peroxide generator and target appear the most feasible, although it must be noted that there is still only limited evidence for either being a major pathway. Site-localized reactions of H_2O_2 can occur, but only when there is a highly reactive target that restricts diffusion, and a membrane would not normally be a sufficient barrier. Therefore, site-localized generation of H_2O_2 does not necessarily result in site-localized action. There are documented examples of peroxiredoxins, glutathione peroxidases, or their redox partners acting as sensors for H_2O_2, but it is also a possibility that metalloproteins might have a role. It is also worth considering whether initiation of redox signals need necessarily involve H_2O_2, as although thiol changes are well documented, evidence for H_2O_2 is often less convincing. However, further elucidation of the relevance of these mechanisms requires further studies that couple identification of specific cellular targets of H_2O_2 with rigorous methodology for its detection.

REFERENCES

Antunes, F., & Cadenas, E. (2000). Estimation of H_2O_2 gradients across biomembranes. *FEBS Letters*, *475*, 121–126.

Baker, L. M., & Poole, L. B. (2003). Catalytic mechanism of thiol peroxidase from Escherichia coli. Sulfenic acid formation and overoxidation of essential CYS61. *The Journal of Biological Chemistry*, *278*, 9203–9211.

Bakhmutova-Albert, E. V., Yao, H., Denevan, D. E., & Richardson, D. E. (2010). Kinetics and mechanism of peroxymonocarbonate formation. *Inorganic Chemistry*, *49*, 11287–11296.

Belousov, V. V., Fradkov, A. F., Lukyanov, K. A., Staroverov, D. B., Shakhbazov, K. S., Terskikh, A. V., et al. (2006). Genetically encoded fluorescent indicator for intracellular hydrogen peroxide. *Nature Methods*, *3*, 281–286.

Bienert, G. P., Moller, A. L., Kristiansen, K. A., Schulz, A., Moller, I. M., Schjoerring, J. K., et al. (2007). Specific aquaporins facilitate the diffusion of hydrogen peroxide across membranes. *The Journal of Biological Chemistry*, *282*, 1183–1192.

Bienert, G. P., Schjoerring, J. K., & Jahn, T. P. (2006). Membrane transport of hydrogen peroxide. *Biochimica et Biophysica Acta*, *1758*, 994–1003.

Burkitt, M. J., & Wardman, P. (2001). Cytochrome *c* is a potent catalyst of dichlorofluorescin oxidation: Implications for the role of reactive oxygen species in apoptosis. *Biochemical and Biophysical Research Communications*, *282*(1), 329–333.

Carballal, S., Radi, R., Kirk, M. C., Barnes, S., Freeman, B. A., & Alvarez, B. (2003). Sulfenic acid formation in human serum albumin by hydrogen peroxide and peroxynitrite. *Biochemistry*, *42*, 9906–9914.
Chevion, M. (1988). A site-specific mechanism for free radical induced biological damage: The essential role of redox-active transition metals. *Free Radical Biology & Medicine*, *5*, 27–37.
Claiborne, A., Miller, H., Parsonage, D., & Ross, R. P. (1993). Protein-sulfenic acid stabilization and function in enzyme catalysis and gene regulation. *The FASEB Journal*, *7*, 1483–1490.
Cox, A. G., Winterbourn, C. C., & Hampton, M. B. (2009). Mitochondrial peroxiredoxin involvement in antioxidant defence and redox signalling. *The Biochemical Journal*, *425*, 313–325.
Cox, A. G., Peskin, A. V., Paton, L. N., Winterbourn, C. C., & Hampton, M. B. (2009). Redox potential and peroxide reactivity of human peroxiredoxin 3. *Biochemistry*, *48*, 6495–6501.
D'Autreaux, B., & Toledano, M. B. (2007). ROS as signalling molecules: Mechanisms that generate specificity in ROS homeostasis. *Nature Reviews. Molecular Cell Biology*, *8*, 813–824.
Davies, M. J., Hawkins, C. L., Pattison, D. I., & Rees, M. D. (2008). Mammalian heme peroxidases: From molecular mechanisms to health implications. *Antioxidants and Redox Signalling*, *10*, 1199–1234.
Delaunay, A., Pflieger, D., Barrault, M. B., Vinh, J., & Toledano, M. B. (2002). A thiol peroxidase is an H_2O_2 receptor and redox-transducer in gene activation. *Cell*, *111*, 471–481.
Ferrer-Sueta, G., Manta, B., Botti, H., Radi, R., Trujillo, M., & Denicola, A. (2011). Factors affecting protein thiol reactivity and specificity in peroxide reduction. *Chemical Research in Toxicology*, *24*, 434–450.
Fisher, A. B. (2011). Peroxiredoxin 6: A bifunctional enzyme with glutathione peroxidase and phospholipase A(2) activities. *Antioxidants and Redox Signalling*, *15*, 831–844.
Flohe, L. (1982). Glutathione peroxidase brought into focus. In W. A. Pryor (Ed.), *Free radicals in biology*. *5*, (pp. 223–254). New York: Academic Press.
Flohe, L. (2010). Changing paradigms in thiology from antioxidant defense toward redox regulation. *Methods in Enzymology*, *473*, 1–39.
Flohe, L., Toppo, S., Cozza, G., & Ursini, F. (2011). A comparison of thiol peroxidase mechanisms. *Antioxidants and Redox Signalling*, *15*, 763–780.
Forman, H. J., Maiorino, M., & Ursini, F. (2010). Signaling functions of reactive oxygen species. *Biochemistry*, *49*, 835–842.
Foyer, C. H., & Noctor, G. (2009). Redox regulation in photosynthetic organisms: Signaling, acclimation, and practical implications. *Antioxidants and Redox Signalling*, *11*, 861–905.
Gallogly, M. M., Starke, D. W., & Mieyal, J. J. (2009). Mechanistic and kinetic details of catalysis of thiol-disulfide exchange by glutaredoxins and potential mechanisms of regulation. *Antioxidants and Redox Signalling*, *11*, 1059–1081.
Goldstein, S., Meyerstein, D., & Czapski, G. (1993). The Fenton reagents. *Free Radical Biology & Medicine*, *15*, 435–445.
Hall, A., Karplus, P. A., & Poole, L. B. (2009). Typical 2-Cys peroxiredoxins—Structures, mechanisms and functions. *The FEBS Journal*, *276*, 2469–2477.
Hall, A., Parsonage, D., Poole, L. B., & Karplus, P. A. (2010). Structural evidence that peroxiredoxin catalytic power is based on transition-state stabilization. *Journal of Molecular Biology*, *402*, 194–209.
Haque, A., Andersen, J. N., Salmeen, A., Barford, D., & Tonks, N. K. (2011). Conformation-sensing antibodies stabilize the oxidized form of PTP1B and inhibit its phosphatase activity. *Cell*, *147*, 185–198.

Hara-Chikuma, M., Chikuma, S., Sugiyama, Y., Kabashima, K., Verkman, A. S., Inoue, S., et al. (2012). Chemokine-dependent T cell migration requires aquaporin-3-mediated hydrogen peroxide uptake. *The Journal of Experimental Medicine, 209*, 1743–1752.

Hugo, M., Turell, L., Manta, B., Botti, H., Monteiro, G., Netto, L. E., et al. (2009). Thiol and sulfenic acid oxidation of AhpE, the one-cysteine peroxiredoxin from Mycobacterium tuberculosis: Kinetics, acidity constants, and conformational dynamics. *Biochemistry, 48*, 9416–9426.

Imlay, J. A. (2003). Pathways of oxidative damage. *Annual Review of Microbiology, 57*, 395–418.

Imlay, J. A. (2008). Cellular defenses against superoxide and hydrogen peroxide. *Annual Review of Biochemistry, 77*, 755–776.

Jang, S., & Imlay, J. A. (2007). Micromolar intracellular hydrogen peroxide disrupts metabolism by damaging iron-sulfur enzymes. *The Journal of Biological Chemistry, 282*, 929–937.

Kagan, V. E., Tyurin, V. A., Jiang, J., Tyurina, Y. Y., Ritov, V. B., Amoscato, A. A., et al. (2005). Cytochrome *c* acts as a cardiolipin oxygenase required for release of proapoptotic factors. *Nature Chemical Biology, 1*, 223–232.

Kalyanaraman, B., Darley-Usmar, V., Davies, K. J., Dennery, P. A., Forman, H. J., Grisham, M. B., et al. (2012). Measuring reactive oxygen and nitrogen species with fluorescent probes: Challenges and limitations. *Free Radical Biology & Medicine, 52*, 1–6.

Karlsson, M., Kurz, T., Brunk, U. T., Nilsson, S. E., & Frennesson, C. I. (2010). What does the commonly used DCF test for oxidative stress really show? *The Biochemical Journal, 428*, 183–190.

Kemp, M., Go, Y. M., & Jones, D. P. (2008). Nonequilibrium thermodynamics of thiol/disulfide redox systems: A perspective on redox systems biology. *Free Radical Biology & Medicine, 44*, 921–937.

Kurz, T., Gustafsson, B., & Brunk, U. T. (2006). Intralysosomal iron chelation protects against oxidative stress-induced cellular damage. *The FEBS Journal, 273*, 3106–3117.

Lambeth, J. D. (2004). NOX enzymes and the biology of reactive oxygen. *Nature Review of Immunology, 4*, 181–189.

Lancaster, J. R., Jr. (1996). Diffusion of free nitric oxide. *Methods in Enzymology, 268*, 31–50.

Lee, S. P., Hwang, Y. S., Kim, Y. J., Kwon, K. S., Kim, H. J., Kim, K., et al. (2001). Cyclophilin a binds to peroxiredoxins and activates its peroxidase activity. *The Journal of Biological Chemistry, 276*, 29826–29832.

Leto, T. L., Morand, S., Hurt, D., & Ueyama, T. (2009). Targeting and regulation of reactive oxygen species generation by Nox family NADPH oxidases. *Antioxidants and Redox Signalling, 11*, 2607–2619.

Maghzal, G. J., Krause, K. H., Stocker, R., & Jaquet, V. (2012). Detection of reactive oxygen species derived from the family of NOX NADPH oxidases. *Free Radical Biology & Medicine, 53*, 1903–1918.

Meyer, A. J., & Dick, T. P. (2010). Fluorescent protein-based redox probes. *Antioxidants and Redox Signalling, 13*, 621–650.

Miller, E. W., Albers, A. E., Pralle, A., Isacoff, E. Y., & Chang, C. J. (2005). Boronate-based fluorescent probes for imaging cellular hydrogen peroxide. *Journal of the American Chemical Society, 127*, 16652–16659.

Murphy, M. P., Holmgren, A., Larsson, N. G., Halliwell, B., Chang, C. J., Kalyanaraman, B., et al. (2011). Unraveling the biological roles of reactive oxygen species. *Cell Metabolism, 13*, 361–366.

Nagy, P., Karton, A., Betz, A., Peskin, A. V., Pace, P., O'Reilly, R. J., et al. (2011). Model for the exceptional reactivity of peroxiredoxins 2 and 3 with hydrogen peroxide: A kinetic and computational study. *The Journal of Biological Chemistry, 286*, 18048–18055.

Nagy, P., & Winterbourn, C. C. (2010). Redox chemistry of biological thiols. In J. C. Fishbein (Ed.), *Advances in molecular toxicology* (pp. 183–222). New York: Elsevier.

Nelson, K. J., Klomsiri, C., Codreanu, S. G., Soito, L., Liebler, D. C., Rogers, L. C., et al. (2010). Use of dimedone-based chemical probes for sulfenic acid detection methods to visualize and identify labeled proteins. *Methods in Enzymology*, *473*, 95–115.

Nelson, K. J., Knutson, S. T., Soito, L., Klomsiri, C., Poole, L. B., & Fetrow, J. S. (2011). Analysis of the peroxiredoxin family: Using active-site structure and sequence information for global classification and residue analysis. *Proteins*, *79*, 947–964.

Neumann, C. A., Cao, J., & Manevich, Y. (2009). Peroxiredoxin 1 and its role in cell signaling. *Cell Cycle*, *8*, 4072–4078.

O'Donnell-Tormey, J., Nathan, C. F., Lanks, K., DeBoer, C. J., & de la Harpe, J. (1987). Secretion of pyruvate. An antioxidant defense of mammalian cells. *The Journal of Experimental Medicine*, *165*, 500–514.

Peskin, A. V., Low, F. M., Paton, L. N., Maghzal, G. J., Hampton, M. B., & Winterbourn, C. C. (2007). The high reactivity of peroxiredoxin 2 with H_2O_2 is not reflected in its reaction with other oxidants and thiol reagents. *The Journal of Biological Chemistry*, *282*(16), 11885–11892.

Rhee, S. G., Bae, Y. S., Lee, S. R., & Kwon, J. (2000). Hydrogen peroxide: A key messenger that modulates protein phosphorylation through cysteine oxidation. *Science's STKE*, *2000*, PE1.

Rhee, S. G., Woo, H. A., Kil, I. S., & Bae, S. H. (2012). Peroxiredoxin functions as a peroxidase and a regulator and sensor of local peroxides. *The Journal of Biological Chemistry*, *287*, 4403–4410.

Saitoh, M., Nishitoh, H., Fujii, M., Takeda, K., Tobiume, K., Sawada, Y., et al. (1998). Mammalian thioredoxin is a direct inhibitor of apoptosis signal-regulating kinase (ASK) 1. *The EMBO Journal*, *17*, 2596–2606.

Salmeen, A., Andersen, J. N., Myers, M. P., Meng, T. C., Hinks, J. A., Tonks, N. K., et al. (2003). Redox regulation of protein tyrosine phosphatase 1B involves a sulphenyl-amide intermediate. *Nature*, *423*, 769–773.

Schrader, M., & Fahimi, H. D. (2006). Peroxisomes and oxidative stress. *Biochimica et Biophysica Acta*, *1763*, 1755–1766.

Stadtman, E. R. (1990). Metal ion catalysed oxidation of proteins: Biochemical mechanism and biological consequences. *Free Radical Biology & Medicine*, *9*, 315–325.

Stone, J. R. (2004). An assessment of proposed mechanisms for sensing hydrogen peroxide in mammalian systems. *Archives of Biochemistry and Biophysics*, *422*, 119–124.

Sutton, H. C., & Winterbourn, C. C. (1989). On the participation of higher oxidation states of iron and copper in Fenton-type reactions. *Free Radical Biology & Medicine*, *6*, 53–60.

Tavender, T. J., Springate, J. J., & Bulleid, N. J. (2010). Recycling of peroxiredoxin IV provides a novel pathway for disulphide formation in the endoplasmic reticulum. *The EMBO Journal*, *29*, 4185–4197.

Tonks, N. K. (2005). Redox redux: Revisiting PTPs and the control of cell signaling. *Cell*, *121*, 667–670.

Toppo, S., Flohe, L., Ursini, F., Vanin, S., & Maiorino, M. (2009). Catalytic mechanisms and specificities of glutathione peroxidases: Variations of a basic scheme. *Biochimica et Biophysica Acta*, *1790*, 1486–1500.

Trindade, D. F., Cerchiaro, G., & Augusto, O. (2006). A role for peroxymonocarbonate in the stimulation of biothiol peroxidation by the bicarbonate/carbon dioxide pair. *Chemical Research in Toxicology*, *19*, 1475–1482.

Trujillo, M., Ferrer-Sueta, G., Thomson, L., Flohe, L., & Radi, R. (2007). Kinetics of peroxiredoxins and their role in the decomposition of peroxynitrite. *Sub-Cellular Biochemistry*, *44*, 83–113.

Truong, T. H., & Carroll, K. S. (2012). Redox regulation of epidermal growth factor receptor signaling through cysteine oxidation. *Biochemistry*, *51*, 9954–9965.

Truong, T. H., Garcia, F. J., Seo, Y. H., & Carroll, K. S. (2011). Isotope-coded chemical reporter and acid-cleavable affinity reagents for monitoring protein sulfenic acids. *Bioorganic & Medicinal Chemistry Letters*, *21*, 5015–5020.
Ursini, F., Maiorino, M., Brigelius-Flohe, R., Aumann, K. D., Roveri, A., Schomburg, D., et al. (1995). Diversity of glutathione peroxidases. *Methods in Enzymology*, *252*, 38–53.
Vasquez-Vivar, J., Denicola, A., Radi, R., & Augusto, O. (1997). Peroxynitrite-mediated decarboxylation of pyruvate to both carbon dioxide and carbon dioxide radical anion. *Chemical Research in Toxicology*, *10*, 786–794.
Veal, E. A., Findlay, V. J., Day, A. M., Bozonet, S. M., Evans, J. M., Quinn, J., et al. (2004). A 2-Cys peroxiredoxin regulates peroxide-induced oxidation and activation of a stress-activated MAP kinase. *Molecular Cell*, *15*, 129–139.
Walling, C. (1975). Fenton's reagent revisited. *Accounts of Chemical Research*, *8*, 125–131.
Wardman, P. (2007). Fluorescent and luminescent probes for measurement of oxidative and nitrosative species in cells and tissues: Progress, pitfalls, and prospects. *Free Radical Biology & Medicine*, *43*, 995–1022.
Winterbourn, C. C. (1995). Toxicity of iron and hydrogen peroxide: The Fenton reaction. *Toxicology Letters*, *82–83*, 969–974.
Winterbourn, C. C. (2008). Reconciling the chemistry and biology of reactive oxygen species. *Nature Chemical Biology*, *4*, 278–286.
Winterbourn, C. C. (2012). Biological chemistry of reactive oxygen species. In C. Chatgilialoglu & A. Studer (Eds.), *Encyclopedia of radicals in chemistry, biology & materials: Vol. 3*. (pp. 1259–1282). Chichester: Wiley.
Winterbourn, C. C., & Hampton, M. B. (2008). Thiol chemistry and specificity in redox signaling. *Free Radical Biology & Medicine*, *45*, 549–561.
Winterbourn, C. C., Hampton, M. B., Livesey, J. H., & Kettle, A. J. (2006). Modeling the reactions of superoxide and myeloperoxidase in the neutrophil phagosome: Implications for microbial killing. *The Journal of Biological Chemistry*, *281*, 39860–39869.
Winterbourn, C. C., & Metodiewa, D. (1999). Reactivity of biologically important thiol compounds with superoxide and hydrogen peroxide. *Free Radical Biology & Medicine*, *27*(3–4), 322–328.
Wood, Z. A., Poole, L. B., & Karplus, P. A. (2003). Peroxiredoxin evolution and the regulation of hydrogen peroxide signaling. *Science*, *300*, 650–653.
Yu, Z., Persson, H. L., Eaton, J. W., & Brunk, U. T. (2003). Intralysosomal iron: A major determinant of oxidant-induced cell death. *Free Radical Biology & Medicine*, *34*, 1243–1252.
Zhang, H., Joseph, J., Gurney, M., Becker, D., & Kalyanaraman, B. (2002). Bicarbonate enhances peroxidase activity of Cu,Zn-superoxide dismutase. Role of carbonate anion radical and scavenging of carbonate anion radical by metalloporphyrin antioxidant enzyme mimetics. *The Journal of Biological Chemistry*, *277*, 1013–1020.
Zielonka, J., Sikora, A., Hardy, M., Joseph, J., Dranka, B. P., & Kalyanaraman, B. (2012). Boronate probes as diagnostic tools for real time monitoring of peroxynitrite and hydroperoxides. *Chemical Research in Toxicology*, *25*, 1793–1799.

CHAPTER TWO

Reactive Oxygen Species in the Activation of MAP Kinases

Yong Son*, Sangduck Kim†, Hun-Taeg Chung‡, Hyun-Ock Pae§,1

*Department of Anesthesiology and Pain Medicine, Wonkwang University School of Medicine, Iksan, Republic of Korea

†Department of Ophthalmology, Wonkwang University School of Medicine, Iksan, Republic of Korea

‡Department of Biological Science, University of Ulsan, Ulsan, Republic of Korea

§Department of Microbiology and Immunology, Wonkwang University School of Medicine, Iksan, Republic of Korea

[1]Corresponding author: e-mail address: hopae@wku.ac.kr

Contents

1. Introduction 28
2. Reactive Oxygen Species 29
   2.1 Cellular antioxidants 30
   2.2 Superoxide and hydrogen peroxide 30
   2.3 ROS in cell signaling 32
3. Mitogen-Activated Protein Kinases 33
   3.1 Activation of ERK pathway 35
   3.2 Activation of p38 MAPK pathway 36
   3.3 Activation of JNK pathway 36
   3.4 Inactivation of MAPK pathways by MKPs 37
   3.5 Cross-talk relationship between the members of MAPK pathways 38
4. Roles of ROS in MAPK Activation 39
   4.1 Roles of ROS in ASK1 activation 40
   4.2 Roles of ROS in transactivation of growth factor receptors 42
   4.3 Roles of ROS in expression of MKPs 43
5. Summary 44
Acknowledgment 45
References 45

Abstract

There are three well-defined subgroups of mitogen-activated protein kinases (MAPKs): the extracellular signal-regulated kinases (ERKs), the c-Jun N-terminal kinases (JNKs), and the p38 MAPKs. Three subgroups of MAPKs are involved in both cell growth and cell death, and the tight regulation of these pathways, therefore, is paramount in determining cell fate. MAPK pathways have been shown to be activated not only by receptor ligand interactions but also by different stressors placed on the cell. MAPK phosphatases (MKPs) dephosphorylate and deactivate MAPKs. Reactive oxygen species (ROS), such as

Methods in Enzymology, Volume 528
ISSN 0076-6879
http://dx.doi.org/10.1016/B978-0-12-405881-1.00002-1

hydrogen peroxide, have been reported to activate ERKs, JNKs, and p38 MAPKs, but the mechanisms by which ROS can activate these kinases are unclear. Oxidative modifications of MAPK signaling proteins and inactivation and/or degradation of MKPs may provide the plausible mechanisms for activation of MAPK pathways by ROS, which will be reviewed in this chapter.

1. INTRODUCTION

The partially reduced metabolites of oxygen molecules (O_2), which are often referred to as "reactive oxygen species" (ROS) due to their higher reactivities relative to oxygen, play critical roles for the determination of cell fate by eliciting a wide variety of cellular responses, such as proliferation, differentiation and cell death, depending on cell types, cellular contexts, and amounts and duration of ROS generation (Matsuzawa & Ichijo, 2008). In many cases, low concentrations of ROS may enhance cell survival and proliferation, whereas high concentrations of ROS may cause cell death. Many intracellular signaling pathways involved in ROS-induced cellular responses have been shown to be regulated by the intracellular reduction–oxidation (redox) state, which depends on the balance between the levels of oxidizing and reducing equivalents. Excessively generated ROS is generally counteracted by ubiquitously expressed antioxidant molecules and enzymes (Birben, Sahiner, Sackesen, Erzurum, & Kalayci, 2012). Once the generation of ROS exceeds the capacity of the antioxidant proteins, cells suffer "oxidative stress" (OS), which causes severe dysfunction or death of cells.

Mitogen-activated protein kinases (MAPKs) are evolutionarily conserved regulators that mediate signal transduction and play essential roles in various cellular processes such as cell growth, differentiation, development, cell cycle, survival, and cell death (Ravingerová, Barancík, & Strnisková, 2003). MAPK cascades are well organized as modular pathways in which activation of upstream kinases by cell surface receptors and signaling molecules leads to sequential activation of an MAPK module (Keshet & Seger, 2010). Upon activation, MAPKs can phosphorylate a variety of intracellular targets, including transcription factors, nuclear pore proteins, membrane transporters, cytoskeletal elements, and other protein kinases (Kyriakis & Avruch, 2012). MAPK phosphatases (MKPs) provide a negative regulatory network that acts to modulate the duration, magnitude, and spatiotemporal profile of MAPK activities in response to both physiological and pathological stimuli (Boutros, Chevet, &

Metrakos, 2008). Individual MKPs may exhibit either exquisite specificity toward a single MAPK isoform or be able to regulate multiple MAPK pathways in a single cell or tissue. In animal cells, MAPKs and MKPs have been shown to be programmed to be activated in response to various stresses, including ROS-mediated OS, and to interact with one another for proper regulation of various cellular processes.

Studies have demonstrated that ROS can induce or mediate the activation of the MAPK pathways (McCubrey, Lahair, & Franklin, 2006). A number of cellular stimuli that induce ROS production in parallel can activate MAPK pathways in multiple cell types (Torres & Forman, 2003). The prevention of ROS accumulation by antioxidants blocks MAPK activation after cell stimulation with cellular stimuli (McCubrey et al., 2006; Torres & Forman, 2003), indicating the involvement of ROS in activation of MAPK pathways. Moreover, direct exposure of cells to exogenous hydrogen peroxide, to mimic OS, leads to activation of MAPK pathways (Dabrowski, Boguslowicz, Dabrowska, Tribillo, & Gabryelewicz, 2000; Ruffels, Griffin, & Dickenson, 2004). The mechanisms by which ROS can activate the MAPK pathways, however, is not well understood, largely due to a lack of information regarding the fundamental roles of ROS in activation of MAPK pathways. Thus, this chapter will mainly consider what is currently known about possible roles of ROS in activation of MAPK pathways.

2. REACTIVE OXYGEN SPECIES

Atmospheric oxygen (O_2) is usually nonreactive to organic molecules, unless it absorbs sufficient energy to reverse the spin of one of its unpaired electrons and is converted into singlet oxygen ($^1O_2$). However, a subsequent reduction of an oxygen molecule can occur in the animal cells, thereby leading to ROS generation (Hensley, Robinson, Gabbita, Salsman, & Floyd, 2000). For instance, one-electron reduction of oxygen molecule leads to the formation of the free radical superoxide ($O_2^{\bullet -}$), while two-electron reduction leads to hydrogen peroxide (H_2O_2) which is not a free radical because all its electrons are paired. In the presence of trace amounts of cellular iron, superoxide and hydrogen peroxide will interact with one another to form the highly reactive hydroxyl radical ($^{\bullet}OH$) that can interact with all biological molecules and may generate alkoxyl ($RO^{\bullet}$) and peroxyl ($ROO^{\bullet}$) radicals and organic hydroperoxides (ROOH). ROS are a group of free radicals, reactive molecules, and ions that are derived from O_2, but biologically most of them are derived from either superoxide and/or hydrogen peroxide, which, thus, will be further discussed below, alone with the brief comments on the cellular antioxidants and some roles of ROS in cell signaling.

2.1. Cellular antioxidants

All ROS are extremely harmful to organisms at high concentrations; ROS can pose a threat to cells by causing peroxidation of lipids, oxidation of proteins, damage to nucleic acids, enzyme inhibition, and activation of apoptotic pathway, and ultimately lead to cell death (Chang & Chuang, 2010). Continuous exposure to ROS from numerous sources has led the cell and the entire organism to develop defense mechanisms for protection against potential cytotoxicity of ROS. In fact, both superoxide and hydrogen peroxide are only moderately reactive to organic molecules, partly because of their elimination by the antioxidant enzymes, including superoxide dismutase (SOD), catalase (CAT), and glutathione (GSH) peroxidase (GSH Px), and/or nonenzymatic antioxidants, including vitamin C (ascorbic acid), vitamin E (tocopherol), flavonoids, and GSH (Valko et al., 2007). Whereas SOD can reduce superoxide to hydrogen peroxide, both CAT and GSH Px can reduce hydrogen peroxide to water (H_2O) (Fig. 2.1). Unfortunately, animal cells have no enzymatic mechanism for elimination of hydroxyl radical, and, thus, its excess production can eventually lead to cell death. Because the formation of hydroxyl radical is dependent on superoxide and hydrogen peroxide, its formation is subject to inhibition by SOD and CAT. When the level of ROS exceeds the antioxidant defense mechanisms, a cell is said to be in a state of OS; in other words, OS is defined as a persistent imbalance between the production of ROS and antioxidant defenses (Chang & Chuang, 2010; Valko et al., 2007).

2.2. Superoxide and hydrogen peroxide

Virtually all cells in the human body have apparatuses necessary for ROS generation, and are therefore capable of producing ROS in various amounts. In all animal cells that contain mitochondria, superoxide has been shown to be formed in mitochondria electron transfer chain via univalent reduction of oxygen (Richter et al., 1995). However, mitochondria are capable of scavenging superoxide by virtue of Cu/Zn-SOD and Mn-SOD and the resulting hydrogen peroxide by virtue of CAT and GSH Px and may be equipped with powerful means to eliminate toxic ROS generated inside the mitochondria; otherwise, mitochondria should be damaged by

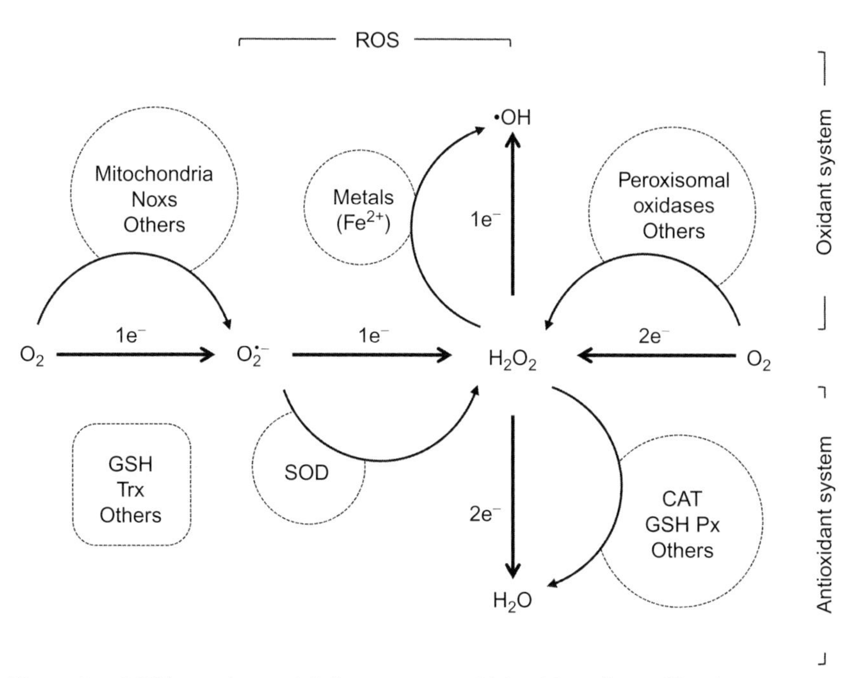

Figure 2.1 ROS formation and defenses against ROS. ROS are formed by the interaction of ionizing radiation with biological molecules, an unavoidable by-product of cellular respiration, a reaction between superoxide and hydrogen peroxide, and enzymes, including Noxs and peroxisomal oxidases. Cells have a variety of defenses against the harmful effects of ROS. These include antioxidant enzymes, such as SOD and CAT, as well as several small antioxidant molecules, such as GSH and Trx.

accumulated toxic ROS. Superoxide is also produced by activated phagocytes as a part of their mechanisms to combat bacteria and other invaders (Katsuyama, Matsuno, & Yabe-Nishimura, 2012). Following stimulation, phagocytes, such as neutrophils and macrophages, undergo a respiratory burst through activation of nicotinamide adenine dinucleotide phosphate (NADPH) oxidase (Katsuyama et al., 2012). This NADPH oxidase (Nox), which is now renamed Nox2, is one of the enzymes which catalyze superoxide production by the one-electron reduction of oxygen using NADPH as the electron donor. The produced superoxide serves as a precursor of microbicidal ROS, including hydrogen peroxide, hydroxyl radical, and hypochlorous acid. It is needless to say that the production of ROS should occur at restricted sites, because they are also toxic, not only to the invading microbes but also to the surrounding tissues. Therefore, Nox2 is expected to be activated exclusively on the phagosomal membrane after formation of the phagosome, a membrane-bound cytoplasmic vesicle within the phagocyte containing the phagocytized materials, such as microbes (Bylund, Brown, Movitz, Dahlgren, & Karlsson, 2010). Additionally, ROS that escape the phagosome may be neutralized by antioxidant enzymes, including SOD and CAT, and nonenzymatic antioxidants located around the phagosome. Because Noxs are activated only when needed, they are considered to be the major source of ROS, and, thus, further discussed later.

Superoxide is not thought to cross membranes appreciably, because of its charge and reactivity, so most of the superoxide produced within an organelle is going to stay there, prior to consumption. The most superoxide is assumed to be converted (by SOD or spontaneously in phagosome) to hydrogen peroxide. There are some enzymes that can also produce hydrogen peroxide directly or indirectly. Although hydrogen peroxide is considered a nonradical oxygen metabolite, it can cause damage to the cell at a relatively low concentration. It is freely dissolved in aqueous solution and can easily penetrate biological membranes. Consequently, it can cause oxidative damage far from the site of formation. Because hydrogen peroxide is the only ROS that can diffuse over larger distances within the cell and is relatively stable compared to other ROS, it has received particular attention as a signaling molecule involved in the regulation of specific biological processes (Muller-Delp, Gurovich, Christou, & Leeuwenburgh, 2012). The deleterious chemical effects of hydrogen peroxide can be divided into the categories of direct activity, originating from its oxidizing properties, and indirect activity, in which it serves as a source for more deleterious species, such as hydroxyl radicals. Direct activities of hydrogen peroxide include degradation of heme proteins; release of iron; inactivation of enzymes; and oxidation of DNA, lipids, thiol groups, and keto acids (Kohen & Nyska, 2002).

2.3. ROS in cell signaling

Besides their roles as destructive agents, ROS may also participate in a diverse array of biological processes, including normal cell growth, induction, and maintenance of the transformed state, apoptosis, and cellular senescence. How can ROS trigger such divergent responses? It is most likely that ROS may act as intracellular messengers or alter the intracellular redox state and/or the protein structure and function by modifying critical amino acid residues. Among ROS, hydrogen peroxide is a small, uncharged, freely diffusible molecule that can be synthesized and destroyed rapidly in response to external stimuli (Muller-Delp et al., 2012). These features meet all the important criteria for an intracellular messenger. Indeed, accumulating evidence supports the notion that hydrogen peroxide is a ubiquitous intracellular messenger (Rhee et al., 2005).

The cytosol is normally maintained under strong reducing conditions, which is accomplished by the redox-buffering capacity of intracellular thiols, primarily GSH and thioredoxin (Trx). The levels of reduced and oxidized forms of GSH and Trx are maintained by the activity of GSH reductase and Trx reductase, respectively. Both these thiol redox systems can reduce intracellular ROS, and thus counteract OS. The evidence that ROS regulate transcription factors, such as nuclear factor-κB and activation protein 1 (AP-1), through the modulation of cellular redox state is already convincing (Sen & Packer, 1996). However, in many cases, it still remains unclear whether the observed alteration in the redox state is the cause or merely the consequence of the subsequent cellular events.

Because ROS can induce oxidation of all molecules in animal cells, proteins that can be sensitively and reversibly oxidized by ROS may be candidates for mediating the signaling function of ROS. Because there are so many mechanisms for induction of protein oxidation and all of the amino acids can become oxidatively modified, there may be numerous different types of protein oxidative modification. One of the best described oxidation-susceptible amino acid residues is the cysteine residue (—SH) (Miki & Funato, 2012). All ROS can induce oxidative modification of cysteine residues. Oxidation of cysteine residues leads to the formation of disulfide bond (—S—S—), sulfenyl moiety (—SOH), sulfinyl moiety (—SO_2H), and sulfonyl moiety (—SO_3H) (Fig. 2.2; Miki & Funato, 2012). Such alterations may alter the activity of an enzyme if the critical cysteine is located within its catalytic domain or the ability of a transcription factor to bind DNA if it is located within its DNA-binding motif. It should be noted that the animal cells possess protein repair pathways to rescue oxidized proteins and restore their functions. If these repair processes fail, oxidized proteins may function abnormally. It is most likely that oxidative protein modifications may be specific and reversible in the cellular systems.

3. MITOGEN-ACTIVATED PROTEIN KINASES

MAPKs are members of a major intracellular signal transduction pathway that has been demonstrated to play an important role in various

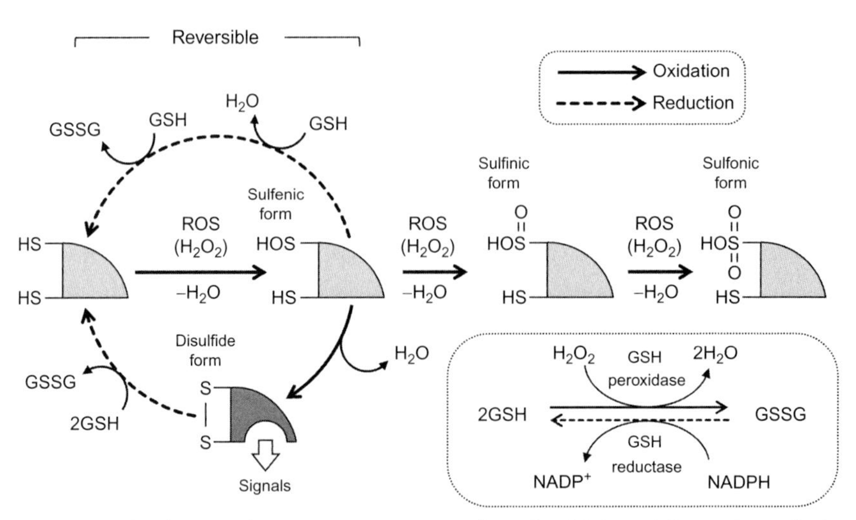

Figure 2.2 Oxidation of protein cysteine residues by ROS and formation of glutathionylated proteins. The cysteine moiety of a protein is oxidized to become the sulfenyl moiety. The sulfenyl moiety can form a disulfide bond with another cysteine moiety. The sulfenyl moiety and disulfide bond can be reduced by various antioxidants in cells, such as GSH; thus, a protein can be sensitively and reversibly oxidized by ROS, and generate cellular signals. The sulfenyl moiety can be further oxidized to become the sulfinyl and sulfonyl moiety, which cannot be reduced under the normal intracellular system.

physiological processes (Plotnikov, Zehorai, Procaccia, & Seger, 2011). The activation of an MAPK employs a core three-kinase cascade consisting of an MAPK kinase kinase (MAP3K or MAPKKK), which phosphorylates and activates an MAPK kinase (MAP2K, MEK, or MKK), which then phosphorylates and increases the activity of one or more MAPKs (Fig. 2.3). In mammalian cells, there are three well-defined subgroups of MAPKs: the extracellular signal-regulated kinases (ERKs, including ERK-1 and ERK-2 isoforms), the c-Jun N-terminal kinases (JNKs, including JNK-1, JNK-2, and JNK-3 isoforms), and the p38 MAPKs (including p38α, p38β, p38γ, and p38δ isoforms). Obviously, the three subgroups of MAPKs (i.e., ERKs, JNKs, and p38 MAPKs) are involved in both cell growth and cell death, and the tight regulation of these pathways, therefore, is paramount in determining cell fate (Plotnikov et al., 2011). The deleterious consequences of sustained activation of MAPK pathways may include excessive production of MAPK-regulated genes, uncontrolled proliferation, and unscheduled cell death. MAPKs can be activated by a wide variety of different stimuli, but in general, ERK-1 and ERK-2 are preferentially activated in

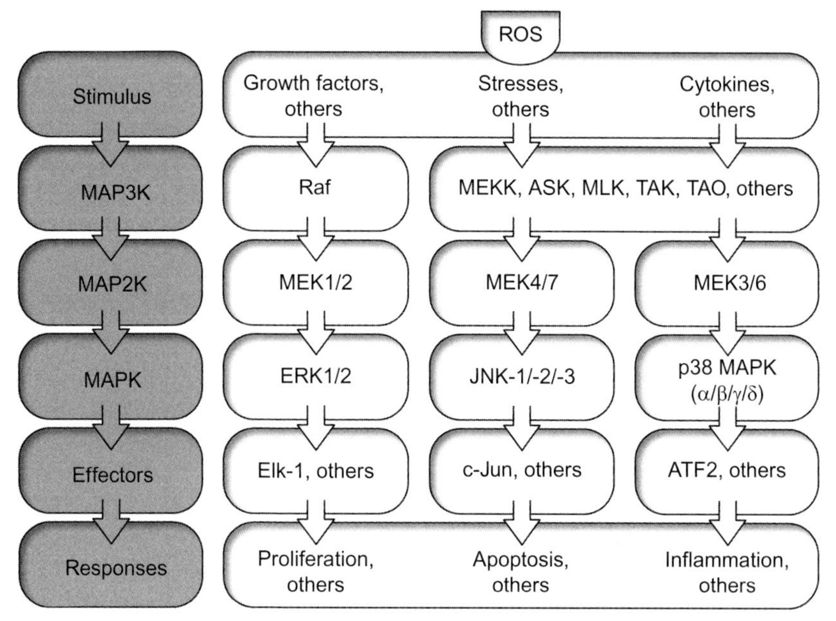

Figure 2.3 MAPK cascades. MAPK signaling pathways mediate intracellular signaling initiated by extracellular or intracellular stimuli. MAP3Ks phosphorylate MAP2Ks, which in turn phosphorylate MAPKs. Activated MAPKs phosphorylate various substrate proteins (e.g., transcription factors), resulting in regulation of various cellular activities (e.g., proliferation, inflammation, and apoptosis). Activation by MAPK signaling cascades is achieved either through a series of binary interactions among the kinase components or through formation of a multiple kinase complex. It should be noted that ROS can activate ERK, JNK, and p38 MAPK pathways.

response to growth factors, while the JNKs and p38 MAPKs are more responsive to stress stimuli, as briefly discussed below.

3.1. Activation of ERK pathway

The ERK pathway is activated mainly by growth factors such as epidermal growth factor (EGF) and platelet-derived growth factor (PDGF; Stadler, 2005). Briefly, ligation of growth factor receptors with their ligands can transmit activating signals to the Raf/MEK/ERK cascade through different isoforms of the small GTP-binding protein Ras (Friday & Adjei, 2008). Ras, a membrane-bound protein, is activated through the exchange of bound GDP to GTP. Activated Ras, then, recruits cytoplasmic Raf (MAP3K) to the cell membrane for activation. Activated Raf binds to and phosphorylates the dual specificity kinase MEK1/2, which, in turn, phosphorylates ERK1/2

(Friday & Adjei, 2008). Activated ERK1/2 can translocate to the nucleus, where it activates several transcription factors (Plotnikov et al., 2011). Activated ERK1/2 can also phosphorylate several cytoplasmic and nuclear kinases (Plotnikov et al., 2011). ROS have been shown to activate growth factor receptors in the absence of the growth factor receptor ligands (Verbon, Post, & Boonstra, 2012). EGF receptor is one of the receptor tyrosine kinases (RTKs) that are most commonly activated through ligand-induced dimerization or oligomerization and generally involved in the regulation of cell proliferation, survival, migration, and differentiation (Verbon et al., 2012). Interestingly, ROS have recently been found to activate the EGF receptor even in the absence of its ligand (León-Buitimea et al., 2012), which is now referred to as "receptor transactivation."

3.2. Activation of p38 MAPK pathway

The p38 MAPKs are described as stress-activated protein kinases because they are primarily activated through extracellular stresses and cytokines, such as tumor necrosis factor-α (TNF-α) and interleukin-1β (IL-1β), and consequently have been extensively studied in the field of inflammation (Yong, Koh, & Moon, 2009). The p38 MAPKs are known to be activated by MEK3 (or MKK3) and MEK6 (or MKK6). MKK6 can phosphorylate the four p38 MAPK family members, while MKK3 phosphorylates p38α, p38γ, and p38δ, but not p38β. They are subject to activation by one of the several upstream MAP3Ks, these being ASK1 (apoptosis signal-regulating kinase 1), DLK1 (dual-leucinezipper-bearing kinase 1), TAK1 (transforming growth factor β-activated kinase 1), TAO (thousand-and-one amino acid) 1 and 2, TPL2 (tumor progression loci 2), MLK3 (mixed lineage kinase 3), MEKK3 (MEK kinase 3) and MEKK4, and ZAK1 (leucine zipper and sterile-α motif kinase 1; Cuadrado & Nebreda, 2010). The diversity of MAP3Ks and their regulatory mechanisms may provide the ability to respond to a wide range of stimuli and to integrate p38 MAPK activation with other signaling pathways. Once activated by MKK3 or MKK6, p38 MAPKs can potentiate the downstream signals or activate other proteins directly (Cuadrado & Nebreda, 2010). It should be noted that some MAP3Ks that trigger p38 MAPK activation can also activate the JNK pathway.

3.3. Activation of JNK pathway

The JNKs, also named stress-activated protein kinases, were discovered originally by their ability to phosphorylate the N-terminal transactivating domain of the transcription factor c-Jun (Davies & Tournier, 2012); however, they are now known to have the ability to phosphorylate other target proteins. Like the p38 MAPKs, the initial signaling pathway starts with the activation of several MAP3Ks, which are the same as those activating p38 MAPK pathway. Two MEK family members, MEK4 (or MKK4) and MEK7 (or MKK7), have been implicated in phosphorylation of JNKs (Davies & Tournier, 2012). A number of different MAP3Ks can activate MKK4 and MKK7, suggesting that a wide range of stimuli can activate this MAPK pathway. These include MEKK1, 2, 3, and 4; MLK; and ASK1. Several transcription factors are downstream proteins activated by JNKs, especially c-Jun (Karin & Gallagher, 2005). This transcription factor is one of members of the AP-1 family of transcription factors (Karin & Gallagher, 2005). Once activated by phosphorylation, c-Jun can form heterodimers or homodimers with other factors and can induce the transcription of many genes (Karin & Gallagher, 2005). Research into the molecular mechanisms of ROS-mediated activation of JNK and p38 MAPK pathways has focused on redox-sensitive proteins, such as Trx and glutaredoxin (Grx; Katagiri, Matsuzawa, & Ichijo, 2010). It is well known that ROS oxidizes Trx to dissociate from ASK-1 for its activation, resulting in the activation of JNK and p38 pathways (Katagiri et al., 2010).

3.4. Inactivation of MAPK pathways by MKPs

The inactivation of MAPK pathways has been shown to be essential for cell physiology for cells to remain responsive to stimuli and to prevent potential deleterious effects of prolonged stimulation of these pathways. As mentioned above, MAPK pathways are activated through their phosphorylation. Thus, the dephosphorylation of MAPKs by phosphatases may be the most efficient mode of negative regulation. A number of protein phosphatases that are known to deactivate MAPKs include dual specificity phosphatases, protein serine/threonine phosphatases, and protein tyrosine phosphatases (Boutros et al., 2008; Patterson, Brummer, O'Brien, & Daly, 2009). A group of dual specificity protein phosphatases that are responsible primarily for dephosphorylation of MAPKs are often referred to as MKPs (Boutros et al., 2008). Since MKPs dephosphorylate MAPKs on their regulatory residues, aberrant regulation of MAPK activity may arise through defective regulation of the MKPs; this may emphasize the importance of balance between the phosphorylating MAPKs and dephosphorylating MKPs in regulating these pathways. The factors that can activate MAPK pathways, such as environmental stresses and growth factor stimulation, can also activate MKP pathways (Boutros et al., 2008; Patterson et al., 2009), supporting the notion that there is a tight and specific control of MAPK activation by MKP activation. In mammalian cells, at least 11 MKP family members have been identified so far: some of them include MKP-1, MKP-2, MKP-3, MKP-4, MKP-5, MKP-7, and MKP-X. According to their subcellular localization, MKPs can be grouped: MKP-1 and MKP-2 are found in the nucleus; MKP-3, MKP-4, and MKP-X are found in the cytoplasm; and MKP-5 and MKP-7 are found in both the nucleus and the cytoplasm (Kondoh & Nishida, 2007; Patterson et al., 2009). These MKPs exhibit distinct biochemical properties with regard to their substrate specificity. MKP-1 and MKP-2 show selectivity for p38 MAPKs and JNKs over ERKs. MKP-3 and MKP-X primarily inactivate ERKs. MKP-5 and MKP-7 show selectivity for JNKs and p38 MAPKs, while MKP-4 inactivates ERKs and p38 MAPKs. MKP-1, the archetype, was initially discovered as a stress-responsive protein phosphatase (Kondoh & Nishida, 2007; Patterson et al., 2009). Since MKP-1 deactivates MAPKs and is robustly induced by stress stimuli that also activate MAPKs, MKP-1 is regarded as an important feedback control mechanism that regulates the MAPK pathways. Compared with other MKPs, MKP-1 has been most closely examined. The activity of MKP-1 may be regulated at multiple levels, including transcriptional induction, protein stabilization, catalytic activation, and acetylation (Boutros et al., 2008). It has been reported that JNK and p38 MAPK pathways are highly activated in MKP-1-deficient mouse embryonic fibroblasts (Wu & Bennett, 2005), supporting that MKP-1 functions as a critical negative regulator during MAPK activation. However, it should be noted that all MKPs may act cooperatively to modulate the MAPK pathways and to orchestrate appropriate cellular responses.

3.5. Cross-talk relationship between the members of MAPK pathways

Until now, for reasons of simplicity and clarity, the axes of the MAPKs have been discussed independently of one another with respect to their signal transduction pathways. However, this is not the case in cellular systems. Relationships between the members of a given MAPK module are not always linear, and there could be both stimulatory and inhibitory interactions within and across MAPK modules. For example, activation of ERK pathway in human alveolar macrophages decreases JNK activity by stabilizing MKP-7 (Monick et al., 2006). Inhibition of p38 MAPK pathway upregulates JNK and ERK activities in M1 cells and in peritoneal macrophages (Hall & Davis, 2002), suggesting that p38 MAPK may regulate JNK and ERK activities. In COS-7 cells, sustained JNK activation blocks ERK activation (Shen et al., 2003), suggesting the existence of a negative cross-talk relationship between the stress-activated JNK pathway and the growth factor-activated ERK pathway. Other members of the MAPKs may also interact with both stimulatory and inhibitory consequences. Overall, it is most likely that the stress-activated JNK and p38 MAPK pathways may inhibit the growth factor-activated ERK pathway.

4. ROLES OF ROS IN MAPK ACTIVATION

Generally, the signaling molecules, when they are needed, can be synthesized and broken down again in specific reactions by enzymes or ion channels, and act to transmit signals from a receptor to a target. For these reasons, ROS produced by Noxs, the membrane-bound multicomponent enzyme complexes that are functionally linked to receptors and present in phagocytes as well as nonphagocytic cells, have been suggested to act as signaling molecules. ROS derived from Noxs, especially hydrogen peroxide, can specifically and reversibly react with proteins, altering their activity, localization, and half-life (Brown & Griendling, 2009). There are seven homologous Nox proteins in animal cells, comprising Nox1–5 and two larger dual oxidases (Maghzal, Krause, Stocker, & Jaquet, 2012). Classic Nox is the phagocytic Nox found in neutrophils. The phagocytic Nox complex consists of the catalytic subunit $gp91^{phox}$ (renamed Nox2) together with the regulatory subunit $p22^{phox}$ which is located in the membrane. The other regulatory components (i.e., $p47^{phox}$, $p40^{phox}$, $p67^{phox}$, and the small GTPase Rac) are normally located in the cytoplasm. Upon stimulation, the cytosolic subunits translocate to the membrane-bound cytochrome complex leading to enzymatic activity (Brown & Griendling, 2009). Among the seven members of the Nox family, Nox1 is structurally and functionally similar to Nox2. All Noxs are able to catalyze the reduction of oxygen to superoxide, but there may be key differences in their activation, subunit composition, localization, and expression (Brown & Griendling, 2009). It has been reported that antisense-mediated suppression of the expression of Mox1, which is now known as Nox1, resulted in the reduction of both ROS production and cell growth, whereas genetic overexpression of Mox1, or Nox1, in NIH3T3 cells increased both ROS production and cell growth (Suh et al., 1999), and this, therefore, strongly supports the notion that ROS derived from Noxs can function as signaling molecules. It is most likely that the signaling molecule action of ROS may depend upon their ability to react with the cysteine residues of a certain group of target proteins. ROS can rapidly oxidize the highly reactive thiol groups to form a disulfide bond. The recovery from this oxidized state back to a fully reduced thiol group is carried out by the Grx and/or the Trx systems, thus suggesting that oxidative protein modifications can be specific and reversible. The oxidized target proteins activate a number of oxidation-sensitive processes that bring about a number of cellular responses, such as gene activation, modulation of ion channels, and the activity of other signaling pathways, including MAPK cascade. In this regard, the well-known roles of ROS as signaling molecules in activation of MAPK pathways are discussed below.

4.1. Roles of ROS in ASK1 activation

Because the JNK and p38 MAPK pathways are generally activated by inflammatory cytokines, such as TNF-α and IL-1β, and a diverse array of cellular stresses, it is not surprising that ROS, as they can induce OS in biological systems, can activate JNK and p38 MAPK pathways in animal cells. Important questions, however, remain concerning how ROS can activate these pathways; is there any signaling intermediate that can sense the redox state and transmit its information to signaling molecules? In other words, is there a redox-sensitive protein that can undergo reversible oxidation/reduction and may switch "on" and "off" depending on the cellular redox state? Most importantly, in the MAPK module, is there any upstream kinase that can sense the ROS signal and translate it to the downstream kinases? An example of the best-characterized redox-sensitive proteins is Trx, which is an antioxidant protein and additionally plays pivotal roles in maintaining intracellular redox balance (Spindel, World, & Berk, 2012). MAP3Ks may also play key roles in redox signaling. Among the MAP3K family, ASK1 has been extensively characterized as an ROS-responsive kinase (Fujisawa, Takeda, & Ichijo, 2007). ASK1 is a MAP3K of the MKK4/MKK7-JNK and MKK3/MKK6-p38 MAPK signaling cascades (Fujisawa et al., 2007). ASK1 possesses a serine/threonine kinase domain in the middle part of the molecule flanked with the N- and C-terminal coiled-coil (CCC) domains. ASK1 is activated by various types of stress, such as ROS, TNF-α, and lipopolysaccharide, among which ROS is the most potent activator of ASK1 (Matsuzawa & Ichijo, 2008). Trx has been identified as a binding protein of ASK1 (Matsuzawa & Ichijo, 2008). Briefly, Trx inhibits the kinase activity of ASK1 by its direct binding to the N-terminal region of ASK1. Trx has a redox-active site, in which two cysteine residues produce the sulfhydryl groups associated with Trx-dependent reducing activity. Whereas the reduced form binds to ASK1 in inactivated cells, ROS, such as hydrogen peroxide, convert Trx to the oxidized form and dissociates Trx from ASK1 (Fujisawa et al., 2007; Matsuzawa & Ichijo, 2008). ASK1, thereby, is activated, and induces the phosphorylation of a critical threonine residue within the kinase domain of ASK1. Because Trx–ASK1 complex functions as a signaling complex competent to ROS-dependent activation of ASK1, it is now designated as "ASK1 signalosome." This signalosome has two important domains for Trx-dependent regulation of ASK1: the CCC domain and N-terminal coiled-coil domain (NCC; Fujino et al., 2007). Trx constitutively disrupts N-terminal homophilic interaction through the NCC domain of ASK1 in inactivated cells. Upon ROS-dependent dissociation of Trx from ASK1, freed ASK1 has been shown to be tightly oligomerized through its NCC domains in addition to the basal interaction through the CCC domain and thereby is activated. In ASK1-deficient cells, ROS-induced sustained activations of JNK and p38 MAPK pathways were dramatically diminished (Nakagawa et al., 2008), suggesting that ASK1 is one of the upstream kinases that can sense the ROS signal and translate it to the downstream kinases.

TNF receptor-associated factor 2 (TRAF2) and TRAF6 were reported to be recruited to the ASK1 signalosome and required for the activation of ASK1 (Fujino et al., 2007). TRAF2 activates the JNK pathway through the association with MAP3Ks, including ASK1. TRAF6 is also a critical regulator of MAPK pathways through the TNF receptor superfamily and Toll/IL-1 receptor family. TNF-α-induced association of ASK1 with TRAF2 and subsequent activation of ASK1 have been shown to depend largely on intracellular ROS production by TNF-α (Noguchi et al., 2005). Consistently, TRAF2 was recruited to the ASK1 signalosome in response not only to TNF-α but also to ROS. Furthermore, TRAF6 was also recruited to the activated ASK1 signalosome upon ROS stimulation (Fujino et al., 2007). TRAF2 and TRAF6 have been shown to be recruited to ASK1 to form the activated ASK1 signalosome, following the ROS-induced dissociation of Trx from the ASK1 signalosome. Interestingly, TNF-α-induced sustained activations of JNK and p38 MAPK pathways were lost in ASK1-deficient cells (Tobiume et al., 2001), suggesting that

TNF-α activates the ROS-dependent sustained JNK and p38 MAPK pathways through ASK1.

4.2. Roles of ROS in transactivation of growth factor receptors

Generally, ERK pathway is activated by growth factors and survival factors, whereas JNK and p38 MAPK pathways are activated by stress stimuli (Plotnikov et al., 2011). Considering their ability to induce OS in cells, ROS have been shown to activate JNK and p38 MAPK pathways rather than the ERK pathway. Nevertheless, a number of studies have demonstrated the ability of ROS to activate the ERK pathway in animal cells (Mehdi, Azar, & Srivastava, 2007). The mechanism(s) for this effect is unclear, and the precise molecular target(s) is unknown. It is most likely that ROS-mediated ERK activation may be an upstream event at the level of growth factor receptors. In fact, many growth factors have cysteine-rich motifs, and they, if not all, may be targets of ROS. There is evidence indicating that accumulation of hydrogen peroxide preceded by depletion of GSH are sequential events leading to phosphorylation of EGF receptor (Meves, Stock, Beyerle, Pittelkow, & Peus, 2001). Similarly, ROS also induce the ligand-independent activation of PDGF receptor (Eyries, Collins, & Khachigian, 2004). Phosphorylation of EGF and PDGF receptors can induce a subsequent activation of the Raf/MEK/ERK signaling pathways (Friday & Adjei, 2008). A study has determined the contribution of EGF receptor to activation of ERK pathway by insulin-like growth factor-I (IGF-I) in vascular smooth muscle cells (Meng, Shi, Jiang, & Fang, 2007). This study showed that IGF-I induced phosphorylation of EGF receptor and ERK activation. AG1478, an EGF receptor inhibitor, inhibited IGF-I-induced phosphorylation of EGF receptor and ERK activation, suggesting that activation of ERK pathway results from EGF receptor activation. IGF-I stimulated ROS production and antioxidants inhibited IGF-I-induced ROS generation and activation of both EGF receptor and ERK, indicating that IGF-I activates ERK pathway through ROS-mediated activation of EGF receptor. It has been reported that ROS generation is essential for EGF receptor transactivation in the endothelin-1 (ET-1)-mediated signaling pathway (Cheng et al., 2003). The increase of ROS specifically inhibited the activity of Src homology 2-containing tyrosine phosphatase (SHP-2) in the early period of ET-1 treatment to facilitate the transient increase of phosphorylation of EGF receptor. After EGF receptor signaling pathway, ET-1 induced the phosphorylation of ERK to promote the proliferation of rat cardiac fibroblasts. In other words, ROS generation is involved in EGF receptor transactivation through the transient oxidization of SHP-2 in ET-1-triggered mitogenic signaling pathway.

4.3. Roles of ROS in expression of MKPs

As mentioned above, the function of MKPs is to dephosphorylate and, therefore, inactivate ERKs, p38 MAPKs, and JNKs. Because MKPs are robustly induced by stress stimuli that can also activate MAPK pathways, it is reasonable to assume that any changes in the protein level of MKPs could lead to inactivation or activation of MAPK pathways, resulting in an increase or decrease in the cellular responses. Indeed, a study has demonstrated that hydrogen peroxide induces MKP-1 expression and activates MAPK pathways, and MKP-1 expression correlates with inactivation of MAPK pathways (Zhou, Liu, & Wu, 2006). Importantly, this study showed that overexpression of MKP-1 renders MCF-7 cells resistant to hydrogen peroxide-induced cell death by inhibiting p38 MAPK and JNK activation, while loss of MKP-1 by downregulation via small interfering RNA (siRNA) or deletion of *mkp-1* gene sensitizes cells to hydrogen peroxide-induced cell death. Another study has also demonstrated that Nox4 attenuates ERK pathway by controlling MKP-1 expression in preadipocytes (Schröder, Wandzioch, Helmcke, & Brandes, 2009). This study showed that whereas downregulation of Nox4 attenuated MKP-1 expression, overexpression of the Nox increased it. Consequently, the basal and growth factor-stimulated phosphorylation or ERK-1/2 was enhanced when Nox4 was downregulated by siRNA, whereas it was attenuated when Nox4 was overexpressed.

Because ROS can alter protein structure and function by oxidizing critical amino acid residues of proteins, it is also reasonable to assume that there could be a catabolic dysfunction of MKPs when exposed to nontoxic doses of ROS. It has been demonstrated that treatment of fibroblasts with TNF-α induces the intracellular accumulation of hydrogen peroxide and inactivates MKPs by oxidation of their catabolic cysteine residue (Kamata et al., 2005). This oxidation leads to sustained activation of JNK and p38 MAPK pathways. It has been also demonstrated that ROS-induced MKP inactivation causes sustained activation of JNK pathway (Hou, Torii, Saito, Hosaka, & Takeuchi, 2008).

It may be difficult to determine if MKPs could be activated or inactivated by ROS in a given cell. Interestingly, a study has investigated the regulation of MKP-1 expression and JNK activation by the induction of light damage that is known to enhance ROS production in ARPE-19 cells (Lornejad-Schäfer, Schäfer, Schöffl, & Frank, 2009). In this study, low light doses upregulated MKP-1 expression in ARPE-19 cells, this being accompanied by inactivation of the JNK pathway. High light doses, however, led to a decrease in the expression of MKP-1, resulting in sustained activation of the JNK pathway. Hence, the paradox in the roles of ROS as "inducers" in MKP expression and as "inhibitors" may be, at least in part, related to differences in the concentrations of ROS.

5. SUMMARY

The old concept regarding the roles of ROS has been shown to be that oxidative modification of proteins represents a detrimental process in which the modified proteins were irreversibly inactivated, leading to cellular dysfunction. The new concept, however, has been shown to be that oxidative protein modifications can be specific and reversible and, thus, may play a key role in normal cellular physiology. The evidence supporting that ROS can

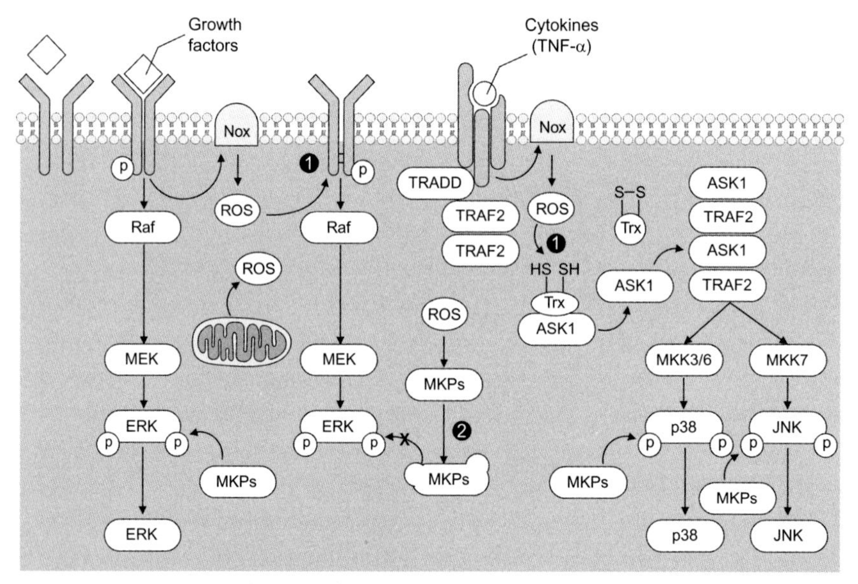

Figure 2.4 Putative mechanisms for ROS-mediated activation of MAPK pathways. ROS are generated from Noxs, which are activated by growth factors, cytokines, and various stresses, and rapidly removed by intracellular antioxidants. ROS, once ROS production exceeds the capacity of the antioxidants, may induce oxidative modification of MAPK signaling proteins, including RTKs and MAP3Ks (see the pathway 1), thereby leading to MAPK activation. ROS may activate MAPK pathways via inhibition and/or degradation of MKPs (see the pathway 2).

activate MAPK pathways at cellular levels is based largely on the following findings: (1) cellular stimuli that are capable of producing ROS can also activate MAPK pathways in a number of different cell types; (2) antioxidants and inhibitors of ROS-producing enzymatic systems block MAPK activation; and (3) exogenous addition of hydrogen peroxide, one of ROS, activates MAPK pathways. As illustrated in Fig. 2.4, the putative mechanisms by which ROS, on the basis of their oxidation potentials, can activate MAPK pathways may include (1) oxidative modifications of MAPK signaling proteins (e.g., RTKs and MAP3Ks) and (2) inactivation of MKPs. Finally, the site of ROS production and the concentration and kinetics of ROS production as well as cellular antioxidant pools and redox state are most likely to be important factors in determining the effects of ROS on activation of MAPK pathways.

ACKNOWLEDGMENT

This chapter was supported by Wonkwang University in 2011.

REFERENCES

Birben, E., Sahiner, U. M., Sackesen, C., Erzurum, S., & Kalayci, O. (2012). Oxidative stress and antioxidant defense. *The World Allergy Organization Journal*, *5*(1), 9–19.

Boutros, T., Chevet, E., & Metrakos, P. (2008). Mitogen-activated protein (MAP) kinase/MAP kinase phosphatase regulation: Roles in cell growth, death, and cancer. *Pharmacological Reviews*, *60*(3), 261–310.

Brown, D. I., & Griendling, K. K. (2009). Nox proteins in signal transduction. *Free Radical Biology & Medicine*, *47*(9), 1239–1253.

Bylund, J., Brown, K. L., Movitz, C., Dahlgren, C., & Karlsson, A. (2010). Intracellular generation of superoxide by the phagocyte NADPH oxidase: How, where, and what for? *Free Radical Biology & Medicine*, *49*(12), 1834–1845.

Chang, Y. C., & Chuang, L. M. (2010). The role of oxidative stress in the pathogenesis of type 2 diabetes: From molecular mechanism to clinical implication. *American Journal of Translational Research*, *2*(3), 316–331.

Cheng, C. M., Hong, H. J., Liu, J. C., Shih, N. L., Juan, S. H., Loh, S. H., et al. (2003). Crucial role of extracellular signal-regulated kinase pathway in reactive oxygen species-mediated endothelin-1 gene expression induced by endothelin-1 in rat cardiac fibroblasts. *Molecular Pharmacology*, *63*(5), 1002–1011.

Cuadrado, A., & Nebreda, A. R. (2010). Mechanisms and functions of p38 MAPK signalling. *The Biochemical Journal*, *429*(3), 403–417.

Dabrowski, A., Boguslowicz, C., Dabrowska, M., Tribillo, I., & Gabryelewicz, A. (2000). Reactive oxygen species activate mitogen-activated protein kinases in pancreatic acinar cells. *Pancreas*, *21*(4), 376–384.

Davies, C., & Tournier, C. (2012). Exploring the function of the JNK (c-Jun *N*-terminal kinase) signalling pathway in physiological and pathological processes to design novel therapeutic strategies. *Biochemical Society Transactions*, *40*(1), 85–89.

Eyries, M., Collins, T., & Khachigian, L. M. (2004). Modulation of growth factor gene expression in vascular cells by oxidative stress. *Endothelium*, *11*(2), 133–139.

Friday, B. B., & Adjei, A. A. (2008). Advances in targeting the Ras/Raf/MEK/Erk mitogen-activated protein kinase cascade with MEK inhibitors for cancer therapy. *Clinical Cancer Research*, *14*(2), 342–346.

Fujino, G., Noguchi, T., Matsuzawa, A., Yamauchi, S., Saitoh, M., Takeda, K., et al. (2007). Thioredoxin and TRAF family proteins regulate reactive oxygen species-dependent activation of ASK1 through reciprocal modulation of the N-terminal homophilic interaction of ASK1. *Molecular and Cellular Biology, 27*(23), 8152–8163.

Fujisawa, T., Takeda, K., & Ichijo, H. (2007). ASK family proteins in stress response and disease. *Molecular Biotechnology, 37*(1), 13–18.

Hall, J. P., & Davis, R. J. (2002). Inhibition of the p38 pathway upregulates macrophage JNK and ERK activities, and the ERK, JNK, and p38 MAP kinase pathways are reprogrammed during differentiation of the murine myeloid M1 cell line. *Journal of Cellular Biochemistry, 86*(1), 1–11.

Hensley, K., Robinson, K. A., Gabbita, S. P., Salsman, S., & Floyd, R. A. (2000). Reactive oxygen species, cell signaling, and cell injury. *Free Radical Biology & Medicine, 28*(10), 1456–1462.

Hou, N., Torii, S., Saito, N., Hosaka, M., & Takeuchi, T. (2008). Reactive oxygen species-mediated pancreatic beta-cell death is regulated by interactions between stress-activated protein kinases, p38 and c-Jun *N*-terminal kinase, and mitogen-activated protein kinase phosphatases. *Endocrinology, 149*(4), 1654–1665.

Kamata, H., Honda, S., Maeda, S., Chang, L., Hirata, H., & Karin, M. (2005). Reactive oxygen species promote TNFalpha-induced death and sustained JNK activation by inhibiting MAP kinase phosphatases. *Cell, 120*(5), 649–661.

Karin, M., & Gallagher, E. (2005). From JNK to pay dirt: Jun kinases, their biochemistry, physiology and clinical importance. *IUBMB Life, 57*(4–5), 283–295.

Katagiri, K., Matsuzawa, A., & Ichijo, H. (2010). Regulation of apoptosis signal-regulating kinase 1 in redox signaling. *Methods in Enzymology, 474*, 277–288.

Katsuyama, M., Matsuno, K., & Yabe-Nishimura, C. (2012). Physiological roles of NOX/NADPH oxidase, the superoxide-generating enzyme. *Journal of Clinical Biochemistry and Nutrition, 50*(1), 9–22.

Keshet, Y., & Seger, R. (2010). The MAP kinase signaling cascades: A system of hundreds of components regulates a diverse array of physiological functions. *Methods in Molecular Biology, 661*, 3–38.

Kohen, R., & Nyska, A. (2002). Oxidation of biological systems: Oxidative stress phenomena, antioxidants, redox reactions, and methods for their quantification. *Toxicologic Pathology, 30*(6), 620–650.

Kondoh, K., & Nishida, E. (2007). Regulation of MAP kinases by MAP kinase phosphatases. *Biochimica et Biophysica Acta, 1773*(8), 1227–1237.

Kyriakis, J. M., & Avruch, J. (2012). Mammalian MAPK signal transduction pathways activated by stress and inflammation: A 10-year update. *Physiological Reviews, 92*(2), 689–737.

León-Buitimea, A., Rodríguez-Fragoso, L., Lauer, F. T., Bowles, H., Thompson, T. A., & Burchiel, S. W. (2012). Ethanol-induced oxidative stress is associated with EGF receptor phosphorylation in MCF-10A cells overexpressing CYP2E1. *Toxicology Letters, 209*(2), 161–165.

Lornejad-Schäfer, M. R., Schäfer, C., Schöffl, H., & Frank, J. (2009). Cytoprotective role of mitogen-activated protein kinase phosphatase-1 in light-damaged human retinal pigment epithelial cells. *Photochemistry and Photobiology, 85*(3), 834–842.

Maghzal, G. J., Krause, K. H., Stocker, R., & Jaquet, V. (2012). Detection of reactive oxygen species derived from the family of NOX NADPH oxidases. *Free Radical Biology & Medicine, 53*(10), 1903–1918.

Matsuzawa, A., & Ichijo, H. (2008). Redox control of cell fate by MAP kinase: Physiological roles of ASK1-MAP kinase pathway in stress signaling. *Biochimica et Biophysica Acta, 1780*(11), 1325–1336.

McCubrey, J. A., Lahair, M. M., & Franklin, R. A. (2006). Reactive oxygen species-induced activation of the MAP kinase signaling pathways. *Antioxidants & Redox Signaling, 8*(9–10), 1775–1789.

Mehdi, M. Z., Azar, Z. M., & Srivastava, A. K. (2007). Role of receptor and nonreceptor protein tyrosine kinases in H_2O_2-induced PKB and ERK1/2 signaling. *Cell Biochemistry and Biophysics*, *47*(1), 1–10.

Meng, D., Shi, X., Jiang, B. H., & Fang, J. (2007). Insulin-like growth factor-I (IGF-I) induces epidermal growth factor receptor transactivation and cell proliferation through reactive oxygen species. *Free Radical Biology & Medicine*, *42*(11), 1651–1660.

Meves, A., Stock, S. N., Beyerle, A., Pittelkow, M. R., & Peus, D. (2001). H_2O_2 mediates oxidative stress-induced epidermal growth factor receptor phosphorylation. *Toxicology Letters*, *122*(3), 205–214.

Miki, H., & Funato, Y. (2012). Regulation of intracellular signalling through cysteine oxidation by reactive oxygen species. *Journal of Biochemistry*, *151*(3), 255–261.

Monick, M. M., Powers, L. S., Gross, T. J., Flaherty, D. M., Barrett, C. W., & Hunninghake, G. W. (2006). Active ERK contributes to protein translation by preventing JNK-dependent inhibition of protein phosphatase 1. *Journal of Immunology*, *177*(3), 1636–1645.

Muller-Delp, J. M., Gurovich, A. N., Christou, D. D., & Leeuwenburgh, C. (2012). Redox balance in the aging microcirculation: New friends, new foes, and new clinical directions. *Microcirculation*, *19*(1), 19–28.

Nakagawa, H., Maeda, S., Hikiba, Y., Ohmae, T., Shibata, W., Yanai, A., et al. (2008). Deletion of apoptosis signal-regulating kinase 1 attenuates acetaminophen-induced liver injury by inhibiting c-Jun N-terminal kinase activation. *Gastroenterology*, *135*(4), 1311–1321.

Noguchi, T., Takeda, K., Matsuzawa, A., Saegusa, K., Nakano, H., Gohda, J., et al. (2005). Recruitment of tumor necrosis factor receptor-associated factor family proteins to apoptosis signal-regulating kinase 1 signalosome is essential for oxidative stress-induced cell death. *The Journal of Biological Chemistry*, *280*(44), 37033–37040.

Patterson, K. I., Brummer, T., O'Brien, P. M., & Daly, R. J. (2009). Dual-specificity phosphatases: Critical regulators with diverse cellular targets. *The Biochemical Journal*, *418*(3), 475–489.

Plotnikov, A., Zehorai, E., Procaccia, S., & Seger, R. (2011). The MAPK cascades: Signaling components, nuclear roles and mechanisms of nuclear translocation. *Biochimica et Biophysica Acta*, *1813*(9), 1619–1633.

Ravingerová, T., Barancík, M., & Strnisková, M. (2003). Mitogen-activated protein kinases: A new therapeutic target in cardiac pathology. *Molecular and Cellular Biochemistry*, *247*(1–2), 127–138.

Rhee, S. G., Kang, S. W., Jeong, W., Chang, T. S., Yang, K. S., & Woo, H. A. (2005). Intracellular messenger function of hydrogen peroxide and its regulation by peroxiredoxins. *Current Opinion in Cell Biology*, *17*(2), 183–189.

Richter, C., Gogvadze, V., Laffranchi, R., Schlapbach, R., Schwezer, M., Suter, M., et al. (1995). Oxidants in mitochondria: From physiology to diseases. *Biochimica et Biophysica Acta*, *1271*(1), 67–74.

Ruffels, J., Griffin, M., & Dickenson, J. M. (2004). Activation of ERK1/2, JNK and PKB by hydrogen peroxide in human SHSY5Y neuroblastoma cells: Role of ERK1/2 in H_2O_2-induced cell death. *European Journal of Pharmacology*, *483*(2–3), 163–173.

Schröder, K., Wandzioch, K., Helmcke, I., & Brandes, R. P. (2009). Nox4 acts as a switch between differentiation and proliferation in pre-adipocytes. *Arteriosclerosis, Thrombosis, and Vascular Biology*, *29*(2), 239–245.

Sen, C. K., & Packer, L. (1996). Antioxidant and redox regulation of gene transcription. *The FASEB Journal*, *10*(7), 709–720.

Shen, Y. H., Godlewski, J., Zhu, J., Sathyanarayana, P., Leaner, V., Birrer, M. J., et al. (2003). Cross-talk between JNK/SAPK and ERK/MAPK pathways: Sustained activation of JNK blocks ERK activation by mitogenic factors. *The Journal of Biological Chemistry*, *278*(29), 26715–26721.

Spindel, O. N., World, C., & Berk, B. C. (2012). Thioredoxin interacting protein: Redox dependent and independent regulatory mechanisms. *Antioxidants & Redox Signaling, 16*(6), 587–596.

Stadler, W. M. (2005). Targeted agents for the treatment of advanced renal cell carcinoma. *Cancer, 104*(11), 2323–2333.

Suh, Y. A., Arnold, R. S., Lassegue, B., Shi, J., Xu, X., Sorescu, D., et al. (1999). Cell transformation by the superoxide-generating oxidase Mox1. *Nature, 401*(6748), 79–82.

Tobiume, K., Matsuzawa, A., Takahashi, T., Nishitoh, H., Morita, K., Takeda, K., et al. (2001). ASK1 is required for sustained activations of JNK/p38 MAP kinases and apoptosis. *EMBO Reports, 2*(3), 222–228.

Torres, M., & Forman, H. J. (2003). Redox signaling and the MAP kinase pathways. *BioFactors, 17*(1–4), 287–296.

Valko, M., Leibfritz, D., Moncol, J., Cronin, M. T. D., Mazur, M., & Telser, J. (2007). Free radicals and antioxidants in normal physiological functions and human disease. *The International Journal of Biochemistry & Cell Biology, 39*(1), 44–84.

Verbon, E. H., Post, J. A., & Boonstra, J. (2012). The influence of reactive oxygen species on cell cycle progression in mammalian cells. *Gene, 511*(1), 1–6.

Wu, J. J., & Bennett, A. M. (2005). Essential role for mitogen-activated protein (MAP) kinase phosphatase-1 in stress-responsive MAP kinase and cell survival signaling. *The Journal of Biological Chemistry, 280*(16), 16461–16466.

Yong, H. Y., Koh, M. S., & Moon, A. (2009). The p38 MAPK inhibitors for the treatment of inflammatory diseases and cancer. *Expert Opinion on Investigational Drugs, 18*(12), 1893–1905.

Zhou, J. Y., Liu, Y., & Wu, G. S. (2006). The role of mitogen-activated protein kinase phosphatase-1 in oxidative damage-induced cell death. *Cancer Research, 66*(9), 4888–4894.

CHAPTER THREE

Hydrogen Peroxide Signaling Mediator in the Activation of p38 MAPK in Vascular Endothelial Cells

Rosa Bretón-Romero, Santiago Lamas[1]

Centro de Biología Molecular "Severo Ochoa" (CSIC-UAM), Madrid, Spain

[1]Corresponding author: e-mail address: slamas@cbm.uam.es

Contents

1. Introduction 50
2. Materials and Methods 51
   2.1 Cell culture 51
   2.2 Flow experiments 52
   2.3 ROS detection. Hydrogen peroxide and superoxide anion measurements 52
   2.4 Enzymatic generation of hydrogen peroxide fluxes. Hydrogen peroxide and glucose/glucose oxidase treatments 54
   2.5 Analysis of •NO production 55
   2.6 Western blot analysis 56
   2.7 Small interfering RNA 57

References 58

Abstract

Substantial evidence suggests that a transient increase of hydrogen peroxide (H_2O_2) behaves as an intracellular messenger able to trigger the activation of different signaling pathways. These include phosphatases, protein kinases, and transcription factors among others; however, most of the studies have been performed using supraphysiological levels of H_2O_2. Reactive oxygen species (ROS) generation occurs under physiological conditions and different extracellular stimuli including cytokines, growth factors, and shear stress are able to produce both low levels of superoxide anion and H_2O_2. Here, we explore the redox-dependent activation of key signaling pathways induced by shear stress. We demonstrate that laminar shear stress (LSS) rapidly promotes a transient generation of H_2O_2 that is necessary for the activation of the stress-activated protein kinase p38 MAPK. We describe p38 MAPK as an early redox sensor in LSS. Our studies show that it is essential for the activation of endothelial nitric oxide synthase, the subsequent nitric oxide generation, and the protection of endothelial function.

Methods in Enzymology, Volume 528
ISSN 0076-6879
http://dx.doi.org/10.1016/B978-0-12-405881-1.00003-3

1. INTRODUCTION

For many years, reactive oxygen species (ROS) were described as unwanted toxic products of cellular metabolism (Rhee, 2006). However, substantial evidence in the past decades has proved that ROS are important signaling molecules (Ray, Huang, & Tsuji, 2012; Rhee, Chang, Bae, Lee, & Kang, 2003). Different redox-active species have distinct biological properties, which include reactivity, half-life, and lipid solubility (D'Autreaux & Toledano, 2007). Among them, only H_2O_2 has in fact been proposed as a physiological second messenger (Rhee et al., 2003). The intrinsic nature of H_2O_2, a small, diffusible, and ubiquitous molecule able to reach and react with different cellular subcomponents, makes it an excellent candidate for acting in intracellular signaling in different cells and tissues (D'Autreaux & Toledano, 2007; Rhee, 1999; Rhee, Bae, Lee, & Kwon, 2000; Rhee et al., 2003). The addition of exogenous H_2O_2 or its intracellular production affects the function of several proteins such as protein kinases, protein phosphatases, transcription factors, phospholipases, ion channels, and G-proteins (Rhee et al., 2000). Intracellular production of H_2O_2 occurs under physiological as well as pathophysiological conditions related to the maintenance of steady vessel wall conditions, and the vascular response to fluid flow. Laminar shear stress (LSS) is associated with the generation of ROS and redox-induced signaling responses (Lehoux, 2006). It can activate several intracellular pathways simultaneously, and the great majority of them converge into MAP kinases cascades, suggesting their role in mechanotransduction (Li, Haga, & Chien, 2005). Activation of p38 MAPK has traditionally been associated with stress response, being also known as stress-activated protein kinase. In fact, ROS have been shown to activate p38 MAPK not only by supraphysiological levels of exogenous H_2O_2 (Ushio-Fukai, Alexander, Akers, & Griendling, 1998) but also by ligands able to elicit intracellular H_2O_2 such as angiotensin II (Ushio-Fukai et al., 1998; Zafari et al., 1998) or thrombin (Kanda, Nishio, Kuroki, Mizuno, & Watanabe, 2001). We have previously described that LSS-induced H_2O_2 increase accounts for the activation of the stress-activated MAP kinase p38, and also highlighted the functional relevance of this redox activation in the atheroprotective role of LSS due to the production of the vasoactive molecule nitric oxide ($^{\bullet}$NO Breton-Romero et al., 2012). We now describe in detail the most important technical and methodological approaches that have allowed us to identify p38 MAPK as an early sensor of the redox state in the vascular response to fluid flow.

2. MATERIALS AND METHODS

2.1. Cell culture

Note: All cells were maintained in a standard humidified incubator (37 °C, 5% CO_2).

Note: Studies were performed after maintaining cells under starvation overnight.

Note: All solutions and equipment in contact with cells must be sterile.

2.1.1 Primary cells

Bovine aortic endothelial cells (BAECs): Aortas were obtained from an authorized slaughterhouse and transported to the laboratory in cold phosphate-buffered saline (PBS) supplemented with antibiotics (100 μg/ml penicillin, 100 U/ml streptomycin, and 50 μg/ml Fungizone). Aortas were cleaned from adipose and connective tissues, and stumps of intercostal vessels were tightly closed. After cleaning aortas with PBS, a sterile solution of 0.03% collagenase (Sigma) in Hank's Balanced Salt Solution (HBSS; Gibco) was dripped onto the luminal surface of the aorta and incubated at 37 °C for 15 min. Aortas were washed with RPMI medium supplemented with 10% fetal bovine serum (FBS; Gibco), and the digest was centrifugated, resuspended in the same medium, and grown on plates coated with 0.2% gelatin (Sigma) in PBS. BAECs were utilized from passage 3–7 and maintained in medium RPMI (Gibco), supplemented with 10% FBS (Gibco) and 1% penicillin–streptomycin (Gibco).

Mouse lung endothelial cells (MLECs): MLECs were isolated from lungs of wild type and null mice of 3–4 weeks old. Animals were sacrificed by cervical dislocation, and lungs were collected in ice-cold HBSS (Gibco). Lung tissue was minced and digested for 1 h at 37 °C in 0.1% collagenase type I (Sigma). The digest was passed through a blunt 19.5-gauge needle and filtered through a 70-μm cell strainer. Cells were plated onto coated plates with MLEC medium containing 50% DMEM (Gibco), 50% F-12 (Invitrogen), 20% FBS, 10 mg/ml Heparin (Sigma), 2 m*M* L-glutamine (Gibco), 1% penicillin–streptomycin, 50 μg/ml endothelial cell growth factor. Cells were subjected to negative and positive sorting with magnetic beads coated with CD16/CD32 or ICAM-2, to obtain a pure population of endothelial cells. MLECs were characterized by their morphology as polygonal cells forming a tight-fitting monolayer as well as by positive immunofluorescent staining with an antibody against factor VIII (von Willebrand factor). MLECs were used from passage 2–4.

Human umbilical vein endothelial cells (HUVECs): The umbilical vein was carefully dilated with a cannula and one end of the cord was clamped with a hemostat. Using a syringe, a solution of 0.1% collagenase type I (Sigma) in HBSS (Gibco) was flushed through the vein. The cord was clamped on the other end with a second hemostat, and collagenase solution was spread by inverting the cord two or three times. After incubation for 30 min at 37 °C, one end of the cord was opened and the digest effluent from the vein was collected into complete medium supplemented with serum. Detached endothelial cells were centrifuged, resuspended in Endothelial Basal Medium-2 (Lonza) supplemented with endothelial growth medium 2 (EGM-2), and grown on plates coated with 0.2% gelatin (Sigma) in PBS. HUVECs were used from passage 2–5.

2.2. Flow experiments

Shear stress experiments were performed using the well-characterized cone-and-plate viscometer model as it was first described and applied to biological systems in the early 1980s (Dewey, Bussolari, Gimbrone, & Davies, 1981). Endothelial cells were grown to confluence monolayers and were exposed to physiological levels of steady LSS (12 dyn/cm^2) along various periods to investigate the effects of shear stress on the endothelium. Briefly, the cone-and-plate system is essentially a rotational viscometer. It consists of a stationary plate (60-mm tissue culture dish) coated with gelatin (0.2% in PBS), and a rotating inverted cone in apposition to this plate but without direct contact. The cone is usually designed with an angle of less than 4°. Endothelial cells were grown to confluence and stimulated by rotating the cone unidirectionally inside the culture medium to produce steady LSS.

2.3. ROS detection. Hydrogen peroxide and superoxide anion measurements

2.3.1 Superoxide radical anion detection ($O_2^{\bullet-}$)

After exposure to either static or laminar flow conditions, endothelial cells were incubated for 20 min at 37 °C in 5% CO_2 humidified air with 5 μ*M* dihydroethidium (Invitrogen). After incubation, cells were trypsinized, centrifuged, and resuspended in PBS for flow cytometry measurement. Superoxide radical anion production was analyzed by flow cytometry using an FL3 channel. *Aconitase* is a member of [4Fe–4S] containing (de)hydratases particularly susceptible to become inactivated by $O_2^{\bullet-}$. To determine aconitase activity, cells were grown in 60-mm diameter culture dishes and exposed to different periods of LSS. After treatment, cells were rinsed with

PBS and scraped in ice-cold PBS. Cells were centrifuged for 5 min at 1500 rpm and resuspended in lysis buffer (50 m*M* Tris–HCl, 2 m*M* sodium citrate and 0.6 m*M* $MnCl_2$, pH 7.4). Lysates were snap-frozen at −80 °C processed after defrost and centrifuged for 20 s at 14,000 rpm. Between 30 and 50 μg were diluted in a buffer made of 50 n*M* Tris–HCl, 5 m*M* sodium citrate, 0.6 m*M* $MnCl_2$, 0.2 m*M* NADPH+, and 2 U isocitrate dehydrogenase. Aconitase activity was measured by following NADPH formation spectrophotometrically at 340 n*M*.

2.3.2 Hydrogen peroxide detection (H_2O_2)

Endothelial cells were transfected with the cytosol-targeted *HyPer* vector (HyPer-cyto, Evrogen) using opti-MEM and lipofectamine 2000 (Invitrogen).

Exposure to LSS was performed 48 h after transfection and fluorescence was measured using an Axiovert 200 (Zeiss) microscope. Image intensities were quantified using Metamorph and ImageJ software (Fig. 3.1).

Intracellular hydrogen peroxide production was also measured via extracellular leakage of H_2O_2 from endothelial cells, using peroxide-linked Amplex Red fluorescence assay. Cells were exposed to either static or shear stress conditions, and after treatment, their supernatants were incubated with 5 μ*M* Amplex red reagent (Molecular Probes) and 0.1 U/ml horseradish peroxidase type II (HRP) for 30 min at 37 °C. Supernatant fraction fluorescences were detected via plate reader at excitation and emission wavelenghts of 571 and 585 nm, respectively. H_2O_2 production was estimated using a standard curve (Fig. 3.2).

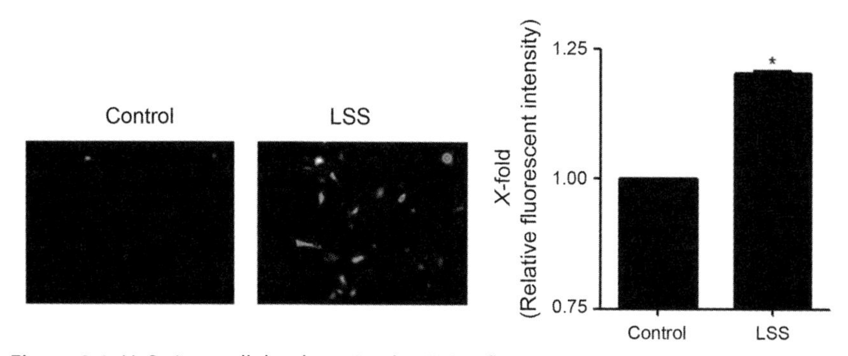

Figure 3.1 H_2O_2 intracellular detection by HyPer fluorescence. BAECs were transfected with the H_2O_2 biosensor pHyPer-Cyto and exposed to laminar shear stress conditions. Image was obtained 5 min after the exposure and fluorescence intensity was detected using confocal microscopy and quantified by Metamorph and ImageJ software. *, $p<0.05$. (See Color Insert.)

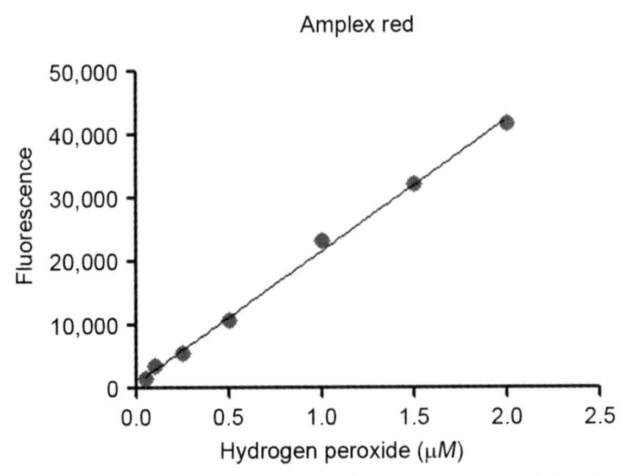

Figure 3.2 Detection of H_2O_2 using Amplex red. Reactions containing 5 μ*M* Amplex red reagent, 0.1 U/ml HRP, and the indicated amounts of H_2O_2 were incubated for 30 min at 37 °C in the dark. Fluorescence was measured with a fluorescence microplate reader at excitation and emission wavelengths of 571 and 585 nm, respectively. (For color version of this figure, the reader is referred to the online version of this chapter.)

2.4. Enzymatic generation of hydrogen peroxide fluxes. Hydrogen peroxide and glucose/glucose oxidase treatments

To generate a continuous flux of H_2O_2, endothelial cells were cultured in Dulbecco's PBS buffer supplemented with 5.6 m*M* glucose and 1 m*M* L-arginine. Different quantities of glucose oxidase (GO) were then added to the extracellular medium. The rate of H_2O_2 production by GO was previously calculated spectrophotometrically using the coupled reaction with HRP and 2,2′-azino-bis(3-ethylbenzothiazoline-6-sulfonic acid) diammonium salt (ABTS) as substrate ($\varepsilon_{415\ \mathrm{nm}} = 3.6 \times 10^4\ M^{-1}\ \mathrm{cm}^{-1}$) in the same buffer used for cell assays, covering a wide range of concentrations of GO (Fig. 3.3).

$$\begin{aligned}
\text{Glucose} + O_2 &\xrightarrow{\text{GO}} \text{Gluconic acid} + H_2O_2 \\
H_2O_2 + \text{HRP} &\longrightarrow \text{Compound I} + H_2O \\
\text{Compound I} + \text{ABTS} &\longrightarrow \text{ABTS}^* + \text{Compound II} \\
\text{Compound II} + \text{ABTS} &\longrightarrow \text{ABTS}^* + \text{HRP}
\end{aligned}$$

2.4.1 Protocol

1. Prepare the calibration solution and warm it at 37 °C (Dulbecco's PBS buffer supplemented with 5.6 m*M* glucose, 1 m*M* L-arginine, 20 μg/ml HRP, and 50 μ*M* ABTS).

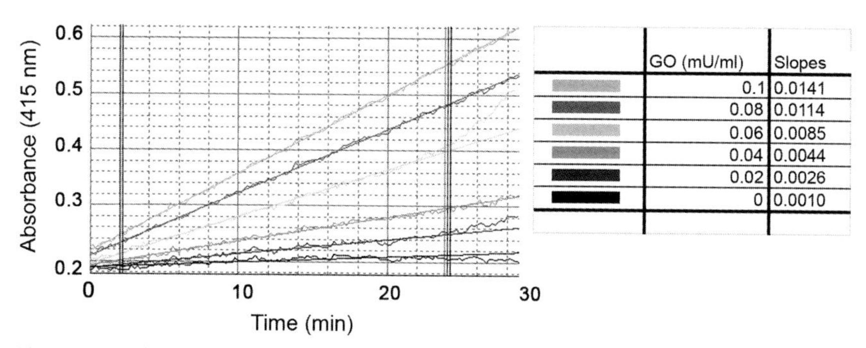

Figure 3.3 Glucose oxidase generation of hydrogen peroxide fluxes. The rate of H_2O_2 production by glucose oxidase was calculated spectrophotometrically in Dulbeccos PBS buffer supplemented with 5.6 m*M* glucose and 1 m*M* L-arginine, 20 μg/ml HRP and 50 μ*M* ABTS. The absorbance of different amounts of glucose oxidase was measured for 20–30 min. Hydrogen peroxide flux generated by each amount of glucose oxidase was determined taking into account the slopes of its kinetic reaction. (For color version of this figure, the reader is referred to the online version of this chapter.)

2. Prepare desired concentrations of GO in the calibration solution.
3. Measure the absorbance of the mixture at 415 nM using a spectrophotometer for 20–30 min.
4. Hydrogen peroxide flux generated by each amount of GO will be determined taking into account the slopes of its kinetic reaction (absorbance vs. time), ABTS ε (ABTS $\varepsilon_{415\ \mathrm{nm}} = 3.6 \times 10^4\ M^{-1}\ \mathrm{cm}^{-1}$) and the fact that one single molecule of hydrogen peroxide per two of ABTS* is produced.

2.5. Analysis of $^{\bullet}$NO production

The production of $^{\bullet}$NO by vascular endothelial cells plays an essential role in normal vascular physiology. In vascular endothelial cells, $^{\bullet}$NO is produced by the type III endothelial NO synthase (eNOS) isoform. To investigate $^{\bullet}$NO production we studied eNOS activation by detecting its active phosphorylation at Ser 1177 by Western blot analysis as described below (Garcia-Cardena et al., 1998) and by measuring $^{\bullet}$NO production using chemiluminescence. The chemiluminescence method detects $^{\bullet}$NO and its oxidation products nitrite (NO_2^-) and nitrate (NO_3^-) in liquid culture medium (Martinez-Ruiz et al., 2005). Hundred microliter aliquots of the supernatants just before and after different treatments were aspirated into a Hamilton syringe and injected through an air-tight septum into a purge vessel containing a reaction mixture, consisting of 45 m*M* potassium iodide (KI) and 10 m*M* iodine (I_2) in glacial acetic acid kept at 60 °C continuously bubbled with nitrogen. Under these conditions, NO_2^- and NO_3^- are converted to NO. The sample/carrier gas mixture passes first through a cooled water jacket, bubbles through 1 *M* NaOH, and is then passed through an inline filter to prevent acid vapors from reaching the NOA reaction, before reaching the $^{\bullet}NO$ analyzer (Sievers NOA, Model 280, Boulder, CO). $^{\bullet}NO$ gas combines with ozone to produce the excited state NO_2^*, which upon decay emits light that passes though a red filter and is detected at 600 nm by a photomultiplier tube inside the NOA. The emitted light is recorded by the instrument software (Feelisch et al., 2002; Hart, Kleinhenz, Dikalov, Boulden, & Dudley, 2005).

2.6. Western blot analysis

2.6.1 Preparation of whole cell lysates

Following experimental treatment, cells were kept on ice, washed once with warm PBS, and then scraped in the presence of ice-cold lysis buffer containing 50 m*M* Tris–HCl, pH 7.4, 150 m*M* NaCl, 1% Nonidet P-40, 0.25% sodium deoxycholate, 1 m*M* EDTA, 2 m*M* Na_3VO_4, 1 m*M* NaF, and protease inhibitor cocktail tablets (Roche, following the instructions provided by the supplier) using a plastic cell scraper. Cell lysates were further homogenized by passing cells through an insulin syringe before centrifuging them at 4° for 20 min at 12,800 rpm.

2.6.2 Protein electrophoresis and Western blotting

The supernatant was collected and protein content was determined by the bicinchoninic acid protein assay kit assay using bovine serum albumin (BSA) as protein standard. Equal amounts of proteins (20–50 μg of protein) were mixed with loading buffer (125 m*M* Tris–HCl, pH 6.8, 10% glycerol, 4% sodium dodecylsulfate (SDS), 5% 2-mercaptoethanol, and 0.01% bromophenol blue) heated for 10 min at 100 °C prior to use and subjected to electrophoresis on SDS polyacrylamide gels (from 7.5% to 15% acrylamide) and separated using the Mini-PROTEAN 3 electrophoresis module (Bio-Rad). The molecular weight of proteins was determined by comparison with the prestained protein marker (Precision Plus Protein Dual Color, Bio-Rad). After protein separation by SDS-PAGE, proteins were transferred to a nitrocellulose membrane (Whatman) by electroblotting. The protein transfer can be achieved either by placing the gel–membrane sandwich between Whatman paper in transfer buffer (semidry transfer) or by complete immersion of a gel–membrane sandwich in a buffer (wet transfer). In this work, semidry transfer was used for proteins with molecular masses of $\leq$100 kDa.

2.6.3 Specific protein staining

After electroblotting, unspecific protein-binding sites were blocked by incubating the membrane in 3% nonfat milk or BSA in Tris-buffered saline (TBS)-Tween [10 m*M* Tris–HCl, pH 7.4, 100 m*M* NaCl, and 0.1% (v/v) Tween 20] for 1 h at room temperature. The membranes were then probed with primary antibodies at room temperature for a couple of hours or overnight at 4 °C [anti-Akt (Cell Signaling, 1:1000), anti-phospho Akt (Cell Signaling, 1:1000), anti-mouse β-actin antibody (Sigma, 1:15,000), anti-c-myc (Santa Cruz Biotechnology, 1:1000), anti-eNOS antibody (BD Transduction Laboratories, 1:2000), anti-phospho eNOS Ser 1177 (Cell Signaling, 1:1000), anti-phospho eNOS Ser 637 (BD Transduction Laboratories, 1:1000), anti-phospho eNOS Thr 495 (Cell Signaling, 1:1000), anti-p38 MAPK (Cell Signaling, 1:1000), and anti-phospho p38 MAPK Thr 180/Tyr 182 (Cell Signaling, 1:1000)]. Membranes were washed with TBS-Tween three times (5 min each) and then incubated with secondary antibody for 1 h at room temperature, followed by three additional washing steps. Proteins were visualized by Li-Cor Odyssey Infrared Imaging System (Biosciences).

2.7. Small interfering RNA

Small interfering RNA (siRNA) transfection assay was performed to downregulate particular genes of interest. We designed siRNA duplex oligonucleotides targeting constructs for the specific knockdown of the proteins of our interest. Constructs and negative control siRNA were purchased from Ambion.

2.7.1 Protocol

1. Opti-MEM (Gibco) was used to dilute lipofectamine 2000 (0.15%, v/v; Invitrogen) and siRNA (30–40 n*M*) separately for 5 min.
2. Both dilutions were combined, and the mixture was incubated for 25 min at room temperature.
3. Endothelial cells were transfected at a confluence around 80% in Opti-MEM medium.
4. After 5 h of transfection, the medium was removed and replaced by culture medium supplemented with 10% FBS (Gibco) and 1% penicillin–streptomycin (Gibco).
5. After 24 or 48 h at 37 °C, the experimental approach was completed.

The following sense sequences were used:

Protein	Sequence
p38α MAPK	5′-GGUCUCUGGAGGAAUUCAAtt-3′
NOX4	5′-ACUGAGGUACAGCUGAAUGtt-3′

REFERENCES

Breton-Romero, R., Gonzalez de Orduna, C., Romero, N., Sanchez-Gomez, F. J., de Alvaro, C., Porras, A., et al. (2012). Critical role of hydrogen peroxide signaling in the sequential activation of p38 MAPK and eNOS in laminar shear stress. *Free Radical Biology & Medicine*, *52*(6), 1093–1100.

D'Autreaux, B., & Toledano, M. B. (2007). ROS as signalling molecules: Mechanisms that generate specificity in ROS homeostasis. *Nature Reviews. Molecular Cell Biology*, *8*(10), 813–824.

Dewey, C. F., Jr., Bussolari, S. R., Gimbrone, M. A., Jr., & Davies, P. F. (1981). The dynamic response of vascular endothelial cells to fluid shear stress. *Journal of Biomechanical Engineering*, *103*(3), 177–185.

Feelisch, M., Rassaf, T., Mnaimneh, S., Singh, N., Bryan, N. S., Jourd'Heuil, D., et al. (2002). Concomitant S-, N-, and heme-nitros(yl)ation in biological tissues and fluids: Implications for the fate of NO in vivo. *The FASEB Journal*, *16*(13), 1775–1785.

Garcia-Cardena, G., Fan, R., Shah, V., Sorrentino, R., Cirino, G., Papapetropoulos, A., et al. (1998). Dynamic activation of endothelial nitric oxide synthase by Hsp90. *Nature*, *392*(6678), 821–824.

Hart, C. M., Kleinhenz, D. J., Dikalov, S. I., Boulden, B. M., & Dudley, S. C., Jr. (2005). The measurement of nitric oxide production by cultured endothelial cells. *Methods in Enzymology*, *396*, 502–514.

Kanda, Y., Nishio, E., Kuroki, Y., Mizuno, K., & Watanabe, Y. (2001). Thrombin activates p38 mitogen-activated protein kinase in vascular smooth muscle cells. *Life Sciences*, *68*(17), 1989–2000.

Lehoux, S. (2006). Redox signalling in vascular responses to shear and stretch. *Cardiovascular Research*, *71*(2), 269–279.

Li, Y. S., Haga, J. H., & Chien, S. (2005). Molecular basis of the effects of shear stress on vascular endothelial cells. *Journal of Biomechanics*, *38*(10), 1949–1971.

Martinez-Ruiz, A., Villanueva, L., Gonzalez de Orduna, C., Lopez-Ferrer, D., Higueras, M. A., Tarin, C., et al. (2005). S-nitrosylation of Hsp90 promotes the inhibition of its ATPase and endothelial nitric oxide synthase regulatory activities. *Proceedings of the National Academy of Sciences of the United States of America*, *102*(24), 8525–8530.

Ray, P. D., Huang, B. W., & Tsuji, Y. (2012). Reactive oxygen species (ROS) homeostasis and redox regulation in cellular signaling. *Cellular Signalling*, *24*(5), 981–990.

Rhee, S. G. (1999). Redox signaling: Hydrogen peroxide as intracellular messenger. *Experimental & Molecular Medicine*, *31*(2), 53–59.

Rhee, S. G. (2006). Cell signaling. H_2O_2, a necessary evil for cell signaling. *Science*, *312*(5782), 1882–1883.

Rhee, S. G., Bae, Y. S., Lee, S. R., & Kwon, J. (2000). Hydrogen peroxide: A key messenger that modulates protein phosphorylation through cysteine oxidation. *Science's STKE*, *2000*(53), pe1.

Rhee, S. G., Chang, T. S., Bae, Y. S., Lee, S. R., & Kang, S. W. (2003). Cellular regulation by hydrogen peroxide. *Journal of the American Society of Nephrology*, *14*(8 Suppl. 3), S211–S215.

Ushio-Fukai, M., Alexander, R. W., Akers, M., & Griendling, K. K. (1998). p38 Mitogen-activated protein kinase is a critical component of the redox-sensitive signaling pathways activated by angiotensin II. Role in vascular smooth muscle cell hypertrophy. *The Journal of Biological Chemistry*, *273*(24), 15022–15029.

Zafari, A. M., Ushio-Fukai, M., Akers, M., Yin, Q., Shah, A., Harrison, D. G., et al. (1998). Role of NADH/NADPH oxidase-derived H_2O_2 in angiotensin II-induced vascular hypertrophy. *Hypertension*, *32*(3), 488–495.

CHAPTER FOUR

In Vivo Imaging of Nitric Oxide and Hydrogen Peroxide in Cardiac Myocytes

Juliano L. Sartoretto[1], Hermann Kalwa[1], Natalia Romero, Thomas Michel[2]

Cardiovascular Division, Department of Medicine, Brigham and Women's Hospital, Harvard Medical School, Boston, Massachusetts, USA

[1]These authors contributed equally to this work.

[2]Corresponding author: e-mail address: thomas_michel@harvard.edu

Contents

1. Introduction 62
2. Isolation and Culture of Adult Mouse Ventricular Cardiac Myocytes 63
3. Live Cell Imaging of Cardiac Myocytes 65
4. Imaging Intracellular NO with Cu_2(FL2E) Dye 66
5. Production and *In Vivo* Expression of Lentivirus Expressing the HyPer2 H_2O_2 Biosensor 68
   5.1 Detailed protocol 70
   5.2 Materials needed for virus production 71
6. Imaging Intracellular H_2O_2 in Cardiac Myocytes and Endothelial Cells Expressing HyPer2 76

Acknowledgments 77

References 77

Abstract

Nitric oxide (NO) and hydrogen peroxide (H_2O_2) are synthesized within cardiac myocytes, and both molecules play key roles in modulating cardiovascular responses. However, the interconnections between NO and H_2O_2 in cardiac myocyte signaling have not been properly understood. Adult mouse cardiac myocytes represent an informative model for the study of receptor-modulated signaling pathways involving reactive oxygen species and reactive nitrogen species. However, these cells typically survive for only 1–2 days in culture, and the limited abundance of cellular protein undermines many biochemical analyses. We have exploited chemical sensors and biosensors for use in *in vivo* imaging studies of H_2O_2 and NO in adult cardiac myocytes. Here we describe detailed methods for the isolation of cardiac myocytes suitable for imaging studies. We also present our methods for the generation of recombinant lentiviral preparations encoding the H_2O_2 biosensor HyPer2 that permit analysis of intracellular H_2O_2 levels using fluorescence microscopy in living cardiac myocytes following tail vein

Methods in Enzymology, Volume 528
ISSN 0076-6879
http://dx.doi.org/10.1016/B978-0-12-405881-1.00004-5

injection and in cultured endothelial cells following infection. We also describe our protocols for using the NO chemical sensor Cu_2(FL2E) in living adult mouse cardiac myocytes to study the effects of agonist-modulated H_2O_2 production on NO synthesis. Using these techniques, we have demonstrated that receptor-stimulated increases in intracellular H_2O_2 modulate NO levels in living cardiac myocytes. These and similar approaches may facilitate a broad range of studies in other terminally differentiated cells that involve the interaction of NO- and H_2O_2-regulated signaling responses.

1. INTRODUCTION

Cardiac myocytes are critical determinants of the heart's contractile function (Bers, 2000; Brady, 1991). Nitric oxide (NO) is a free radical signaling molecule that has been shown to play key roles in modulating cardiac responses (Balligand, Feron, & Dessy, 2009). Both the endothelial (eNOS) and the neuronal (nNOS) isoforms of NO synthase are robustly expressed in cardiac myocytes. eNOS and nNOS are Ca^{2+}-calmodulin-dependent enzymes that can be regulated by diverse extracellular stimuli acting both via cell surface receptors and by mechanochemical signals. NO synthases are also modulated by diverse protein kinases and phosphoprotein phosphatases (Dudzinski, Igarashi, Greif, & Michel, 2006). At least some of these phosphorylation pathways are influenced by cellular oxidants. Under specific pathophysiological conditions, NOS enzymes can be "uncoupled" from NO synthesis, leading to the synthesis of superoxide anion instead of NO. Clearly, there is a complex interplay between NO and oxidant pathways in cardiac cells (Maron & Michel, 2012).

Cell-derived reactive oxygen species (ROS) oxidize a broad array of biomolecules and are implicated in pathological states ranging from neurodegeneration to atherosclerosis (Stocker & Keaney, 2004; Storz, 2006). However, not all effects of ROS are deleterious: endogenously generated ROS have been implicated in posttranslational protein modifications that subserve critical physiological roles in cellular signaling (Rudolph & Freeman, 2009). H_2O_2 is one such ROS that has been identified as a key signaling molecule in many cell types (Cai, 2005; D'Autreaux & Toledano, 2007; Sartoretto, Kalwa, Pluth, Lippard, & Michel, 2011; Sartoretto et al., 2012). Whereas the physiological role of NO in the heart has been extensively characterized, the physiological role of H_2O_2 is less well understood, and much remains to be learned about the interplay between H_2O_2 and the reactive nitrogen species (RNS) in cardiac myocytes. Diverse cell surface receptor-modulated pathways activate eNOS, and yet other extracellular stimuli enhance H_2O_2 synthesis, but the relationships between NO and H_2O_2 in cardiac myocyte signaling are incompletely understood. We have recently shown that H_2O_2 is a critical intracellular mediator that modulates eNOS phosphorylation and enzyme activation in adult cardiac myocytes (Sartoretto et al., 2011, 2012).

The isolation and analysis of primary adult ventricular cardiac myocytes are of particular importance due to the lack of a representative and reliable cell line for studying the redox signaling pathways that modulate cardiac myocyte function. Moreover, the analysis of cardiac responses in transgenic mouse models involving ROS and RNS requires the isolation and culture of primary adult mouse cardiac myocytes. While adult mouse cardiac myocytes represent an informative model for studies involving ROS and NO, the isolation of suitable cells and their subsequent culture under physiological conditions for *in vivo* imaging studies can be problematic. The challenge for us was to adapt established adult cardiac myocyte isolation protocols (Liao & Jain, 2007; O'Connell, Rodrigo, & Simpson, 2007; Schluter & Schreiber, 2005) to permit the rapid and reproducible isolation of healthy cardiac myocytes suitable for *in vivo* imaging analyses. Moreover, the cells need to survive overnight without a significant loss of response to agonists. Freshly isolated myocytes need to be suitable for both biochemical studies (such as measurement of protein phosphorylation) and imaging studies to quantitate endogenous levels of ROS or RNS. Below we describe our protocols for the isolation and culture of adult mouse cardiac myocytes to detect receptor-modulated changes in H_2O_2 or NO levels in living cells by the use of the HyPer2 biosensor or of the Cu_2(FL2E) chemical sensor, respectively. We also describe the production of recombinant lentiviruses expressing HyPer2 and the method used to infect adult mice with the HyPer2 lentivirus. These methods facilitate studies of the dynamic regulation of NO and H_2O_2 metabolism in cardiac myocytes as a basis for understanding the interplay of ROS and RNS in cardiac physiology and pathophysiology.

2. ISOLATION AND CULTURE OF ADULT MOUSE VENTRICULAR CARDIAC MYOCYTES

Protocols for cardiac myocyte isolation have been previously published in detail (Liao & Jain, 2007; O'Connell et al., 2007; Schluter & Schreiber, 2005), and we have slightly modified these methods to improve the yield of cells suitable for cell imaging experiments, as described below.

For the detection of H_2O_2 and NO in living cardiac myocytes, we typically isolate cells from 8- to 10-week-old mice. The animals are placed in a glass desiccator jar and lightly anesthetized with isoflurane for 10–20 s prior to intraperitoneal injection of heparin (100 U/mL, 0.3 mL). We found that isoflurane anesthesia prior to heparin injection improves the yield of viable cardiac cells, possibly reflecting decreased stress on the animals. Ten minutes following heparin injection (the amount of time needed to effect full anticoagulation) the animals are sacrificed with isoflurane (1–2 min). Immediately after the animal dies, the peritoneal cavity and chest are opened with scissors, and the heart is exposed and lifted with forceps. Next, the descending thoracic aorta and common carotid arteries are cut, and the heart is placed in a 100-mm dish containing "perfusion buffer" (0.6 m*M* KH_2PO_4; 14.7 m*M* KCl; 0.6 m*M* Na_2HPO_4; 4.6 m*M* $NaHCO_3$; 120 m*M* NaCl; 1.2 m*M* $MgSO_4 \cdot 7H_2O$; 5.5 m*M* glucose; 30 m*M* taurine; 10 m*M* 2,3-butanedione monoxime; 10 m*M* HEPES; adjust pH to 7.0). Visualized using a dissecting microscope, the aorta is cannulated above the aortic valve and below the carotid arteries using a 25-gauge needle with a blunted tip. In order to permit perfusion of the heart, a small clip (straight micro clip, 10 mm length, 1.5 mm width, Roboz Company, Rockville, MD) is placed between the carotid arteries, and the aorta is tied below the clip to the cannula using cotton thread. The time from removal of the heart to the beginning of perfusion should not exceed 1 min; practice is essential.

The isolated heart is next perfused through the aorta using a peristaltic pump at a rate of 3–4 mL/min with perfusion buffer warmed to 37 °C for at least 5 min or until all the blood is removed. After the heart is completely free of blood, the next step is the enzymatic digestion of the heart. Collagenase type 2 (Worthington Biochemical) is added to the perfusion buffer at a final concentration of 2.4 mg/mL (now called "digestion buffer"). After 2 min of perfusion with digestion buffer, $CaCl_2$ is added at a final concentration of 37.5 μ*M* and the perfusion is continued for another 8–10 min. The time of heart digestion may vary from heart to heart, depending on the age, sex, and weight of the animals; digestion time also varies between different lots of collagenase type 2. A proper digestion is confirmed by a spongy, swollen, and pale appearance of the heart. Once the heart is thoroughly digested (the heart turns pallid, distended, and soft), the heart is removed from the cannula and placed in a 60 mm dish with 2 mL cardiac perfusion buffer containing bovine calf serum (10%, v/v) to stop the collagenase digestion. The ventricles are immediately torn into ~10 small pieces using forceps (micro dissecting tweezers, 45° angled, tip 0.05 mm × 0.01 mm, Roboz Company, Rockville, MD), and the solution is pipetted up and down using a 1-mL pipette until it becomes a cell suspension. The cell suspension is transferred to a 50-mL conical tube attached to a cell strainer (100 μm nylon) to separate dissociated cells from undigested pieces of heart tissue. Next, the cells are allowed to sediment by gravity for 10–15 min. For the final steps, calcium is gradually reintroduced to the cells. After the cells sediment, the supernatant is removed and the cells are resuspended in cardiac perfusion buffer containing 10% bovine calf serum. Three solutions with increasing concentrations of calcium (100, 400, and 900 μ*M*) are added sequentially to the cells. Between each step, the cells are allowed to sediment by gravity for at least 15 min.

After gradually reintroducing calcium to the cells, the cardiac myocytes are plated in laminin (10 μg/mL)-coated 35 mm petri dishes (containing 10 mm Microwell #0 cover glasses) in "plating medium," which consists of Minimum Essential Medium with Hank's Balanced Salt Solution, supplemented with calf serum (10%, v/v), 2,3-butanedione monoxime (10 m*M*), penicillin–streptomycin (100 units/mL), glutamine (2 m*M*), and ATP (2 m*M*). Cardiac myocytes used for determining H_2O_2 and NO production should be plated at about 50% confluence. For *in vivo* analyses of cardiac myocyte production of H_2O_2 or NO, the quality of the cells isolated is far more important than the quantity of cells obtained per isolation. Viable cardiac myocytes have a rod-shaped appearance under the microscope, and moribund myocytes are round; the percentage of rod-shaped to round myocytes should be greater than 60%; if not, the preparation should be discarded and the myocyte isolation repeated.

After the cells attach to the coverslip (1 h), the plating medium is changed to "culture medium" consisting of Minimum Essential Medium with Hank's Balanced Salt Solution, supplemented with bovine serum albumin (1 mg/mL), penicillin/streptomycin (100 units/mL), glutamine (2 m*M*), 2,3-butanedione monoxime (10 m*M*), insulin (5 μg/mL), transferrin (5 μg/mL), and selenium (5 ng/mL). The cells can be studied immediately or may be cultured overnight at 37 °C in a humidified incubator at 2% CO_2. Before performing live cell imaging experiments, it is critical to use fresh medium one or two times to wash away dead cells and other cellular debris after the cells have properly settled, to avoid imaging artifacts.

3. LIVE CELL IMAGING OF CARDIAC MYOCYTES

Live cell imaging of cardiac myocytes infected with the HyPer2 H_2O_2 biosensor or loaded with the Cu_2(FL2E) NO chemical sensor requires the use of specialized equipment to maintain cells under optimal culture conditions and to minimize thermal drift. For these imaging experiments, cells are maintained at 37 °C, 2% CO_2 in a humidified environment using a Tokai on-stage incubator mounted on an Olympus IX81 inverted fluorescence microscope equipped with a mercury arc lamp (Sartoretto et al., 2011). Whatever specific apparatus is chosen for live cell imaging, it is critical to use an environmental chamber that minimizes fluctuations in temperature, humidity, and CO_2 levels. For control of the microscope and for data acquisition and analysis, we used Metamorph software, which we found to work as a versatile platform for a wide range of imaging experiments. Of course, other combinations of hardware and software designed for live cell imaging may also be used. A detailed description of Metamorph scripting, filter selection, and control of background fluorescence can be found in the papers of Aoki and Matsuda (2009) and Pase, Nowell, and Lieschke (2012). These two informative publications also contain detailed descriptions on data analysis. Here we will focus on describing the specific considerations for using cardiac myocytes in live cell imaging experiments.

The photosensitivity of cardiac myocytes mandates that phototoxic stress during cell imaging be reduced by keeping the exposure time as short as possible and by minimizing the measurement frequency. Depending on the time course of the response being analyzed, it is usually sufficient to obtain an image every few seconds. We also use a medium level transmission neutral density filter (6–12%) to further reduce phototoxicity. If high spatial resolution is not necessary, "binning" of the data procured by the camera chip can decrease the exposure time needed to obtain a good signal-to-noise ratio. In our experiments, we used symmetrical 2 × binning as previously described (Aoki & Matsuda, 2009). To minimize background fluorescence from the imaging solutions, we use phenol red-free medium, and routinely evaluate other medium components for intrinsic fluorescence. Importantly, cardiac myocytes have a very high natural fluorescence in the green and red parts of the spectrum, and it is critical to acquire the baseline fluorescence characteristics of the cells of interest. This baseline then serves as a reference to detect expression of the biosensor and/or loading of the cell with the chemical sensor used for the actual experiment.

4. IMAGING INTRACELLULAR NO WITH Cu_2(FL2E) DYE

Details of the synthesis, purification, and validation of Cu_2(FL2E) have been described in detail (McQuade, Pluth, & Lippard, 2010). We obtained this reagent from Professor Stephen Lippard at MIT; Cu_2(FL2E) is now commercially available from Strem Chemicals (Newburyport, MA). The Cu_2(FL2E) is solubilized in DMSO and can be stored at − 80 °C for no more than 2 weeks; the Cu_2(FL2E) powder is more stable, and can be stored at 4 °C for several months. Before loading the cells with Cu_2(FL2E), the plating medium is removed and replaced with Tyrode's solution (140 m*M* NaCl; 5 m*M* KCl; 10 m*M* HEPES; 5.5 m*M* glucose; 1 m*M* $CaCl_2$; 1 m*M* $MgCl_2$; adjust pH to 7.4). Cells are loaded at 37 °C and 2% CO_2 with 5 μ*M* Cu_2(FL2E) for 2 h in Tyrode's solution, washed twice with warm Tyrode's solution, and then imaged. The culture dishes are placed in an on-stage incubator (e.g., Tokai, Tokyo, Japan) on a suitable fluorescence microscope; we use an Olympus IX81 inverted microscope equipped with an UPlan 40X/1.3 oil objective in a low-volume glass-covered recording chamber. Fluorescence signals are analyzed by using a Hamamatsu Orca CCD camera (Hamamatsu, Tokyo, Japan) at 470 nm and processed using Metamorph. Viable rod-shaped cardiac myocytes with rectangular ends are selected by differential interface contrast imaging and then subjected to fluorescence imaging following agonist treatments. Variations in levels of basal fluorescence between experimental preparations are observed because of differences in cellular loading of the NO dye from day to day. The signal from the NO sensor Cu_2(FL2E) is analyzed as the slope of the fluorescence increase seen following the addition of agonist or vehicle. Analysis is done with a standard FITC/GFP exciter emitter block at 480 nm (Semrock GFP 3035-B-OMF-zero). Figure 4.1 shows results of an imaging

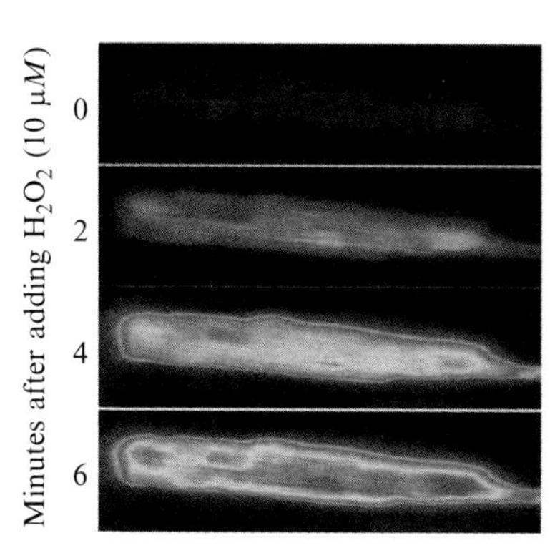

Figure 4.1 Effects of exogenous H_2O_2 on cardiac myocyte NO synthesis. Adult mouse cardiac myocytes were loaded with the NO dye Cu_2(FL2E), and then treated with hydrogen peroxide (H_2O_2, 10 μ*M*) and analyzed by fluorescence microscopy. Fluorescence images obtained at varying times after adding H_2O_2 are shown, as indicated. (See Color Insert.)

experiment using cardiac myocytes loaded with the Cu_2(FL2E) chemical sensor for NO. As can be seen, the addition of a physiological concentration of H_2O_2 (10 μM) promotes a prompt increase in the fluorescence signal, indicative of an increase in intracellular NO in response to H_2O_2. This robust response of the chemical sensor permits the analysis of the signaling pathways that connect H_2O_2 and NO. For example, the experiment shown in Fig. 4.2 indicates that H_2O_2-promoted NO synthesis in cardiac myocytes can be blocked by pretreating the cells with pharmacological concentrations of the clinically important calcium channel blocker nifedipine. As the Cu_2(FL2E) NO chemical sensor becomes more widely available, we expect that this reagent will become even more broadly used to probe NO pathways in diverse tissues and disease states.

5. PRODUCTION AND *IN VIVO* EXPRESSION OF LENTIVIRUS EXPRESSING THE HyPer2 H_2O_2 BIOSENSOR

Detection of H_2O_2 using the HyPer2 biosensor has facilitated studies of ROS metabolism in a broad range of cells transfected with plasmids encoding this informative biosensor (Belousov et al., 2006; Pase et al., 2012). However, analyses of HyPer2 biosensor responses are more challenging to study in cardiac myocytes and in other cell types that are difficult to transfect and/or are unstable in cell culture. Recombinant lentiviruses represent powerful tools for the delivery of transgenes into dividing and nondividing cells *in vitro* as well as *in vivo*. Lentiviruses do not usually elicit a severe systemic immune reaction in mice and are generally safe and well tolerated by the animal. Recombinant proteins encoded by lentiviral constructs can be studied in cardiac myocytes and other cells that are not suitable for prolonged cell culture by using the lentivirus-infected mouse as a "living incubator" to allow viral protein expression in a broad range of tissues. The key element for this approach in our studies is the production of a concentrated and stabilized solution of the lentivirus with a viral titer that is high enough to permit efficient viral infection of target tissues, leading to levels of recombinant HyPer2 protein expression that are sufficient for H_2O_2 detection. To briefly summarize our approach, the coding sequence of HyPer2 was cloned into the pWPXL lentiviral expression plasmid downstream of the EF1-α promoter. Recombinant vesicular stomatitis virus-glycoprotein pseudo-typed lentivirus particles were generated in HEK293-T cells by transfection of the envelope:packaging:transgene plasmids at a 1:1:1.5 ratio

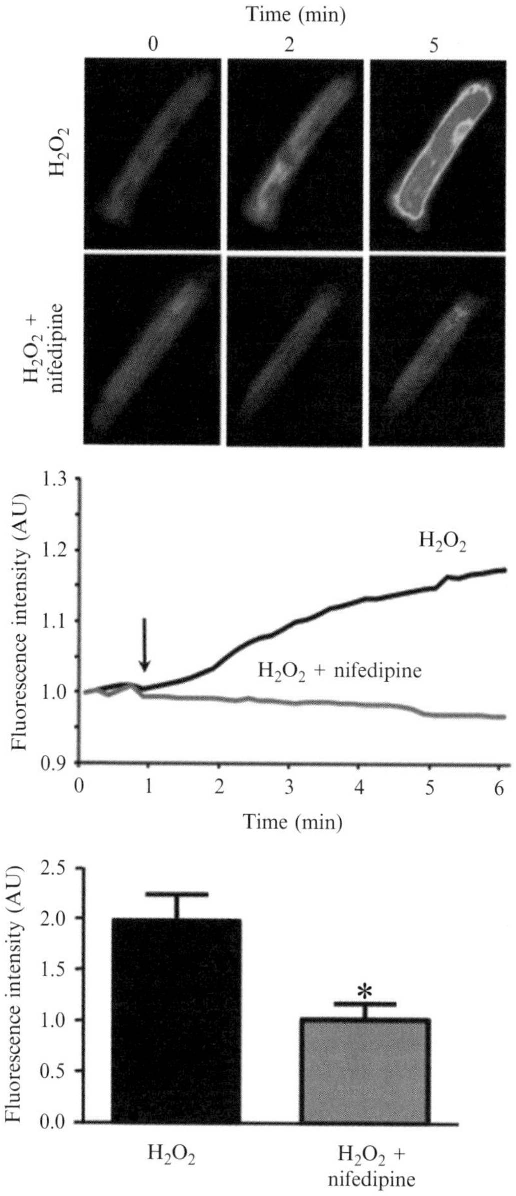

Figure 4.2 Nifedipine treatment abrogates H_2O_2-promoted NO synthesis in cardiac myocytes. Mouse cardiac myocytes were loaded with the NO chemical sensor Cu_2(FL2E), and then treated with nifedipine (100 μM) or vehicle followed by hydrogen peroxide (H_2O_2, 10 μM) treatment. The upper panel shows representative fluorescence images at 0, 2, and 5 min following treatments as indicated. The middle panel shows representative fluorescence tracings of single cells treated with H_2O_2 or H_2O_2 in the presence of nifedipine. The lower panel shows the results of pooled data analyzed from greater than three independent experiments; $^*p < 0.05$. *Adapted with modifications from Sartoretto et al. (2012).* (See Color Insert.)

with Lipofectamine (Invitrogen) according to the manufacturer's protocol. The viral titer was determined with Lenti-X GoStix (Clontech), and virus particles were concentrated by polyethylene glycol precipitation with PEG-it solution (SBI Bioscience), according to the manufacturer's protocol. The virus pellet is resuspended in PBS and used immediately for tail vein injections. For the purposes of cell infection, the lentiviral titer does not need to be quite so high, and the viral suspension may be stored at −80 °C with nominal loss of efficacy.

Below we describe our protocols for construction and injection of the HyPer2 lentivirus, and for the analysis of H_2O_2 responses in cardiac myocytes.

5.1. Detailed protocol

The protocol is based on the second-generation lentivirus system developed by Trono (Zufferey, Nagy, Mandel, Naldini, & Trono, 1997). Viruses are generated by the simultaneous transfection of three plasmids: one consisting of an HIV-1 backbone into which the cDNA of interest is cloned; a second plasmid encoding the vesicular stomatitis virus glycoprotein; and a third plasmid expressing the (VSV-G) envelope protein (which determines the viral tropism). This lentiviral system is well suited to generate and deliver the required viral titers. The plasmids are transfected into a suitable producer cell line, typically HEK293-T cells. After transfection, the viral particles are secreted into the cell culture supernatant. This supernatant is then purified, enriched, and stabilized by precipitation with polyethylene glycol. This viral suspension is ready for injection into various animal models or for the infection of cells in culture. The system separates the viral genome into three independent plasmids, thereby ensuring safety and modularity, generating only replication-deficient viral particles; these procedures are performed in Level 2 biological containment facilities. This is a well-established method that has been widely used.

The three lentivirus construction plasmids are all available from Addgene; the HyPer2 plasmid is from Evrogen. The pWPXL plasmid vector comes with an eGFP insert; methods for exchanging the eGFP insert for the gene of interest can be found in standard molecular biology protocol reference. The vector comes in two easy to confound variations (pWPXL and pWPXLd); the difference is in the loxP acceptor sides flanking the gene of interest in the pWPXLd plasmid. This might render the construct unsuitable for the transduction of inducible knockout mice. The inverted terminal repeat regions in this plasmid are prone to bacterial recombination.

Therefore it is not advisable to use classical bacterial strains (e.g., DH5α) for cloning and plasmid preparation, but instead use bacteria designed for the propagation of unstable inserts (e.g., Invitrogen "One Shot Stbl3," or other similar strains). High viral titers are required for efficient expression following mouse tail vein injection: each mouse receives 250 μL of a freshly made virus suspension containing at least 10^8 infectious particles per mL (pfu/mL). To reach these high levels, it is important to implement optimization and quality control procedures along the viral production process.

Time line

The overall production time is 6 days from start to finish, with a total working time of approximately 45 min per day.

Day 1: seeding of producer cells (HEK293-T)

Day 2: transfection of lentiviral constructs

Day 3: visual assessment of transfection efficiency

Day 4: virus collection

Day 5: virus collection

Day 6: virus concentration and injection

5.2. Materials needed for virus production

5.2.1 Plasmid preparation

The viral system can be requested via Addgene (www.addgene.org)

pWPXL (Plasmid 12258), for carrying the transgene of interest

psPAX2 (Plasmid 12260) HIV helper genome

pMD2.G (Plasmid 12259) VSV-G

One Shot® Stbl3 Chemically Competent *Escherichia coli* (Invitrogen C7373-03)

HiSpeed Plasmid Maxi Kit (Qiagen 12662)

5.2.2 Cell culture

HEK293-T cells (ATCC cell line. # CRL-11268)

DMEM high glucose with glutamate (Invitrogen 11995073)

Lipofectamine 2000 (Invitrogen 11668019)

Optimem (Invitrogen 31985070)

5.2.3 Virus titration, concentration, and infection

Lenti-X GoStix (Clontech 631243)

PEG-it Virus Precipitation Solution (SBI LV810A-1)

Hexadimethrine bromide (H9268 Sigma)

5.2.4 *Lentivirus production*

5.2.4.1 Day 1: Seeding HEK293-T cells

Seed HEK293-T cells by splitting a subconfluent 10 cm plate (p100) 1:5. Pooling the supernatant of 10 virus-producing plates usually generates enough material for one experiment. The seeding should be done ~24 h before transfection into 10 cm plates, each with 8 mL complete growth medium (DMEM high glucose, 10% FBS) and incubated at 37 °C, 5% CO_2.

5.2.4.2 Day 2: Transfection

Before starting transfection of the HEK293-T cells, change to fresh media and check the level of cell confluency under the tissue culture microscope. The cells should be subconfluent (~90% confluent). DNA and Lipofectamine solutions should be prepared separately in 1.5 mL test tubes with a total volume of 50 μL per plate for each transfection. The following plasmids are combined: 1 μg psPax (gag-pol expresser); 1 μg pMDG (VSV-G expresser); and 1.5 μg pWPXL (containing the cDNA of interest). Optimem is then added to a total of 50 mL. For each transfection reaction, add 40 μL Optimem to a second tube and add 10 μL Lipofectamine 2000 into the center of the tube without touching the sides. Do not vigorously vortex or pipet the mix, as this may decrease the transfection efficiency. Incubate the solutions for 15 min at room temperature, then add the DNA to the Optimem/Lipofectamine mix, and stir with the pipet tip until a homogenous suspension is formed. After incubating at room temperature for 15 min, add the transfection mix to the cells in a dropwise manner while agitating the plates by orbital shaking. After 6 h, change the medium a second time and incubate cultures overnight at 37 °C, 10% CO_2.

5.2.4.3 Day 3: Visual assessment of transfection efficiency

Approximately 24 h after transfection, check the efficiency of viral infection via fluorescence microscopy, which permits identification of fluorescent (HyPer2-expressing) cells as well as showing the formation of cell aggregates, which reflect viral infection of the HEK293-T cells. If the infection efficiency is not high enough (fewer than three fluorescent syncytial cell aggregates per field of view at 20 ×), it is unlikely that an adequate viral titer will be obtained; in this case, it is best to discard the cells and start over, possibly with new plasmid preparations.

5.2.4.4 Day 4/Day 5: Virus collection

Harvest the lentivirus by pooling the supernatant from day 4 and day 5. Collect the media into 50 mL tubes, and check the viral titer using Lenti-X

GoStix (Clontech) according to the manufacturer's protocol. If this test is negative, discard the supernatant and start over. If the titer quality is sufficient, remove cellular debris by centrifugation (1500 × *g* for 15 min) and add 10 mL of cold PEG-it solution to the supernatant. Mix well, and refrigerate the supernatant at 4 °C for up to 48 h.

5.2.4.5 Day 6: Virus concentration and injection

Centrifuge supernatant/PEG-it mixture at 1500 × *g* for 30 min at 4 °C. After centrifugation, the virus particles appear as a beige or white pellet at the bottom of the tube. Carefully aspirate the supernatant and resuspend the pellets in 1/100 of the original volume using cold sterile PBS. Pool the virus suspensions and add hexadimethrine bromide solution to a final concentration of 10 μg/mL. Load individual syringes with 250 μL of virus suspension and proceed with the injections, as reviewed below. After adding the hexadimethrine bromide, perform the tail vein injection in less than 30 min. Do not freeze the viral supernatants for subsequent use in tail vein injections, as this may result in a reduction of viral titer. Obviously, this method does not allow time for quantitative viral titer determination, but the semiquantitative assessment of viral titer using Lenti-X GoStix is sufficient. Because it is important to proceed with the injections shortly after the viral enrichment process, a detailed analysis of the viral titer by serial dilution or qPCR is not feasible. We do not use supernatants giving a weak fluorescence signal on day 3 or a negative signal in the "Quicktest." It is better to discard the cells and start over than work with a suboptimal virus stock. For purposes of troubleshooting and quality control, it is useful to save an aliquot of the injected virus solution for quantitative titer determination. For infection of tissue culture cells (when the viral titer does not need to be so high), the virus can be frozen at −80 °C and stored for up to 6 months before use. The key to a high titer virus solution is optimal and healthy HEK293-T cells. Even slight variations in culture conditions can lead to significant fluctuations of virus quality and quantity. Split the cells three times a week at a ratio of around 1:4–1:6, making sure that cells are completely trypsinized to disrupt cell aggregates. Cells should be monitored for mycoplasma infection regularly, as mycoplasma can interfere with transfection efficiency.

For the creation of stable cell lines expressing lentivirus-encoded HyPer2, we infected p100 dishes of HEK293-T cells with the HyPer2 lentivirus ($\sim 5 \times 10^6$ pfu), prepared as described above. After 6 h, media was changed for DMEM containing 10% FBS without phenol red, and cells were incubated for 24 h before splitting. With this procedure, we generated a stable cell line of HEK293-T cells expressing the HyPer2 biosensor; levels of HyPer2 protein expression are stable through multiple passages, and cells can be frozen down for storage and the thawed cells still retain robust HyPer2 expression. To validate the response of the HyPer2 cell line to H_2O_2, HEK293-T cells stably expressing HyPer2 biosensor were plated in 24-well dishes precoated with polyethylenimine (25 μg/mL) to improve cell attachment. Before starting measurements, media was replaced with dPBS containing 5 m*M* glucose. Fluorescence emission at 520 nm was measured using a fluorescence plate reader (Omega Fluostar, BMG) with 400 and 485 nm excitation filters (Fig. 4.3A). After a period of base line stabilization (due to photoactivation of the YFP fluorophore in HyPer2), the enzyme glucose oxidase (Calbiochem, cat # 345386) was added to the extracellular media to generate fluxes of H_2O_2 (0–15 μ*M*/min). Glucose oxidase activity was determined using commercially available colorimetric kits (e.g., Cayman Chemical). The increase in HyPer2 ratiometric fluorescence (485 nm/400 nm) corresponds to the oxidation of HyPer2, and initial slopes correlate well with H_2O_2 fluxes.

We also utilized HyPer2 lentivirus to infect bovine aortic endothelial cells (BAEC) using the same procedures as described above. BAEC expressing Hyper2 were plated in gelatin-coated 35 mm petri dishes containing 10 mm Microwell #0 coverglasses. After incubating for 4 h in phenol red-free DMEM with 10% FBS to allow cell attachment, the cell media was replaced with Hank's Buffered Salt Solution and live cell imaging was obtained as described before (Pase et al., 2012). Figure 4.3B shows imaging results obtained in BAEC encoding HyPer2 biosensor after the addition of hydrogen peroxide (H_2O_2, 100 μ*M*) to the extracellular media. Once the baseline is established, the addition of H_2O_2 results in a robust increase in fluorescence, monitored and quantitated as described above. BAEC expressing HyPer2 can be used to study the mechanisms of H_2O_2 modulated signaling pathways in endothelial cells. Because BAEC are primary cells, the lentivirus-infected cells are only used for one or two passages following infection with the HyPer2 lentivirus.

For infection of cardiac tissues, the HyPer2 lentivirus is infused through the tail vein (250 μL of 10^8 pfu/mL) of adult male mice (8–10 weeks old). One or two weeks after injection of virus (or saline control), mice are euthanized, and cardiac myocytes are isolated, cultured, and analyzed as described above. Although mouse tail vein injection is a commonly used technique, it takes practice to perform it correctly. Great care has to be taken to confirm

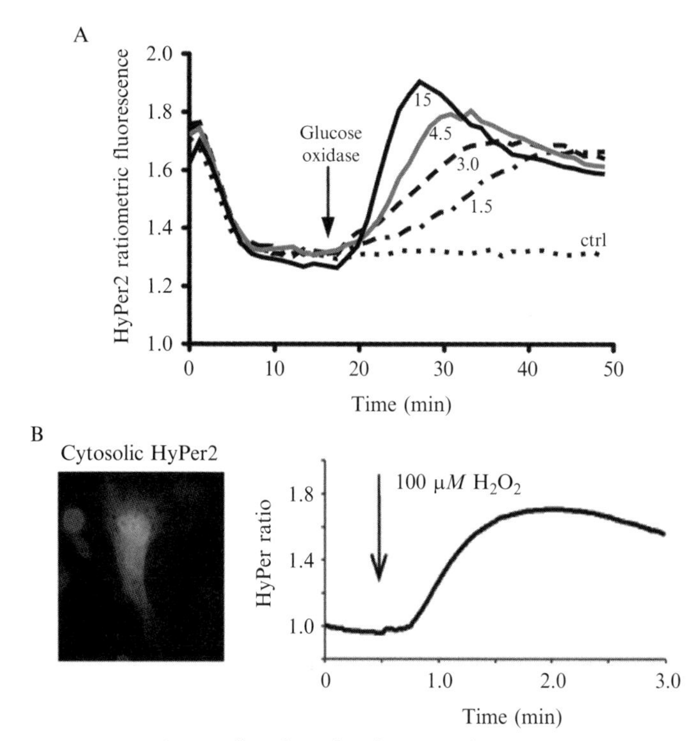

Figure 4.3 Imaging H_2O_2 in cells infected with Hyper2 lentivirus. (A) The results of cellular imaging of HEK293-T cells stably expressing the HyPer2 biosensor. Fluorescence emission at 520 nm was measured using a fluorescence plate reader with 400 and 485 nm excitation filters. After stabilization of the base line, the glucose oxidase was added to the extracellular media to generate H_2O_2 flux (0–15 μ*M*/min), as described in the text. (B) Results obtained in BAEC infected with HyPer2 lentivirus and treated with H_2O_2 (100 μ*M*); images were taken every 5 s. The image shows HyPer2 fluorescence in untreated BAEC that had been infected with the HyPer2 lentivirus; the graph shows a time course of the ratiometric fluorescence change (YFP_{500}/YFP_{420} ratio), revealing a rapid increase in HyPer2 oxidation after adding H_2O_2. (See Color Insert.)

that the injection is indeed into the tail vein (there is an informative demonstration of the tail vein injection technique on line at http://www.youtube.com/watch?v=8MDcyardkmw). Mice have two lateral tail veins, visible along either side of the tail. The mouse is placed in a mouse tail illuminator/restrainer (Braintree Scientific) and the lentivirus is injected using an insulin syringe with a 27-gauge needle. Lentivirus may be diluted with sterile saline to a final volume of 200–250 μL. While injecting, there should be negligible resistance. Two weeks after injection of HyPer2, mouse tissues can be harvested and analyzed for HyPer2 protein expression by probing immunoblots with an antibody against GFP (e.g., the polyclonal GFP antibody from Cell Signaling); the HyPer2 protein migrates at $M_r = 52$ kDa.

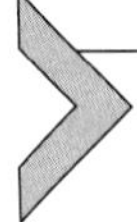

6. IMAGING INTRACELLULAR H_2O_2 IN CARDIAC MYOCYTES AND ENDOTHELIAL CELLS EXPRESSING HyPer2

Lentiviral infection by mouse tail vein injection of the HyPer2 lentivirus leads to a heterogeneous level of HyPer2 expression in cardiac myocytes. In a typical cardiac myocyte preparation following HyPer2 tail vein injection, only approximately 5% of the myocytes express detectable HyPer2. This low proportion of HyPer2-positive cells is further confounded by the high background fluorescence present in cardiac myocytes. The intensity level of this background fluorescence has to be determined first; only cardiac myocytes showing at least twice this fluorescence level are selected for imaging. Following initiation of the live cell imaging protocol, it is important to wait until the baseline signal stabilizes: the first ~30 s of acquired data usually cannot be analyzed because of an optical artifact that arises from the initial photoconversion of YFP (seen as a rapid drop in fluorescence), as has been previously described (Aoki & Matsuda, 2009). The individual imaging conditions (e.g., exposure time and choice of neutral density filters) depend on the specific microscope setup being used. The paper by Pase et al. (2012) contains an extensive description of these considerations. Figure 4.4 shows imaging results obtained in cardiac myocytes isolated from mice after tail vein injection with the HyPer2 lentivirus (Sartoretto et al., 2011). Changes in cell-derived fluorescence were analyzed after treating the cells with hydrogen peroxide (H_2O_2, 10 μ*M*), angiotensin-II (ANG-II, 500 n*M*), or isoproterenol (ISO, 100 n*M*). HyPer2 fluorescence increases after the addition of H_2O_2, serving as a key positive control. ANG-II also promotes a significant increase in HyPer2 fluorescence, but there was no significant increase in HyPer2 fluorescence following the addition of ISO (Fig. 4.4). Both of these agonists promote NOS activation (Sartoretto et al., 2009, 2011), but H_2O_2 plays a differential role in receptor signaling: ANG-II-promoted eNOS activation is dependent on H_2O_2, whereas beta adrenergic receptor activation of nNOS is independent of changes in intracellular H_2O_2. These and similar experimental approaches can be used to probe H_2O_2-dependent signaling pathways in cardiac myocytes and other terminally differentiated cells.

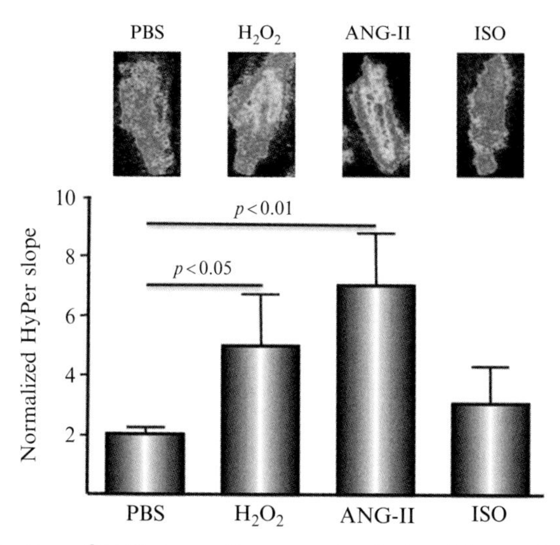

Figure 4.4 Detection of H_2O_2 in cardiac myocytes isolated from mice infected with lentivirus expressing the HyPer2 biosensor. Adult mice were injected via tail vein with lentivirus expressing the HyPer2 H_2O_2 biosensor ($\sim 10^9$ pfu); 2 weeks later the mice were euthanized, and cardiac myocytes were isolated and analyzed. The bar graph shows pooled data from three independent experiments, in which the H_2O_2 response is quantitated as the slope of the fluorescence signal in arbitrary units (AU) measured between $t=0$ and $t=5$ min after the addition of 10 μM of H_2O_2, 500 nM of ANG-II (angiotensin-II), or 100 nM of ISO (isoproterenol). $^*p<0.05$ compared to PBS-treated cells. Also presented are representative HyPer2 images shown in isolated cardiac myocytes treated as displayed. The HyPer2 H_2O_2 image is determined as the YFP_{500}/YFP_{420} excitation ratio; the gray scale is adjusted to improve contrast. *Adapted with modifications from Sartoretto et al. (2011).* (For color version of this figure, the reader is referred to the online version of this chapter.)

ACKNOWLEDGMENTS

This work was supported in part by National Institutes of Health Grants HL46457, HL48743, and GM36259 (to T. M.); by postdoctoral fellowships from the Fonds National de Recherche, Luxembourg and the American Heart Association (to H. K.); and by a Pew Latin American Fellowship (to N. R.).

REFERENCES

Aoki, K., & Matsuda, M. (2009). Visualization of small GTPase activity with fluorescence resonance energy transfer-based biosensors. *Nature Protocols*, *4*(11), 1623–1631.

Balligand, J. L., Feron, O., & Dessy, C. (2009). eNOS activation by physical forces: From short-term regulation of contraction to chronic remodeling of cardiovascular tissues. *Physiological Reviews*, *89*(2), 481–534.

Belousov, V. V., Fradkov, A. F., Lukyanov, K. A., Staroverov, D. B., Shakhbazov, K. S., Terskikh, A. V., et al. (2006). Genetically encoded fluorescent indicator for intracellular hydrogen peroxide. *Nature Methods*, *3*(4), 281–286.

Bers, D. M. (2000). Calcium fluxes involved in control of cardiac myocyte contraction. *Circulation Research*, *87*(4), 275–281.
Brady, A. J. (1991). Mechanical properties of isolated cardiac myocytes. *Physiological Reviews*, *71*(2), 413–428.
Cai, H. (2005). NAD(P)H oxidase-dependent self-propagation of hydrogen peroxide and vascular disease. *Circulation Research*, *96*(8), 818–822.
D'Autreaux, B., & Toledano, M. B. (2007). ROS as signalling molecules: Mechanisms that generate specificity in ROS homeostasis. *Nature Reviews Molecular Cell Biology*, *8*(10), 813–824.
Dudzinski, D. M., Igarashi, J., Greif, D., & Michel, T. (2006). The regulation and pharmacology of endothelial nitric oxide synthase. *Annual Review of Pharmacology and Toxicology*, *46*, 235–276.
Liao, R., & Jain, M. (2007). Isolation, culture, and functional analysis of adult mouse cardiomyocytes. *Methods in Molecular Medicine*, *139*, 251–262.
Maron, B. A., & Michel, T. (2012). Subcellular localization of oxidants and redox modulation of endothelial nitric oxide synthase. *Circulation Journal*, *76*(11), 2497–2512.
McQuade, L. E., Pluth, M. D., & Lippard, S. J. (2010). Mechanism of nitric oxide reactivity and fluorescence enhancement of the NO-specific probe CuFL1. *Inorganic Chemistry*, *49*(17), 8025–8033.
O'Connell, T. D., Rodrigo, M. C., & Simpson, P. C. (2007). Isolation and culture of adult mouse cardiac myocytes. *Methods in Molecular Biology*, *357*, 271–296.
Pase, L., Nowell, C. J., & Lieschke, G. J. (2012). In vivo real-time visualization of leukocytes and intracellular hydrogen peroxide levels during a zebrafish acute inflammation assay. *Methods in Enzymology*, *506*, 135–156.
Rudolph, T. K., & Freeman, B. A. (2009). Transduction of redox signaling by electrophile-protein reactions. *Science Signaling*, *2*(90), re7.
Sartoretto, J. L., Jin, B. Y., Bauer, M., Gertler, F. B., Liao, R., & Michel, T. (2009). Regulation of VASP phosphorylation in cardiac myocytes: Differential regulation by cyclic nucleotides and modulation of protein expression in diabetic and hypertrophic heart. *American Journal of Physiology. Heart and Circulatory Physiology*, *297*(5), H1697–H1710.
Sartoretto, J. L., Kalwa, H., Pluth, M. D., Lippard, S. J., & Michel, T. (2011). Hydrogen peroxide differentially modulates cardiac myocyte nitric oxide synthesis. *Proceedings of the National Academy of Sciences of the United States of America*, *108*(38), 15792–15797.
Sartoretto, J. L., Kalwa, H., Shiroto, T., Sartoretto, S. M., Pluth, M. D., Lippard, S. J., et al. (2012). Role of Ca^{2+} in the control of H_2O_2-modulated phosphorylation pathways leading to eNOS activation in cardiac myocytes. *PLoS One*, 7(9), e44627.
Schluter, K. D., & Schreiber, D. (2005). Adult ventricular cardiomyocytes: Isolation and culture. *Methods in Molecular Biology*, *290*, 305–314.
Stocker, R., & Keaney, J. F., Jr. (2004). Role of oxidative modifications in atherosclerosis. *Physiological Reviews*, *84*(4), 1381–1478.
Storz, P. (2006). Reactive oxygen species-mediated mitochondria-to-nucleus signaling: A key to aging and radical-caused diseases. *Science's STKE*, *2006*(332), re3.
Zufferey, R., Nagy, D., Mandel, R. J., Naldini, L., & Trono, D. (1997). Multiply attenuated lentiviral vector achieves efficient gene delivery in vivo. *Nature Biotechnology*, *15*(9), 871–875.

CHAPTER FIVE

Methods for Studying Oxidative Regulation of Protein Kinase C

Rayudu Gopalakrishna[1], Thomas H. McNeill, Albert A. Elhiani, Usha Gundimeda

Department of Cell and Neurobiology, Keck school of Medicine, University of Southern California, Los Angeles, California, USA

[1]Corresponding author: e-mail address: rgopalak@usc.edu

Contents

Abstract

The protein kinase C (PKC) family of isoenzymes may be a crucial player in transducing H_2O_2-induced signaling in a wide variety of physiological and pathophysiological processes. PKCs contain unique structural features that make them highly susceptible to oxidative modification. Depending on the site of oxidation and the extent to which it is modified, PKC can be either activated or inactivated by H_2O_2. The N-terminal regulatory domain contains zinc-binding, cysteine-rich motifs that are readily oxidized by H_2O_2. When oxidized, the autoinhibitory function of the regulatory domain is compromised, and as a result, PKC is activated in a lipid cofactor-independent manner. The C-terminal catalytic domain contains several reactive cysteine residues, which when oxidized with a higher concentration of H_2O_2 leads to an inactivation of PKC. Here, we

Methods in Enzymology, Volume 528
ISSN 0076-6879
http://dx.doi.org/10.1016/B978-0-12-405881-1.00005-7

describe the methods used to induce oxidative modification of purified PKC isoenzymes by H_2O_2 and the methods to assess the extent of this modification. Protocols are given for isolating oxidatively activated PKC isoenzymes from cells treated with H_2O_2. Furthermore, we describe the methods used to assess indirect regulation of PKC isoenzymes by determining their cytosol to membrane or mitochondrial translocation and tyrosine phosphorylation of PKCδ in response to sublethal levels of H_2O_2. Finally, as an example, we describe the methods used to demonstrate the role of H_2O_2-mediated cell signaling of PKCε in green tea polyphenol-induced preconditioning against neuronal cell death caused by oxygen–glucose deprivation and reoxygenation, an *in vitro* model for cerebral ischemic/reperfusion injury.

1. INTRODUCTION

Although the oxidant hydrogen peroxide (H_2O_2) was previously considered to be cytotoxic and thought to induce cell death, it is becoming clear that at sublethal concentrations it serves as a second messenger that mediates the actions of a variety of agents including growth factors and tumor promoters (Rhee, Chang, Bae, Lee, & Kang, 2003). Low concentrations of H_2O_2 also induce cell signaling leading to a cellular adaptation to higher concentrations of oxidants, which would otherwise be cytotoxic. Similar to other second messengers such as cAMP, cGMP, Ca^{2+}, and diacylglycerol (DAG) which have specific cellular targets such as protein kinases, specific cellular targets for H_2O_2 that mediate its signaling are coming into light (Rhee, 2004). Among these targets is a protein kinase C (PKC) family of isoenzymes that may be important players in transducing H_2O_2-induced signaling in wide variety of physiological and pathophysiological processes (Gopalakrishna & Jaken, 2000).

Conventional PKCs (α, β, and γ) are Ca^{2+} dependent and are activated by a second messenger DAG (Griner & Kazanietz, 2007; Newton, 1997; Nishizuka, 1992; Parker & Murray-Rust, 2004). Conversely, novel PKCs (δ, ε, η, and θ) are Ca^{2+} independent but activated by DAG. Atypical PKCs (ζ and λ/ι) require neither Ca^{2+} nor DAG for expressing optimal activity. PKC-binding proteins direct these isoenzymes to various subcellular compartments (Mochly-Rosen, 1995; Poole, Pula, Hers, Crosby, & Jones, 2004). Although most cells express more than one type of PKC, differences among the isoenzymes with respect to activation conditions and subcellular locations suggest that individual PKC isoenzymes mediate distinct cellular

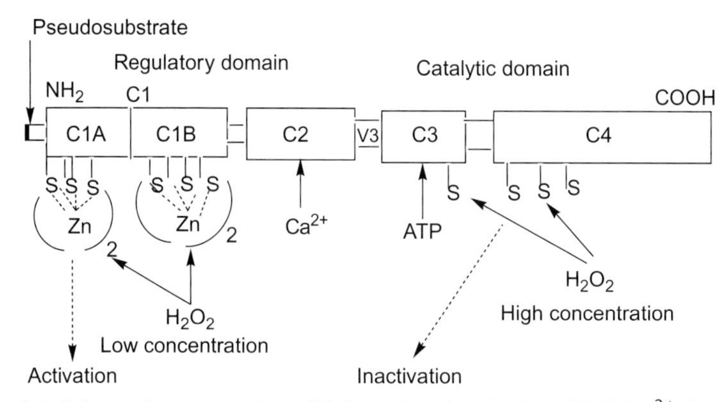

Figure 5.1 Schematic presentation of H_2O_2-induced activation of PKC (Ca^{2+}-dependent isoenzymes). At low concentrations, H_2O_2 reacts with zinc thiolates present in the regulatory domain and activates PKC. At high concentrations, H_2O_2 also reacts with cysteine residues present in the catalytic domain and inactivates the kinase.

processes. For example, in most cell types, PKCε is a promitogenic and prosurvival kinase, whereas PKCδ is an antiproliferative and proapoptotic kinase (Griner & Kazanietz, 2007). Therefore, an inactivation of PKCε and/or PKCδ could lead to cell death.

All PKCs consist of N-terminal regulatory domains and C-terminal catalytic domains (Fig. 5.1). In the absence of lipid coactivators as in the resting cells, PKC assumes an inactive conformation. This is maintained by an intramolecular interaction between an autoinhibitory sequence (the pseudosubstrate) in the regulatory domain and substrate-binding region of the catalytic domain (Gopalakrishna & Jaken, 2000). In this inactive state, PKC is mainly present in the cytosol. Activation is triggered by receptor-mediated stimulation of phospholipase C, which generates DAG. DAG binding to PKC regulatory domain induces its translocation to the plasma membrane, nucleus, and mitochondria (Blumberg et al., 1994; Steinberg, 2008).

Tumor promoters such as phorbol esters mimic DAG activation of PKCs by binding to the same sites within the regulatory domain, thereby inducing similar conformational changes within the regulatory domain (Blumberg et al., 1994). The DAG/phorbol ester-binding sites have been mapped to two pairs of zinc fingers in the regulatory domain (C1A and C1B). Each zinc finger contains six regularly spaced cysteine residues that fold a structure that coordinates two zinc atoms (Kazanietz et al., 1995). The high number of cysteine residues in the zinc finger of the regulatory domain makes this an attractive target for redox regulation.

Both regulatory and catalytic domains of PKC are susceptible to oxidative modification by H_2O_2 (Gopalakrishna & Jaken, 2000). At lower concentrations, H_2O_2 selectively modifies the isolated PKC at its regulatory domain, which can be monitored by the loss of phorbol ester binding. As a consequence, the kinase becomes cofactor independent, presumably due to the oxidant-induced loss of autoinhibition through the regulatory domain. This type of modification which converts the Ca^{2+}/lipid-activated form of PKC (cofactor-dependent) to a cofactor-independent form is also seen in H_2O_2-treated cells. This is particularly intriguing considering that H_2O_2 mimics several actions of phorbol esters in many cell types. On the contrary, oxidative modification of cysteine residues in the catalytic domain leads to inactivation of PKC. An inactivation of PKC isoenzymes, particularly PKCε, has been implicated in cell death (Griner & Kazanietz, 2007).

In H_2O_2-treated cells, PKCδ is activated by phosphorylation at its Tyr311 located in the hinge region between the regulatory and catalytic domains of the kinase (Konishi et al., 2001). This is supported by the fact that purified PKCδ is activated in a cofactor-independent manner by direct phosphorylation at Tyr311 by Src family kinases. However, in a latter study, Kikkawa and his associates have elegantly demonstrated that the mutant PKCδ where Tyr311 is replaced with phenylalanine is recovered as an active form as the wild-type PKCδ from the H_2O_2-treated COS-7 cells (Umada-Kajimoto, Yamamoto, Matsuzaki, & Kikkawa, 2006). Steinberg and her associates have shown an activation of PKCδ by its tyrosine phosphorylation as well as by its oxidative modification in cardiomyocytes (Rybin et al., 2004).

The oxidative activation of PKC was observed in many cases (Gopalakrishna & Anderson, 1989; Knapp & Klann, 2000; Zabouri & Sossin, 2002). Redox regulation of PKC has been implicated in a wide variety of physiological and pathophysiological processes (Giorgi et al., 2010). Some examples include its role in tumor promotion, long-term potentiation, neuritogenesis, and ischemic preconditioning in both the heart and the brain (Gopalakrishna, Gundimeda, Schiffman, & McNeill, 2008; Gopalakrishna & Jaken, 2000; Klann, Roberson, Knapp, & Sweatt, 1998; Perez-Pinzon, Dave, & Raval, 2005).

We recently observed the role of oxidatively activated PKCε in green tea polyphenol (GTPP)-induced preconditioning against cell death induced by oxygen–glucose deprivation (OGD) (Gundimeda et al., 2012). Polyphenol epigallocatechin-3-gallate (EGCG) present in GTPP binds to the 67-kDa laminin receptor (67LR) and triggers cell signaling that leads to activation of NADPH oxidase, which then generates sublethal levels of ROS. ROS, particularly H_2O_2, in turn, activates PKCε by directly inducing its oxidative modification. In addition, H_2O_2 indirectly induces activation and membrane/mitochondrial translocation of PKCε through generation of phospholipid-hydrolysis products. Mitochondrial association of the cell survival isoenzyme PKCε likely protects cells from cell death induced by oxygen–glucose deprivation/reoxygenation (OGD/R). Our previous studies have shown an activation of ERK1/2 by ROS in GTPP-treated cells as well as an activation of ERK by PKC in oxidant-treated cells triggering downstream nuclear events (Gopalakrishna et al., 2008; Gundimeda, McNeill, Schiffman, Hinton, & Gopalakrishna, 2010). Based on these observations, we propose that GTPP induces the expression of antioxidant enzymes, which may precondition cells against OGD/R-induced cell death. Thus, it is possible that activation of PKC by sublethal levels of H_2O_2 during preconditioning may induce an endogenous antioxidant response and thereby protects neuronal cells from oxidative injury caused by lethal levels of ROS during OGD/R.

Here, we describe the methods used to induce oxidative modification of purified PKC isoenzymes by H_2O_2 to determine oxidative activation of PKC isoenzymes occurring in cells treated with H_2O_2, and to assess indirect regulation of PKC isoenzymes in response to sublethal levels of H_2O_2. Finally, as an example, we describe the methods used to demonstrate the role of H_2O_2-induced redox signaling of PKC in GTPP-induced preconditioning against neuronal cell death induced by OGD/R. Emphasis is given to α, β, γ, δ, and ε isoenzymes.

2. MATERIALS

2.1. Reagents and antibodies

Catalase–polyethylene glycol (PEG), xanthine, xanthine oxidase, copper–zinc superoxide dismutase (SOD), catalase, EGCG, aprotinin, leupeptin, pepstatin A, *N*-acetyl-L-cysteine, phosphatidylcholine, phosphatidylserine, 1,2-diolein, and hydrogen peroxide were from Sigma. *N*-Acetyl-D-cysteine was from Research Organics. [γ-^{32}P]ATP (specific activity 20 Ci/mmol) was from MP Biochemicals. [20-^{3}H]phorbol 12,13-dibutyrate (specific activity 20 Ci/mmol) was from Dupont NEN. Cyto-Toxo-One Homogeneous Membrane Integrity assay kit was from Promega. 2′,7′-Dichlorofluorescin diacetate (DCFDA) was obtained from Molecular Probes. Diphenyleneiodonium (DPI) and VAS2870 were obtained from Calbiochem. PKC-specific substrate peptide corresponding to a neurogranin amino acid sequence (residues 28–43) was synthesized at the core facility of Norris Comprehensive Cancer Center.

Decaffeinated extract of GTPP, which was standardized to contain 97% polyphenols and nearly 70% catechins, was obtained from Pharmanex. The typical preparation contained the following polyphenols expressed as percentage of original weight of GTPP preparation: EGCG (36%), ECG (15%), EC (7%), and EGC (3%) (21). Similar composition was also reported with another batch of GTPP (52), suggesting a high degree of consistency in the composition of this GTPP preparation.

Anti-67LR (MLuC5) mouse monoclonal antibodies, anti-PKCδ rabbit polyclonal antibodies, anti-PKCε rabbit polyclonal antibodies, anti-PKCε mouse monoclonal antibodies, mouse IgM, and Protein-A/G Plus-agarose were obtained from Santa Cruz Biotechnology. Anti-PKCα mouse monoclonal antibodies were from Millipore. Mouse monoclonal antiphosphotyrosine antibodies (clone 4G10) were from Upstate Biotechnology.

2.2. Buffers

Buffer A: 10 m*M* Tris–HCl, pH 7.4/1 m*M* EGTA or 1 m*M* $CaCl_2$/2 μ*M* leupeptin/0.15 μ*M* pepstatin A/0.25 m*M* ATP/10 m*M* $MgCl_2$.

Buffer B: 20 m*M* Tris–HCl, pH 7.4/1 m*M* EDTA/0.5 m*M* phenylmethylsulfonyl fluoride/2 μ*M* leupeptin/0.15 μ*M* pepstatin A/1% Igepal CA-630 (v/v).

Buffer C: 20 m*M* Tris–HCl, pH 7.4/1 m*M* EDTA/10 m*M* 2-mercaptoethanol/1 μ*M* leupeptin/0.15 μ*M* pepstatin A.

Buffer D: 20 m*M* Tris–HCl, pH 7.4/1 m*M* EDTA/1% Triton X-100/10 m*M* 2-mercaptoethanol/10 m*M* sodium fluoride/1 m*M* sodium vanadate/0.5 m*M* phenylmethylsulfonyl fluoride/1 μ*M* leupeptin/0.15 μ*M* pepstatin A.

Buffer E: 20 m*M* Tris–HCl, pH 7.4/1 m*M* EDTA/0.15 *M* NaCl/1 μ*M* leupeptin/0.15 μ*M* pepstatin A/1% Igepal CA-630.

Buffer F: 20 m*M* Tris–HCl, pH 7.4/1 m*M* EDTA/0.1 *M* NaCl/1 μ*M* leupeptin/0.15 μ*M* pepstatin A/0.1 μ*M* microcystin-LR.

Buffer G: 10 m*M* Tris–HCl, pH 7.4/250 m*M* sucrose/1 m*M* EDTA/0.5 m*M* phenylmethylsulfonyl fluoride/leupeptin (10 μg/ml)/aprotinin (10 μg/ml).

3. DIRECT OXIDATIVE MODIFICATION OF PKC ISOENZYMES BY H_2O_2

3.1. Modification of purified PKC isoenzymes by H_2O_2

PKC isoenzymes α, β, γ, δ, and ε are purified from rat brains (Gundimeda et al., 2008). In some cases, a mixture of Ca^{2+}-dependent isoenzymes α, β, and γ (specific activity 890 units/mg protein) is used, as Ca^{2+}-dependent hydrophobic chromatography can purify these isoenzymes to homogeneity in high amounts (Gopalakrishna & Anderson, 1989). If these isoenzymes are purified by other conventional methods, it is important to remove protease calpain from the enzyme preparation as described previously (Gopalakrishna, Chen, & Gundimeda, 1995). If trace amounts of proteases are present in the preparations of enzyme used for oxidative modification experiments, the oxidatively modified PKC may be degraded. It is also beneficial to use various protease inhibitors.

Initially, thiol agents are removed from PKC preparation by dialysis or PD-10 Sephadex (Pharmacia) gel filtration columns. The purified isoenzymes are taken into a buffer A in a total volume of 0.25 ml. These samples are brought to 30 °C in a water bath. Oxidative modification is initiated by adding various concentrations of H_2O_2, and the samples are incubated for 5–20 min. Excess H_2O_2 is removed from the treated samples by subjecting them to a centrifuge column technique (Gopalakrishna et al., 1995), and both PKC activity and phorbol ester binding are then determined.

In the original method, we reported that the reaction was arrested by the addition of dithiothreitol (Gopalakrishna & Anderson, 1989). Thus, the modification observed is thiol-irreversible. Therefore, higher concentration of H_2O_2 (4 m*M*) is required for this modification. In the current protocol, stopping the reaction by removing H_2O_2, thiol reversible modification is seen to a greater extent than irreversible modification. This modification requires a lower concentration (<1 m*M*) of H_2O_2. Both regulatory and catalytic domains are susceptible to H_2O_2-induced oxidative modification. H_2O_2 at 1 m*M* concentration decreased phorbol ester binding and increased lipid cofactor-independent activity by nearly 50–60% of that is seen when assayed with cofactors. To prevent modification of the catalytic domain, ATP and Mg^{2+} are required. However, the catalytic activity is also lost when incubated with high (5 m*M*) concentration of H_2O_2.

There is also a substrate preference for oxidatively activated PKC isoenzymes. The commercial type III-S preparation of histone (a mixture of histone H1 and high mobility group proteins) served as a better substrate than pure histone H1 for oxidatively activated PKC (α, β, and γ mixture). A synthetic peptide corresponding to a neurogranin amino acid sequence (residues 28–43) is found to be a good substrate for oxidatively activated PKC isoenzymes.

3.1.1 Measuring cofactor-independent activity of PKC

The assays of PKC are carried out in 96-well plates with fitted filtration discs made of a Durapore membrane (Gopalakrishna, Chen, Gundimeda, Wilson, & Anderson, 1992). Samples (25 μl) to be assayed for PKC activity are pipetted into four wells. Fifty microliters of 20 m*M* EGTA is then added to the first two wells containing PKC samples. To the remaining two wells containing the PKC samples, 50 μl of a sonicated lipid mixture (100 μg/ml phosphatidylserine, 10 μg/ml diolein) is added. The reaction is started by adding a mixture containing 50 m*M* Tris–HCl (pH 7.5)/25 m*M* $MgCl_2$/ 0.8 m*M* $CaCl_2$/0.25 m*M* [γ-^{32}P]ATP (three million cpm)/histone H1 (0.25 mg/ml)/2 μ*M* leupeptin. The samples in a total volume of 125 μl are incubated at 30 °C for 5 min. After arresting the reaction with trichloroacetic acid and ultrafiltration, the radioactivity associated with the protein retained on the membrane is determined. In certain cases, histone H1 is replaced with 5 μ*M* neurogranin substrate peptide (residues 25–43), and the same reactions are carried out in regular 96-well plates without filtration discs. The reaction is arrested with 10 μl of 1 *M* phosphoric acid, the samples are applied to Whatman P81 paper (2 × 2 cm), and the papers are washed four times with 75 m*M* phosphoric acid. Radioactivity retained in the washed paper is then counted.

As the PKC assay use histone H1 is a relatively simple, this substrate is used for determining the activity of conventional PKC isoenzymes (α, β, γ). Histone H1, however, is not a good substrate for measuring the activity of novel and atypical isoenzymes. Conversely, the neurogranin (residues 25–43) is a potent substrate peptide and appropriate for determining the phosphotransferase activity of all PKC isoenzymes (Gonzalez, Klann, Sessoms, & Chen, 1993). Therefore, we use this substrate for determining the activity of all PKC isoenzymes in cell extracts.

3.1.2 Assessing modification of the regulatory domain by measuring phorbol ester binding

As the zinc-coordinating cysteine-rich region is required for phorbol ester binding, this binding assay may indicate the functional integrity of the cysteine-rich region in the regulatory domain. Phorbol ester binding to the H_2O_2-modified PKC is carried out with a multiwell filtration approach using [^{3}H]phorbol 12,13-dibutyrate (PDBu) as a ligand (Gopalakrishna et al., 1992). Briefly, 25 μl of purified PKC is incubated with 20 m*M* Tris–HCl, pH 7.4/0.6 m*M* $CaCl_2$/0.15 μ*M* leupeptin/0.06 μ*M* pepstatin A/20 n*M* [^{3}H]PDBu [20 Ci/mmol; 200,000 dpm]/phenylmethylsulfonyl fluoride-treated bovine serum albumin (0.1 mg/ml)/bovine γ-globulin (0.1 mg/ml) in the wells of the 96-well filtration plate with fitted Durapore membrane disks at 30 °C for 10 min. The ligand-bound PKC is absorbed to DEAE-Sephadex beads, and the beads are then filtered and washed in the same multiwells with 20 m*M* Tris–HCl, pH 7.5. The radioactivity associated with the DEAE-Sephadex beads retained on the filter is then counted.

3.2. Oxidative modification of PKC in cells by exogenous H_2O_2

3.2.1 Protocol for isolation of oxidatively modified PKC isoenzymes from cells

In the original report, the concentration of H_2O_2 needed to induce the oxidative activation of PKC in cells was relatively high (1–5 m*M*) (Gopalakrishna & Anderson, 1989). However, concentration as low as 100–300 μ*M* was found to be sufficient in order to induce the oxidative activation of PKC in J774A.1 macrophage cells (Kaul, Gopalakrishna, Gundimeda, Choi, & Forman, 1998). Differences in effective concentrations are partly due to protocol differences (i.e., lower concentrations are effective in experiments with serum-starved cells in a balanced salt solution). Variations among cell types are most likely due to differences in levels of reductases that reverse the oxidative modification.

Protocol: Various cell lines are grown in the 100-mm Petri dishes in the minimal essential medium (MEM) containing 10% fetal calf serum. At confluency, the medium is changed to serum-free medium, and the cells are incubated for overnight. The cells then are treated with 1 m*M* H_2O_2 for a given period of time. The treated cells are washed three times with Tris–HCl, pH 7.4/0.15 *M* NaCl and Dounce homogenizer in 3.5 ml of buffer B. It is important to use protease inhibitors during the isolation of oxidatively modified PKC from H_2O_2-treated cells and to measure the enzyme as quickly as possible. The homogenates are centrifuged at 13,000 × *g* for 15 min, and the supernatants are used to isolate PKC. To test whether the oxidative modification is reversible, cell homogenization is carried out as mentioned above in buffer B, and the homogenate is divided into two aliquots. To one of these aliquots, 2-mercaptoethanol is added to a final concentration of 10 m*M*. The samples with and without 2-mercaptoethanol are separately processed while maintaining the difference in 2-mercaptoethanol in all buffers used for PKC isolation by DEAE-cellulose chromatography.

Cell extract (2 ml) is applied to a 0.5-ml column of DEAE-cellulose (DE-52) previously equilibrated with buffer C. After washing the column with 2.5 ml of buffer C, the bound native form of PKC is eluted with 1.25 ml 0.1 *M* NaCl in buffer C (referred to as peak A). The oxidatively modified and constitutively active PKC isoenzymes are then eluted with 1.25 ml of 0.25 *M* NaCl in buffer C (peak B). The peak B fraction also has protein phosphatase 2A. Therefore, it is important to add its inhibitor microcystin LR (0.1 μ*M*) to this fraction in order to reliably measure the kinase activity of oxidatively modified PKC. In this method, synthetic neurogranin peptide (residues 28–43), a specific substrate for PKC, is used for determining protein kinase activity. Furthermore, this cofactor-independent protein kinase activity elevated in peak B is inhibited by pan-PKC-specific inhibitor bisindolylmaleimide (BIM). Therefore, this cofactor-independent protein kinase activity increase in peak B is most likely due to activation of PKC, rather than to activation of other protein kinases.

3.2.2 Protocol for measuring inactivation of the PKC regulatory domain by determining phorbol ester binding in intact cells

Cells are grown in 6-well plates. The medium is changed to a serum-free medium, and then, cells are treated with various concentrations of H_2O_2 for a given period of time. Then, 37.5 n*M* [^{3}H]PDBu (0.25 μCi) is added to the medium. To determine nonspecific binding, 10 μ*M* unlabeled PDBu is included with a radiolabel. After incubation for 30 min, cells are washed four times with ice-cold Tris–HCl, pH 7.4/0.15 *M* NaCl and lysed with 0.2 *M* NaOH. The radioactivity present in the cell extract is then determined. The specific binding is calculated by subtracting the nonspecific binding from the observed total binding (38).

4. INDIRECT CELLULAR REGULATION OF PKC ISOENZYMES BY SUBLETHAL LEVELS OF H_2O_2

4.1. H_2O_2-induced cytosol-to-membrane translocation of PKC

The second type of activation of PKC in H_2O_2-treated cells is indirect and is caused by membrane/mitochondrial association of PKC capable of being extracted with detergents. The membrane association of PKC is presumably caused by phospholipid-hydrolysis products. ROS activate phospholipid-hydrolyzing enzymes such as phospholipases C and D (Min, Kim, & Exton, 1998; Natarajan, Vepa, Verma, & Scribner, 1996). This results in the release of second messengers, such as DAG, that induce the binding of PKC to the membrane.

Protocol: Various cell types are grown in 100-mm Petri dishes in MEM supplemented with 10% fetal calf serum. At confluency, the cells are changed to serum-free medium and treated with various agents for a given period of time. Treated cells are washed with Tris–HCl, pH 7.4/0.15 *M* NaCl and homogenized in buffer C loose fitting glass–glass Dounce homogenizer (20 strokes). The homogenate is centrifuged at 13,000 × *g* for 10 min, and the supernatant is used as a soluble fraction. The particulate fractions are resuspended in buffer C with 1% Igepal CA-630 (v/v) by Dounce homogenizing (5 strokes). The suspension is left on ice for 5 min and then centrifuged at 13,000 × *g* for 10 min; the supernatant is used as a detergent-solubilized membrane fraction. The cytosol and membrane fractions obtained by centrifuging at 100,000 × *g* showed similar results to that seen with soluble and particulate fractions obtained by centrifuging at 13,000 × *g*. Therefore, ultracentrifugation is not needed. Both soluble fractions and particulate fractions are subjected to DEAE-cellulose chromatography: peak A and peak B fractions are eluted from the column and pan-PKC activity is determined as described in Section 3.1.1. The cytosol to membrane translocation of PKC also determined by Western immunoblotting of various PKC isoenzymes in these fractions.

4.2. Activation of PKCδ by H_2O_2-induced tyrosine phosphorylation

Although we included this mode of activation of PKCδ in H_2O_2-treated cells as an indirect mechanism mediated by its tyrosine phosphorylation, it might also be possible that a direct modification of the enzyme by H_2O_2 may partially contribute to this activation. The following protocol may be useful for measuring the H_2O_2-induced tyrosine phosphorylation of PKCδ at all tyrosine residues phosphorylated in H_2O_2-treated cells (Tyr311, Tyr322, and Tyr512) or specifically Tyr311.

Protocol: The H_2O_2-treated cells are washed with 20 m*M* Tris–HCl, pH 7.4/0.15 *M* NaCl and lysed in buffer D. The lysate is centrifuged at 13,000 × *g* for 10 min at 4 °C, and the supernatant (1 ml) is incubated with 1 μg anti-PKCδ rabbit polyclonal antibodies at 4 °C. Then, 30 μl of suspension of protein A/G Plus-agarose beads is added, and the mixture is incubated for additional 2 h with agitation. The beads are then collected by centrifugation at 2000 × *g* for 5 min at 4 °C followed by washing of the beads four times with buffer D with added 0.15 m*M* NaCl. The washed pellet is suspended in SDS sample buffer, boiled for 3 min, and centrifuged at 10,000 × *g* for 3 min. The proteins present in the supernatant are separated by SDS polyacrylamide gel electrophoresis. The electrophoretically separated proteins are transferred to a polyvinylidene fluoride membrane. The membranes are blocked with 5% dry milk and subsequently incubated with mouse monoclonal antiphosphotyrosine antibodies followed by rabbit anti-mouse secondary antibodies conjugated with horseradish peroxidase. The immunoreactive bands are visualized by SuperSignal West Femto Maximum Sensitivity Substrate kit (Pierce). These bands are analyzed by densitometric scanning using the Omega 12IC Molecular Imaging System and Ultraquant software.

The commercially available anti-PKCδ-phosphotyrosine-311 antibodies may be used to directly quantify PKCδ phosphorylation at Tyr-311 by Western immunoblotting without prior immunoprecipitation of PKCδ from the cell extracts.

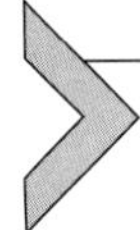

5. H_2O_2-INDUCED SIGNALING IN GTPP-INDUCED PRECONDITIONING FOR CEREBRAL ISCHEMIA

We have previously used various methodologies to determine H_2O_2-induced cell signaling in the action of tamoxifen, catechol, hydroquinone, and neuritogenesis (Gopalakrishna, Chen, & Gundimeda, 1994; Gopalakrishna et al., 2008; Gundimeda, Chen, & Gopalakrishna, 1996). Here, we describe the methodology recently applied to the study of H_2O_2-mediated redox signaling in the action of GTPPs in preconditioning against cell death caused by OGD/R (Gundimeda et al., 2012).

5.1. 67LR-mediated activation of NADPH oxidase

EGCG binds with high affinity to the 67LR, a cell-surface nonintegrin-type receptor (Tachibana, Koga, Fujimura, & Yamada, 2004). The following procedure is used to demonstrate the role of 67LR in mediating the EGCG-induced ROS generation via NADPH oxidase.

Protocol: PC12 cells are grown to confluency in polylysine-coated 96-well plates. The cells are then washed with Krebs–Ringer–HEPES buffer (KRH) and incubated with 10 μ*M* 2′,7′-dichlorofluorescin (DCFDA

Molecular Probes) for 30 min at 37 °C. Then the cells are washed three times with KRH to remove any excess probe. Various concentrations of GTPP or EGCG are added in KRH, and fluorescent intensity is determined using SpectraMax M2e fluorescence microplate reader (Molecular Devices) with excitation at 488 nm and emission at 525 nm. In this context, we determine ROS based on the assumption that H_2O_2 is the major species that oxidizes dichlorofluorescin into a fluorescent compound. Peroxynitrite, however, also oxidizes dichlorofluorescin (Wang & Joseph, 1999). However, pretreating PC12 cells with N^G-nitro-L-arginine methyl ester, a nitric oxide synthase inhibitor, did not decrease the GTPP-induced fluorescence, suggesting that this increase in fluorescence is unlikely to be caused by reactive nitrogen species. To establish the specific role of the 67LR in intracellular H_2O_2 production, PC12 cells are preincubated with anti-67LR antibodies for 2 h. Cells are preloaded with DCFDA and treated with GTPP (0.05 μg/ml). Fluorescence was measured as described above.

DCFDA preloaded cells are incubated with 0.05 μg/ml GTPP along with 10 μ*M* DPI and 25 μ*M* VAS2870 NADPH oxidase inhibitors. The fluorescence intensity is then measured as described above. The decrease in fluorescence is considered as the involvement of NADPH oxidase in GTPP-induced ROS generation.

5.2. Role of endogenous H_2O_2 in GTPP-induced preconditioning

The following methods are useful for establishing the role of H_2O_2 in mediating preconditioning against OGD/R-induced cell death.

5.2.1 Oxygen–glucose deprivation/reoxygenation

Protocol: We are utilizing a modified *in vitro* cell culture model using a rat pheochromocytoma PC12 cell line to mimic ischemia/reperfusion-induced cell death (Tabakman, Lazarovici, & Kohen, 2002). PC12 cells are grown in RPMI medium supplemented with 10% heat-inactivated horse serum, 5% fetal calf serum, 50 units/ml penicillin, and 0.05 mg/ml streptomycin. Poly-L-lysine-coated flasks or Petri dishes are used for cell culture. Briefly, PC12 cells are washed once with glucose-free DMEM previously bubbled with a mixture of 95% nitrogen and 5% CO_2. Cells are kept in this deoxygenated glucose-free medium. The plates are then placed in a modular incubation chamber (Billups-Rothenberg) and flushed with 95% nitrogen/5% CO_2 for 4 min at a flow rate of 10 l/min. The chamber is then sealed and kept in an incubator for 3 h at 37 °C. OGD is terminated by adding glucose to a final concentration of 4.5 mg/ml followed by incubation in a normoxic incubator for 18–20 h (reoxygenation). Control cells are washed with glucose-containing DMEM and incubated in the normoxic incubator for 24 h.

5.2.2 Method for GTPP-induced preconditioning

Initially, we treat PC12 cells with GTPP or EGCG for 24–48 h prior to subjecting them to OGD/R. Although a 24-h pretreatment with GTPP or EGCG showed appreciable protection against OGD/R, a 48-h treatment showed optimal protection. Because of the instability of GTPP constituents in the culture medium, GTPP 0.2 μg/ml is added twice daily to the medium. Due to synergistic interaction and enhanced stability when GTPP constituents are present together (Suganuma et al., 1999), we use unfractionated GTPP mixture instead of pure compounds. However, because EGCG is the major polyphenol responsible for the action of GTPP, we also test this compound to a limited extent. The optimal concentration of EGCG is observed at 2 μ*M*.

5.2.3 Cell death-related assays employed after OGD/R

PC12 cells are grown in 96-well plates. Cellular release of lactate dehydrogenase (LDH) is determined using CytoToxo-ONE Homogeneous Membrane Integrity assay kit (Promega) according to the manufacturer's instructions. Cells are lysed to obtain maximum LDH release values (full-kill numbers). The percent of cell death (% of LDH release) is calculated by dividing the experimental time point by the full-kill values × 100. Caspase-3 activity is determined using tetrapeptide substrate (*N*-acetyl-Asp-Glu-Val-Asp-7-amino-4-methylcoumarin) with an assay kit obtained from BIOMOL. Cells are seeded in 100-mm Petri dishes and allowed to grow for 24 h. The cells are treated with GTPP and subjected to OGD/R. Then, the treated cells are homogenized, and caspase-3 activity is determined fluorometrically according to the manufacturer's instructions. For the apoptosis assay, cells are grown on polylysine-coated chamber slides, treated with GTPP and subjected to ODG/R, and then fixed with 4% paraformaldehyde in phosphate-buffered saline (PBS) for 20 min at room temperature. After rinsing them with PBS, cells are stained with 4′,6-diamidino-2-phenyl-indole (10 μg/ml) for 5 min. The morphology of the nuclei is observed using a fluorescence microscope (Nikon Eclipse TE300) at an excitation wavelength of 345 nm. Apoptotic nuclei are identified by chromatin condensation and fragmentation.

5.2.4 Method to establish the role of H_2O_2 in preconditioning

The cells are pretreated for 18 h with cell-permeable PEG-catalase and then subjected to GTPP-induced preconditioning. Heat-inactivated catalase is used as a negative control. Furthermore, the role of H_2O_2 or other ROS in preconditioning can be established by using an exogenous ROS-generating system comprising of 100 μM xanthine, 1 milliunit/ml of xanthine oxidase, and 40 units/ml SOD by incubating for 3 h. Then the medium is replaced with fresh medium. This step is repeated on the second day. During the incubation of cells with the ROS-generating system, catalase (500 units) is added to determine the role of H_2O_2 in this process. Heat-inactivated catalase is used as a control.

5.3. H_2O_2-induced activation of PKCε in preconditioning

5.3.1 Role of pan-PKC in preconditioning

Protocol: We use two specific inhibitors acting via different mechanisms to determine the role of pan-PKC activity in GTPP-induced preconditioning. Calphostin C is a potent inhibitor of the PKC regulatory domain and induces irreversible inactivation of PKC, whereas BIM is a potent inhibitor of the PKC catalytic site (Gopalakrishna et al., 1995). We test its inactive analog BIM V as a negative control to exclude the possibility of nonspecific inhibition of other enzymes by BIM. Coincubation of PC12 cells with calphostin C or BIM and GTPP for 2 days negates the ability of GTPP to induce preconditioning against OGD/R-induced cell death.

5.3.2 Assessing oxidative activation of pan-PKC in GTPP-treated cells

Oxidative activation of PKC in intact cells is determined by its conversion from cofactor-dependent form to cofactor-independent form (peak A to Peak B conversion) as discussed in Section 3.2.1. However, when GTPP-treated cells are homogenized in a buffer containing 2-mercaptoethanol, pan-PKC activity is mostly eluted in peak A in a cofactor-dependent form, similar to that seen in the control untreated cells. In addition, in cells pretreated with PEG-catalase for 18 h, GTPP treatment does not generate a constitutively active form of PKC. Thus, this method separates the cofactor-dependent form from the constitutively active form which is most likely caused by H_2O_2-induced redox modification of PKC.

In GTPP-treated cells, a cytosol-to-membrane translocation of PKC also occurs which may indicate the indirect activation of PKC by lipid hydrolysis products derived from oxidative regulation of phospholipases as discussed in Section 4.1.

5.3.3 Immunocomplex precipitation assay for PKCε

PKCε is known to play an important role in cell survival and neuroprotection (Wang, Bright, Mochly-Rosen, & Giffard, 2004); therefore, we have focused on PKCε. Cell extracts are prepared in the presence of detergent (1% Igepal CA-630) as described in Section 3.2.1. The extract (1 ml) is incubated in a microfuge tube with 1 μg of anti-PKCε rabbit polyclonal antibodies or anti-PKCε mouse monoclonal antibodies for 1 h at 4 °C. Then, 30 μl of suspension of protein A/G Plus-agarose beads are added and the mixture is incubated for an additional 2 h with agitation. Then the beads are collected by centrifugation at 2000 × *g* for 5 min at 4 °C. The beads are washed twice with buffer E and then additionally washed twice with the same buffer without detergent. The pellet is resuspended in 125 μl of buffer F, and the PKCε activity present in this fraction is determined using neurogranin peptide as a substrate. It is important to determine whether there is any generation of cofactor-independent activity for PKCε in GTPP-treated cells and whether this activity is reversed by 2-mercaptoethanol.

5.3.4 Protocol for determining mitochondrial association of PKCε

Mitochondrial association of PKCε has previously shown to offer neuroprotection (Wang et al., 2004). Mitochondria are isolated by differential centrifugation with a modification of a previously published method (Frezza, Cipolat, & Scorrano, 2007). Briefly, PC12 cells are collected by centrifugation at 600 × *g* for 10 min at 4 °C. The cell pellet is homogenized in buffer G. The homogenate is centrifuged at 600 × *g* for 10 min at 4 °C to remove nuclei and debris. The supernatant is further centrifuged at 9000 × *g* for 15 min at 4 °C, and the resulting mitochondrial pellet is subjected to SDS-PAGE and immunoblotted for PKCε.

5.3.5 Establishing role of PKCε by its overexpression and knockout

We use two different approaches to determine whether there is a correlation between the expression of PKCε in PC12 cells and the extent of GTPP-induced preconditioning against OGD/R-induced cell death. In the first approach, we use previously generated PC12 cells stably transfected with either a metallothionein-driven PKCε expression vector (to overexpress PKCε) or an empty vector (as a control (Gopalakrishna et al., 2008)). Western blot analysis is used to determine the extent of expression of PKCε in these transfectants. Cadmium chloride is used for the optimal expression of PKCε in these transfectants. In the second approach, we suppress the levels of PKCε. PC12 cells are plated in a six-well plate. After 24 h, 50 n*M* PKCε siRNA oligonucleotides (three predesigned Silencer oligonucleotides from Ambion) are transfected into PC12 cells with Lipofectamine 2000 (Life Technologies) according to the manufacturer's instructions. As a negative control, we use Silencer siRNA that did not exhibit homology to any encoding region. The efficiency of transfection and knockout of PKCε is determined by Western immunoblotting.

6. SUMMARY

In this chapter, we focused on describing the protocols applicable to H_2O_2-induced oxidative modification of purified PKC isoenzymes. Methodology is provided for inducing oxidative modification of PKC isoenzymes in intact cells as well as isolation protocols of such modified forms from the H_2O_2-treated cells. We have also described procedures for assessing the indirect regulation of PKC by its translocation from the cytosol to the plasma membrane or mitochondria, as well as PKC regulation by tyrosine phosphorylation. Furthermore, methodology is provided to demonstrate the role of H_2O_2-induced PKC cell signaling in GTPP-induced preconditioning against neuronal cell death caused by OGD/R. In this case, the methods used support the fact that a limited amount of H_2O_2 generated within the cell by receptor-mediated transmembrane signaling is sufficient to induce PKC oxidative activation. We believe that these methodologies are also applicable to growth factors and neurotrophins which induce the generation of H_2O_2 as a second messenger.

ACKNOWLEDGMENTS

This work was supported in part by NIH Grant CA99216 from the National Cancer Institute. We thank Carleen Sarksian, Alan Hung, and David Rayudu for their excellent technical assistance.

REFERENCES

Blumberg, P. M., Acs, G., Areces, L. B., Kazanietz, M. G., Lewin, N. E., & Szallasi, Z. (1994). Protein kinase C in signal transduction and carcinogenesis. *Progress in Clinical and Biological Research*, *387*, 3–19.

Frezza, C., Cipolat, S., & Scorrano, L. (2007). Organelle isolation: Functional mitochondria from mouse liver, muscle and cultured fibroblasts. *Nature Protocols*, *2*(2), 287–295.

Giorgi, C., Agnoletto, C., Baldini, C., Bononi, A., Bonora, M., Marchi, S., et al. (2010). Redox control of protein kinase C: Cell- and disease-specific aspects. *Antioxidants & Redox Signaling*, *13*(7), 1051–1085.

Gonzalez, A., Klann, E., Sessoms, J. S., & Chen, S. J. (1993). Use of the synthetic peptide neurogranin(28-43) as a selective protein kinase C substrate in assays of tissue homogenates. *Analytical Biochemistry*, *215*(2), 184–189.

Gopalakrishna, R., & Anderson, W. B. (1989). Ca^{2+}- and phospholipid-independent activation of protein kinase C by selective oxidative modification of the regulatory domain. *Proceedings of the National Academy of Sciences of the United States of America*, *86*(17), 6758–6762.

Gopalakrishna, R., Chen, Z. H., & Gundimeda, U. (1994). Tobacco smoke tumor promoters, catechol and hydroquinone, induce oxidative regulation of protein kinase C and influence invasion and metastasis of lung carcinoma cells. *Proceedings of the National Academy of Sciences of the United States of America*, *91*(25), 12233–12237.

Gopalakrishna, R., Chen, Z. H., & Gundimeda, U. (1995). Modifications of cysteine-rich regions in protein kinase C induced by oxidant tumor promoters and enzyme-specific inhibitors. *Methods in Enzymology*, *252*, 132–146.

Gopalakrishna, R., Chen, Z. H., Gundimeda, U., Wilson, J. C., & Anderson, W. B. (1992). Rapid filtration assays for protein kinase C activity and phorbol ester binding using multiwell plates with fitted filtration discs. *Analytical Biochemistry*, *206*(1), 24–35.

Gopalakrishna, R., Gundimeda, U., Schiffman, J. E., & McNeill, T. H. (2008). A direct redox regulation of protein kinase C isoenzymes mediates oxidant-induced neuritogenesis in PC12 cells. *The Journal of Biological Chemistry*, *283*(21), 14430–14444.

Gopalakrishna, R., & Jaken, S. (2000). Protein kinase C signaling and oxidative stress. *Free Radical Biology & Medicine*, *28*(9), 1349–1361.

Griner, E. M., & Kazanietz, M. G. (2007). Protein kinase C and other diacylglycerol effectors in cancer. *Nature Reviews. Cancer*, 7(4), 281–294.

Gundimeda, U., Chen, Z. H., & Gopalakrishna, R. (1996). Tamoxifen modulates protein kinase C via oxidative stress in estrogen receptor-negative breast cancer cells. *The Journal of Biological Chemistry*, *271*(23), 13504–13514.

Gundimeda, U., McNeill, T. H., Elhiani, A. A., Schiffman, J. E., Hinton, D. R., & Gopalakrishna, R. (2012). Green tea polyphenols precondition against cell death induced by oxygen-glucose deprivation via stimulation of laminin receptor, generation of reactive oxygen species, and activation of protein kinase Cepsilon. *The Journal of Biological Chemistry*, *287*(41), 34694–34708.

Gundimeda, U., McNeill, T. H., Schiffman, J. E., Hinton, D. R., & Gopalakrishna, R. (2010). Green tea polyphenols potentiate the action of nerve growth factor to induce neuritogenesis: Possible role of reactive oxygen species. *Journal of Neuroscience Research*, *88*(16), 3644–3655.

Gundimeda, U., Schiffman, J. E., Chhabra, D., Wong, J., Wu, A., & Gopalakrishna, R. (2008). Locally generated methylseleninic acid induces specific inactivation of protein kinase C isoenzymes: Relevance to selenium-induced apoptosis in prostate cancer cells. *The Journal of Biological Chemistry*, *283*(50), 34519–34531.

Kaul, N., Gopalakrishna, R., Gundimeda, U., Choi, J., & Forman, H. J. (1998). Role of protein kinase C in basal and hydrogen peroxide-stimulated NF-kappa B activation in the murine macrophage J774A.1 cell line. *Archives of Biochemistry and Biophysics*, *350*(1), 79–86.

Kazanietz, M. G., Wang, S., Milne, G. W., Lewin, N. E., Liu, H. L., & Blumberg, P. M. (1995). Residues in the second cysteine-rich region of protein kinase C delta relevant to phorbol ester binding as revealed by site-directed mutagenesis. *The Journal of Biological Chemistry*, *270*(37), 21852–21859.

Klann, E., Roberson, E. D., Knapp, L. T., & Sweatt, J. D. (1998). A role for superoxide in protein kinase C activation and induction of long-term potentiation. *The Journal of Biological Chemistry*, *273*(8), 4516–4522.

Knapp, L. T., & Klann, E. (2000). Superoxide-induced stimulation of protein kinase C via thiol modification and modulation of zinc content. *The Journal of Biological Chemistry, 275*(31), 24136–24145.

Konishi, H., Yamauchi, E., Taniguchi, H., Yamamoto, T., Matsuzaki, H., Takemura, Y., et al. (2001). Phosphorylation sites of protein kinase C delta in H2O2-treated cells and its activation by tyrosine kinase in vitro. *Proceedings of the National Academy of Sciences of the United States of America, 98*(12), 6587–6592.

Min, D. S., Kim, E. G., & Exton, J. H. (1998). Involvement of tyrosine phosphorylation and protein kinase C in the activation of phospholipase D by H2O2 in Swiss 3T3 fibroblasts. *The Journal of Biological Chemistry, 273*(45), 29986–29994.

Mochly-Rosen, D. (1995). Localization of protein kinases by anchoring proteins: A theme in signal transduction. *Science, 268*(5208), 247–251 [Review] [56 refs].

Natarajan, V., Vepa, S., Verma, R. S., & Scribner, W. M. (1996). Role of protein tyrosine phosphorylation in H2O2-induced activation of endothelial cell phospholipase D. *The American Journal of Physiology, 271*(3 Pt. 1), L400–L408.

Newton, A. C. (1997). Regulation of protein kinase C. *Current Opinion in Cell Biology, 9*(2), 161–167 [Review] [58 refs].

Nishizuka, Y. (1992). Intracellular signaling by hydrolysis of phospholipids and activation of protein kinase C. *Science, 258*(5082), 607–614.

Parker, P. J., & Murray-Rust, J. (2004). PKC at a glance. *Journal of Cell Science, 117*(Pt. 2), 131–132 [Review] [12 refs].

Perez-Pinzon, M. A., Dave, K. R., & Raval, A. P. (2005). Role of reactive oxygen species and protein kinase C in ischemic tolerance in the brain. *Antioxidants & Redox Signaling,* 7(9–10), 1150–1157.

Poole, A. W., Pula, G., Hers, I., Crosby, D., & Jones, M. L. (2004). PKC-interacting proteins: From function to pharmacology. *Trends in Pharmacological Sciences, 25*(10), 528–535.

Rhee, S. G. (2004). Cell signaling. H2O2, a necessary evil for cell signaling. *Science, 312*(5782), 1882–1883.

Rhee, S. G., Chang, T.-S. , Bae, Y. S., Lee, S.-R. , & Kang, S. W. (2003). Cellular regulation by hydrogen peroxide. *Journal of the American Society of Nephrology, 14*(8 Suppl. 3), S211–S215.

Rybin, V. O., Guo, J., Sabri, A., Elouardighi, H., Schaefer, E., & Steinberg, S. F. (2004). Stimulus-specific differences in protein kinase C delta localization and activation mechanisms in cardiomyocytes. *The Journal of Biological Chemistry, 279*(18), 19350–19361.

Steinberg, S. F. (2008). Structural basis of protein kinase C isoform function. *Physiological Reviews, 88*(4), 1341–1378.

Suganuma, M., Okabe, S., Kai, Y., Sueoka, N., Sueoka, E., & Fujiki, H. (1999). Synergistic effects of (–)-epigallocatechin gallate with (–)-epicatechin, sulindac, or tamoxifen on cancer-preventive activity in the human lung cancer cell line PC-9. *Cancer Research, 59*(1), 44–47.

Tabakman, R., Lazarovici, P., & Kohen, R. (2002). Neuroprotective effects of carnosine and homocarnosine on pheochromocytoma PC12 cells exposed to ischemia. *Journal of Neuroscience Research, 68*(4), 463–469.

Tachibana, H., Koga, K., Fujimura, Y., & Yamada, K. (2004). A receptor for green tea polyphenol EGCG. *Nature Structural and Molecular Biology, 11*(4), 380–381.

Umada-Kajimoto, S., Yamamoto, T., Matsuzaki, H., & Kikkawa, U. (2006). The complex formation of PKCdelta through its C1- and C2-like regions in H2O2-stimulated cells. *Biochemical and Biophysical Research Communications, 341*(1), 101–107.

Wang, J., Bright, R., Mochly-Rosen, D., & Giffard, R. G. (2004). Cell-specific role for epsilon- and betaI-protein kinase C isozymes in protecting cortical neurons and astrocytes from ischemia-like injury. *Neuropharmacology, 47*(1), 136–145.

Wang, H., & Joseph, J. A. (1999). Quantifying cellular oxidative stress by dichlorofluorescein assay using microplate reader. *Free Radical Biology & Medicine*, *27*(5–6), 612–616.
Zabouri, N., & Sossin, W. S. (2002). Oxidation induces autonomous activation of protein kinase C Apl I, but not protein kinase C Apl II in homogenates of Aplysia neurons. *Neuroscience Letters*, *329*(3), 257–260.

CHAPTER SIX

p66Shc, Mitochondria, and the Generation of Reactive Oxygen Species

Mirella Trinei[*], Enrica Migliaccio[*], Paolo Bernardi[†], Francesco Paolucci[‡], Piergiuseppe Pelicci[*], Marco Giorgio[*,1]

[*]Department of experimental Oncology, European Institute of Oncology, Milan, Italy
[†]Department of Biomedical Sciences, University of Padova, Padova, Italy
[‡]Department of Chemistry "G. Ciamician", University of Bologna, Bologna, Italy
[1]Corresponding author: e-mail address: marco.giorgio@ieo.eu

Contents

1. Introduction 100
2. The *P66* Gene and Protein 100
3. The Mitochondrial Function of p66Shc 101
4. Preparation of Recombinant p66Shc Protein 102
5. Mitochondrial Swelling Assay 104
6. Mitochondrial ROS Formation by p66Shc 105
   6.1 p66Shc and cytochrome *c* CV 105
   6.2 Mitochondrial ROS formation assay 106
7. Conclusions: Role of p66Shc ROS 107
Acknowledgments 108
References 108

Abstract

Reactive oxygen species (ROS), mainly originated from mitochondrial respiration, are critical inducers of oxidative damage and involved in tissue dysfunction. It is not clear, however, whether oxidative stress is the result of an active gene program or it is the by-product of physiological processes. Recent findings demonstrate that ROS are produced by mitochondria in a controlled way through specialized enzymes, including p66Shc, and take part in cellular process aimed to ensure adaptation and fitness. Therefore, genes generating specifically ROS are selected determinants of life span in response to different environmental conditions.

Methods in Enzymology, Volume 528
ISSN 0076-6879
http://dx.doi.org/10.1016/B978-0-12-405881-1.00006-9

ABBREVIATIONS

CV cyclic voltammetry
PTP permeability transition pore
ROS reactive oxygen species

1. INTRODUCTION

According to the free radical mitochondrial theory of aging, the aging process is due to the oxidative damage caused by reactive oxygen species (ROS) generated mainly during mitochondrial respiration when O_2 may be partially reduced. It is not clear, however, whether oxidative pressure is attributable to active gene programming or is the by-product of critical physiological processes. Indeed, significant electron leakage at specific redox centers during mitochondrial electron transfer chain reactions has been demonstrated to be responsible for a significant fraction of the cellular ROS (Murphy, 2009). However, the emerging picture is that mitochondria generate ROS in a regulated manner, reflecting the metabolic activity of the cell and acting as sensor involved in signal transduction pathways (as examples, see Bell et al., 2007; Hoffman & Brookes, 2009). Moreover, several enzymatic systems associated with mitochondria contribute to oxidative signaling (see Bao et al., 2009 for review on monoamine oxidase and NADPH oxidase). These enzymes are mainly "known" oxidoreductase and include monoamine oxidases and NADPH oxidases that release H_2O_2 although others, such as oxidoreductin 1, directly target sulfide bonds on proteins.

An intriguing example of how the mitochondrial-cell cross talk evolved is represented by the unexpected enzymatic function of p66Shc and phenotype of its mutation in mice, that is, resistance to stresses and longevity. Here, we report on the experiments that revealed the biology of p66Shc, from the electron transfer activity at molecular level to survival in wild at genetic level.

2. THE *P66* GENE AND PROTEIN

The p66Shc protein is the largest isoform encoded by the ShcA locus, mapping in the human chromosome 1 or the mouse chromosome 3. The ShcA locus encodes three isoforms trough two different promoters, one

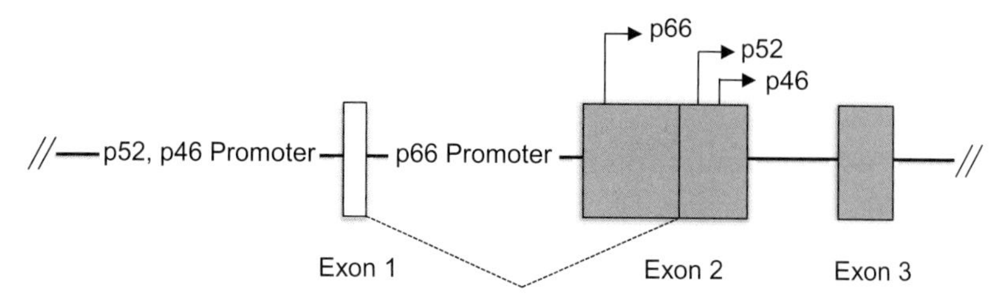

Figure 6.1 Transcription and translation starts of mammalian ShcA isoforms. The ATGs of p66, p52, and p46 ShcA isoforms locate in the second exon of the ShcA locus. The transcription of the mRNA-encoding p52Shc and p46Shc starts from a promoter upstream the first exon that is transcribed but not translated splicing downstream the ATG of p66Shc. The p66Shc isoform is transcribed by another promoter that initiates the transcription upstream the second exon. Then all the three isoforms share the remaining exon–intron structure.

transcribing for the p46 and p52 isoforms, whereas the other one transcribing for the p66 isoform (Fig. 6.1).

The other two isoforms, p52Shc and p46Shc, function as adaptor protein in signal transduction pathways linking different activated receptor tyrosine kinases to the Ras pathway by recruitment of the GRB2/SOS complex. On the contrary, p66Shc is not involved in Ras activation, although p66Shc has the typical domain organization of all members of the Shc family of adaptor proteins, with a carboxy-terminal Src homology type-2 (SH2) domain, an adjacent glycine- and proline-rich region, which has a certain degree of homology with a helix of collagen (collagen homologous region; CH1), a phosphotyrosine-binbing domain and an additional, aminoterminal CH region (CH2). Notably, p66Shc, at variance with the other Shc isoforms, is peculiar of vertebrates being conserved in *Fucu*, *Xenophus*, *Rattus*, *Mus*, and *Homo*, but not in *Saccharomyces*, *Caenorhabditis*, and *Drosophila* (see for review Luzi, Confalonieri, Di Fiore, & Pelicci, 2000; Trinei et al., 2009).

3. THE MITOCHONDRIAL FUNCTION OF p66Shc

The first evidence that p66Shc might possess nonredundant functions with respect to p46 and p52Shc came 15 years ago when two groups (Kao, Waters, Okada, & Pessin, 1997; Migliaccio et al., 1997) independently showed that, in contrast to p52Shc, overexpression of p66Shc was incapable of transforming mouse fibroblasts, did not induce MAPK activation, and had a negative effect on the fos promoter in an a transactivation assay. Then, p66Shc overexpression, contrary to other Shc isoforms, was described to suppress the Ras signaling in several cellular systems (Baldari & Telford, 1999). In 1999, our group reported the preliminary characterization of a knockout mouse for p66Shc (Migliaccio et al., 1999). In striking contrast with ShcA −/− animals, p66Shc null mice developed normally and resulted to be protected from aging-associated diseases, such as atherosclerosis, showing prolonged life span (Berry et al., 2007; Menini et al., 2006; Napoli et al., 2003). Moreover, during the last decade, p66Shc −/− cells were shown to be resistant to apoptosis induced by a variety of different signals, including H_2O_2, UV, staurosporine, taxol, growth factor deprivation, calcium ionophore, osmotic shock, and CD3–CD4 cross-linking (see for review Migliaccio, Giorgio, & Pelicci, 2006), and similarly, different tissues of the p66Shc knockout mice were found to be resistant to apoptosis induced by paraquat (Migliaccio et al., 1999), hypercholesterolemia (Napoli et al., 2003), hyperglycaemia (Menini et al., 2006), immunotoxicity (Su et al., 2012), and ischemia (Carpi et al., 2009).

Consistently with the role proposed for oxidative stress on cell death and aging (Perez-Campo, López-Torres, Cadenas, Rojas, & Barja, 1998), Mouse embryo fibroblasts (MEFs) derived from p66Shc −/− embryos have lower intracellular concentration of ROS, as revealed by the reduced oxidation of ROS-sensitive probes and the reduced accumulation of endogenous markers of oxidative stress (8-oxo-guanosine) (Trinei et al., 2002). Likewise, p66Shc −/− mice have diminished levels of both systemic isoprostane (Napoli et al., 2003) and intracellular (nitrotyrosines, 8-oxo-guanosine) oxidative stress (Trinei et al., 2002).

At mechanistic level, the mechanism of cell death mediated by p66Shc was described always involving the disruption of mitochondrial network, the release of cytochrome *c*, and the activation of caspase cascade (Orsini et al., 2004). Indeed, p66Shc null background is resistant to a variety of stimuli that induce the intrinsic way of apoptosis (see for review Migliaccio et al., 2006). However, substantial evidence that p66Shc induces mitochondrial swelling came from the observation that the p66Shc recombinant protein directly affects mitochondrial integrity.

4. PREPARATION OF RECOMBINANT p66Shc PROTEIN

Expression of eukaryotic proteins in bacteria is a common procedure for basic research and applicative biochemistry. To obtain enough p66Shc (or the p52Shc and p46Shc) protein, we cloned the coding sequences of the human p66Shc cDNA (or the corresponding cDNA of the other two

ShcA isoforms) into pGEX-6P-1 vector (GE Healhcare) was expressed in BL21(DE3)pLysS *Escherichia coli* cells. Bacteria were grown in LB medium containing 50 μg/ml chloramphenicol and 50 μg/ml ampicillin. Protein expression was induced by the addition of 0.5 m*M* IPTG at an OD_{600} of 0.6.

The concentration of cells in the culture has been determined with a spectrophotometer by measuring the amount of 600-nm light scattered by the culture. Time-dependent variations of the registered absorbance (A_{600}), also expressed as optical density $(OD)_{600}$, reflect the growth of cells. Results revealed that p66Shc expression, only, inhibits bacterial growth. The removal of the IPTG in the culture medium of the p66Shc expressing cells restores the normal growth, thereby suggesting a cytostatic property of human p66Shc rather than an effective cytotoxicity (Fig 6.2).

Therefore, to improve p66Shc production, growth was carried on overnight at 18 °C and 150 rpm. Cells were harvested by centrifugation at 4000 × *g* for 20 min, and the obtained bacterial pellet was suspended in lysis buffer (50 m*M* Tris–HCl, pH 7.4, 50 m*M* NaCl, 5% glycerol, 5 m*M* DTT)

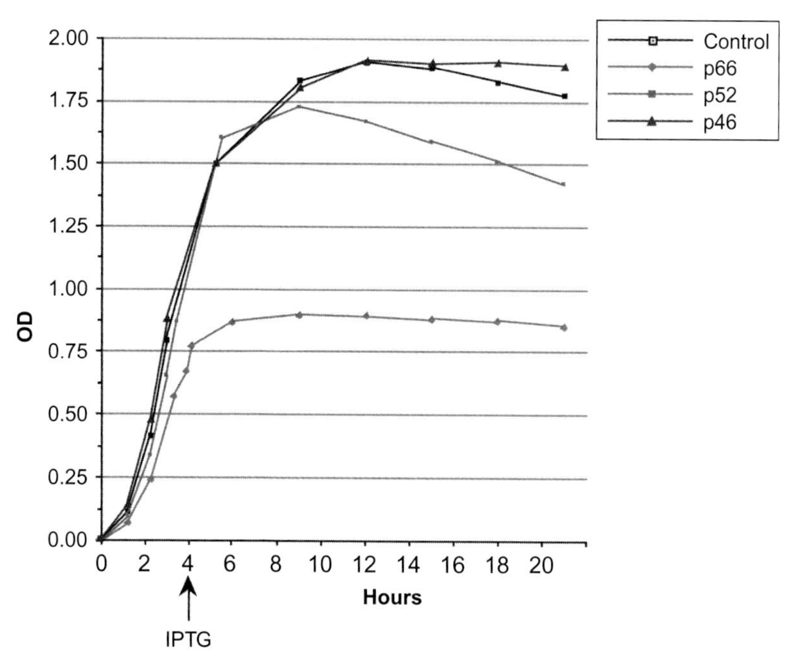

Figure 6.2 Growth of *E. coli* clones expressing the different human ShcA isoforms. Average of eight sample analysis for a representative experiment is reported. (For color version of this figure, the reader is referred to the online version of this chapter.)

plus 1 mg/ml lysozyme and disrupted by sonication. Each step was performed at 4 °C unless stated otherwise.

Cell lysate was cleared by centrifugation at 15,000 × *g* for 30 min, and the supernatant was batch incubated with Glutathione Sepharose 4B beads for 60 min. After two washes in lysis buffer, elution of the fusion protein was performed in the same buffer plus 10 m*M* glutathione. Copurifying DnaK contaminant can be removed from GST by ion-exchange chromatography. For this purpose, ion exchanger RESOURCE Q, Buffer A (50 m*M* Tris–HCl, pH 7.4, 1 m*M* EDTA, 5 m*M* DTT, 25 m*M* NaCl), and Buffer B (Buffer A + 175 m*M* NaCl) were used, with a total elution volume of 200 ml (a fraction size of 10 ml) and a flow rate of 5 ml/min.

The fractions containing GST-p66Shc fusion protein were pooled and subjected to Prescission protease cleavage (one unit/200 µg GST-p66) for 20 h at 4 °C. GST, GST-fusion uncleaved, and GST-Prescission protease were removed using Glutathione beads, and the supernatant was applied again to RESOURCE Q and eluted with 0.5 *M* NaCl in Buffer A. After a Centriprep 30 kDa MWCO step, purified p66Shc protein was obtained using Superdex 200 column with 10 m*M* Tris–HCl, pH 7.4, 100 m*M* NaCl, 1 m*M* EDTA, and 5 m*M* DTT as chromatographic buffer and a flow rate of 0.5 ml/min.

5. MITOCHONDRIAL SWELLING ASSAY

The release of proapoptotic factors from mitochondria is due to the disruption of the organelle integrity (Green & Kroemer, 2004). The opening of a high-conductance channel, the permeability transition pore (PTP), triggers these events. Opening of the PTP provokes an increase in inner membrane permeability to ions and solutes, followed by net water influx toward the mitochondrial matrix, swelling of the organelle, and physical rupture of its outer membrane, with the consequent release of proteins of the intermembrane space, including cytochrome *c* (cyt *c*) (Bernardi, Petronilli, Di Lisa, & Forte, 2001).

To test the effect of p66Shc protein on the swelling response to calcium challenges of isolated mitochondria, fresh mitochondria were purified from WT or p66Shc −/− mouse livers by differential centrifugations as described by Rapino et al. in this volume and assayed for swelling. Briefly, 0.4 mg/ml mitochondrial suspension in 125 m*M* KCl, 10 m*M* MOPS–Tris, 1 m*M* inorganic phosphate, 5 m*M* succinate, and 25 µ*M* Digitonin, pH 7.4 was incubated with [$CaCl_2$] (from 10 to 300 µ*M*) and or different amount of recombinant p66Shc (up to final concentration of 50 μ*M*) in a spectrophotometric cuvette to allow OD_{620} determination. Swollen mitochondria, in fact, decrease the absorbance of the suspension.

Results from this assay revealed that the addition of recombinant p66Shc accelerated the swelling of mitochondria and the simultaneous addition of the PTP inhibitor Cyclosporin A or antioxidant such as *N*-ethylmaleimide prevented the swelling induced by p66Shc, thus suggesting that p66Shc induced mitochondrial permeability transition through oxidative stress (Giorgio et al., 2005).

6. MITOCHONDRIAL ROS FORMATION BY p66Shc

Indeed, ROS induce the opening of the PTP through oxidation-dependent mechanisms and are potent inducers of apoptosis, both in cultured cells and *in vivo* (Petronilli et al., 1994). How p66Shc might stimulate the oxidation of PTP in isolated mitochondria was clear when we could demonstrate the redox properties of recombinant p66Shc protein by cyclic voltammetry (CV).

6.1. p66Shc and cytochrome *c* CV

Briefly, a three-electrode cell with separated compartments was used, composed of a saturated calomel electrode as reference, platinum spiral wire as counter electrode, and a gold disk (2-mm diameter) as working electrode. The gold disk electrode was cleaned with 1, 0.3, and 0.05 μm alumina slurry, sonicated for 5 min in 50% ethanol, and rinsed in water. The surface-assembled monolayer (SAM) was formed by soaking the gold electrode for 5 h in a 1-m*M* solution of 11-mercaptoundecanoic acid in ethanol, rinsed with absolute ethanol and dried under vacuum. To obtain a more compact surface coverage, the modified electrode was electro-activated by performing 20 voltammetric cycles at a sweep rate of 10 mV/s between the range of stability of the alkanethiol film (from 0.7 to −0.3 V). Then, the electrode was immersed again for further 5 h into the alkanethiol solution, rinsed, and finally coated with purified recombinant p66Shc protein by drying a drop of 1.0 mg/ml protein solution (1 m*M* Tris–EDTA, pH 7.0).

Measurements will be performed in phosphate buffer and LiClO4 as supporting electrolyte, at constant ionic strength (100 m*M*) in the presence or absence of different concentrations (approximately up to 1–10 m*M*) of the TCS compounds. Before each experiment, the electrolyte solution will be purged with argon transistor (lower than 1 ppm O_2) and the voltammetric curves will be recorded maintaining an argon blanket over the solution. CV experiments will be carried out at a scan rate in the range between 5 and 50 mV/s with an Autolab Model PGSTAT 30, at 25 °C. Protein midpoint potentials (E1/2) will be calculated as average between the potential of oxidation (anodic peak, Epa) and reduction (cathodic peak, Epc) peaks. CV was also performed in the presence of 100 μ*M* horse-purified bovine cytochrome *c* (Sigma).

In the absence of adsorbed p66Shc protein, the CV curves only display the unperturbed, symmetrical capacitive response to alternate potential scans, typical of SAMs of long-chain alkyl-thiols. Accordingly, background subtraction from the curve results to zero. When the SAM electrode was coated with recombinant p66Shc, the two CV curves were modified by the superimposition of novel oxidation and reduction events, in the regions, respectively, of 100 and −170 mV. Background subtraction from the CV curve resulted in this case in two sharp peaks. The CV response of the SAM electrode coated with recombinant p66Shc was investigated at various potential scan rates and, as expected for surface-confined electroactive species, the intensity of the two peaks was proportional to scan rate (Giorgio et al., 2005).

Exposure of the p66Shc-coated SAM electrode to a 100-μ*M* solution of cyt *c* provoked a dramatic change in the voltammetric response: (i) the reduction/oxidation peaks typical of p66Shc were no longer detected; (ii) a novel and single-electron transfer event was detected, distinct from those of either p66Shc or cyt *c*; (iii) the calculated electrode capacitance was markedly reduced in the presence of cyt *c*; and (iv) the kinetics of electron transfer by p66Shc is accelerated in the presence of cyt *c*.

Together, these results suggest that p66Shc mediates electron transfer reaction with cyt *c* (Giorgio et al., 2005).

6.2. Mitochondrial ROS formation assay

To check whether recombinant p66Shc increases ROS production by isolated mitochondria, we evaluated the oxidation rate of $H_2O_2^-$ or $O_2^{\cdot-}$ sensitive dies, respectively, 2′,7′-dichlorodihydrofluorescein diacetate (H_2DCFDA) and hydroethidin DHE, in the presence of mitoplasts. Briefly, mitoplasts were obtained by hypotonic shock (1:10 dilution in deionized water for 5 min on ice) of isolated mouse liver mitochondria. 0.5 mg/ml mitoplasts and 30 μ*M* H_2DCFDA or 10 μ*M* DHE were sequentially added to a spectrofluorimetric cuvette. Kinetic of fluorescence was registered using a Perkin Elmer Ls55 spectrofluorimeter at 25 °C.

Addition of recombinant p66Shc to succinate or glutamate/malate energized mitoplasts, but not TMPD/ascorbate, increased H_2DFCDA or DHE fluorescence, indicating increased rate of oxidation, indicating that p66Shc is able to generate ROS by acting downstream to reduced cyt *c* (Giorgio et al., 2005; Gertz, Fischer, Wolters, & Steegborn, 2008).

7. CONCLUSIONS: ROLE OF p66Shc ROS

P66Shc mitochondrial function is tightly regulated at multiple levels. Although, the import of p66Shc into mitochondrial intermembrane space is not still understood at a mechanistic level. However, a mechanism that depends on p66Shc posttranslational modifications including serine phosphorylation by stress kinases like Jnk-1 and Pkc-B and prolil-isomerization by Pin-1, allowing p66Shc increase within mitochondria during apoptosis, has been described (Pinton et al., 2007).

A second level of activation of p66Shc mitochondrial function is represented by the effective amount of p66Shc within mitochondrial vesicles. In fact, mitochondrial p66Shc has been observed to associate to a high-molecular-weight complex of about 670 kDa and to the mitochondrial chaperon mtHsp70. Notably, treatment of cells with proapoptotic stimuli such as UVC or H_2O_2 induces the dissociation of this complex and thus the release of monomeric p66Shc free to react with cytochrome *c* (Orsini et al., 2006).

Then, the oxidation of cysteine residues and the oligomerization state of p66Shc have been reported to regulate its redox function within mitochondria (Gertz, Fischer, Leipelt, Wolters, & Steegborn, 2009).

Finally, the total amount of p66Shc seems regulated by transcriptional (Kim et al., 2012) and posttranslational mechanisms. In particular, the half-life of p66Shc has been demonstrated to increase upon apoptotic stimulation, notably in a p53-dependent way, thus linking the proapoptotic activity of p66Shc to the p53 pathway (Trinei et al., 2002).

In fat cells, insulin induces serine 36-specific phosphorylation of p66Shc, thus stimulating p66Shc ROS production, which, in turn, potentiates insulin transduction signaling by inhibiting PTEN phosphatase (Berniakovich et al., 2008). In particular, p66Shc suppresses expression of UCP1 on both mRNA and protein levels in both newborn and adult mice. Therefore, p66Shc behaves like an atypical signal transducer that tunes membrane receptor signaling with the regulation of intracellular redox balance.

As a consequence, p66Shc −/− mice have reduced body weight, due to reduced fat mass of both white and brown adipose tissues (Berniakovich et al., 2008; Tomilov et al., 2011).

Fat has a crucial role in the thermoregulation of mammals. It protects from body heat loss (thermo-insulation) and generates heat for the maintenance of body temperature when animals are exposed to cold (thermogenesis). Notably, p66Shc −/− mice were found to be more sensitive to cold due to the reduced thermal insulation effect of fat pads (Giorgio et al., 2012). Therefore, adaptation to cold as well as optimization of energy storage when food is available, both altered in the lean p66Shc −/− mice, have been proposed as possible evolutionary functions whose fitness pressure preserves the p66Shc gene in mammals.

P66Shc represents a clear example of an antagonistic pleiotropic function, which generates both beneficial and detrimental phenomena in an organism.

ACKNOWLEDGMENTS

This research was supported by the Italian Association for Cancer Research (AIRC) and National Institutes of Health Grant 1P01AG025532-01A1 awarded to PG. P.

REFERENCES

Baldari, C. T., & Telford, J. L. (1999). Lymphocyte antigen receptor signal integration and regulation by the SHC adaptor. *Biological Chemistry*, *380*, 129–134.

Bao, L., Avshalumov, M. V., Patel, J. C., Lee, C. R., Miller, E. W., Chang, C. J., et al. (2009). Mitochondria are the source of hydrogen peroxide for dynamic brain-cell signaling. *Journal of Neuroscience*, *29*, 9002–9010.

Bell, E. L., et al. (2007). The Qo site of the mitochondrial complex III is required for the transduction of hypoxic signaling via reactive oxygen species production. *The Journal of Cell Biology*, *177*, 1029–1036.

Bernardi, P., Petronilli, V., Di Lisa, F., & Forte, M. (2001). A mitochondrial perspective on cell death. *Trends in Biochemical Sciences*, *26*, 112–117.

Berniakovich, I., Trinei, M., Stendardo, M., Migliaccio, E., Minucci, S., Bernardi, P., et al. (2008). P66Shc-generated oxidative signal promotes fat accumulation. *Journal of Biological Chemistry*, *283*, 34283–34293.

Berry, A., Capone, F., Giorgio, M., Pelicci, P. G., de Kloet, E. R., Alleva, E., et al. (2007). Deletion of the life span determinant p66Shc prevents age-dependent increases in emotionality and pain sensitivity in mice. *Experimental Gerontology*, *422*, 37–45.

Carpi, A., Menabò, R., Kaludercic, N., Pelicci, P., Di Lisa, F., & Giorgio, M. (2009). The cardioprotective effects elicited by p66(Shc) ablation demonstrate the crucial role of mitochondrial ROS formation in ischemia/reperfusion injury. *Biochimica et Biophysica Acta*, *1787*, 774–780.

Gertz, M., Fischer, F., Wolters, D., & Steegborn, C. (2008). Activation of the lifespan regulator p66Shc through reversible disulfide bond formation. *Proceedings of the National Academy of Sciences of the United States of America*, *105*, 5705–5709.

Gertz, M., Fischer, F., Leipelt, M., Wolters, D., & Steegborn, C. (2009). Identification of Peroxiredoxin 1 as a novel interaction partner for the lifespan regulator protein p66Shc. *Aging*, *1*, 254–265.

Giorgio, M., Berry, A., Berniakovich, I., Poletaeva, I., Trinei, M., Stendardo, M., et al. (2012). The p66Shc knocked out mice are short lived under natural condition. *Aging Cell*, *11*, 162–168.

Giorgio, M., Migliaccio, E., Orsini, F., Paolucci, D., Moroni, M., Contursi, C., et al. (2005). Electron transfer between cytochrome c and p66Shc generates reactive oxygen species that trigger mitochondrial apoptosis. *Cell*, *122*, 221–233.

Green, D. R., & Kroemer, G. (2004). The pathophysiology of mitochondrial cell death. *Science*, *305*, 626–629.

Hoffman, D. L., & Brookes, P. S. (2009). Oxygen sensitivity of mitochondrial reactive oxygen species generation depends on metabolic conditions. *Journal of Biological Chemistry*, *284*, 16236–16245.

Kao, A. W., Waters, S. B., Okada, S., & Pessin, J. E. (1997). Insulin stimulates the phosphorylation of the 66- and 52-kilodalton Shc isoforms by distinct pathways. *Endocrinology*, *138*, 2474–2480.

Kim, Y. R., Kim, C. S., Naqvi, A., Kumar, A., Kumar, S., Hoffman, T. A., et al. (2012). Epigenetic upregulation of p66shc mediates low-density lipoprotein cholesterol-induced endothelial cell dysfunction. *American Journal of Physiology. Heart and Circulatory Physiology*, *303*, H189–H196.

Luzi, L., Confalonieri, S., Di Fiore, P. P., & Pelicci, P. G. (2000). Evolution of Shc functions from nematode to human. *Current Opinion in Genetics and Development*, *10*, 668–674.

Menini, S., Amadio, L., Oddi, G., Ricci, C., Pesce, C., Pugliese, F., et al. (2006). Deletion of p66Shc longevity gene protects against experimental diabetic glomerulopathy by preventing diabetes-induced oxidative stress. *Diabetes*, *5*, 1642–1650.

Migliaccio, E., Giorgio, M., & Pelicci, P. G. (2006). Apoptosis and aging: Role of p66Shc redox protein. *Antioxidants & Redox Signaling*, *8*, 600–608.

Migliaccio, E., Giorgio, M., Mele, S., Pelicci, G., Reboldi, P., Pandolfi, P. P., et al. (1999). The p66shc adaptor protein controls oxidative stress response and life span in mammals. *Nature*, *402*, 309–313.

Migliaccio, E., Mele, S., Salcini, A. E., Pelicci, G., Lai, K. M., Superti-Furga, G., et al. (1997). Opposite effects of the p52shc/p46shc and p66shc splicing isoforms on the EGF receptor-MAP kinase fos signalling pathway. *EMBO Journal*, *16*, 706–716.

Murphy, M. P. (2009). How mitochondria produce reactive oxygen species. *Biochemical Journal*, *417*, 1–13.

Napoli, C., Martin-Padura, I., de Nigris, F., Giorgio, M., Mansueto, G., Somma, P., et al. (2003). Deletion of the p66Shc longevity gene reduces systemic and tissue oxidative stress, vascular cell apoptosis, and early atherogenesis in mice fed a high-fat diet. *Proceedings of the National Academy of Sciences of the United States of America*, *100*, 2112–2116.

Orsini, F., Migliaccio, E., Moroni, M., Contursi, C., Raker, V. A., Piccini, D., et al. (2004). The life span determinant p66Shc localizes to mitochondria where it associates with mitochondrial heat shock protein 70 and regulates trans-membrane potential. *Journal of Biological Chemistry*, *279*, 25689–25695.

Orsini, F., Moroni, M., Contursi, C., Yano, M., Pelicci, P., Giorgio, M., et al. (2006). Regulatory effects of the mitochondrial energetic status on mitochondrial p66Shc. *Biological Chemistry*, *387*, 1405–1410.

Perez-Campo, R., López-Torres, M., Cadenas, S., Rojas, C., & Barja, G. (1998). The rate of free radical production as a determinant of the rate of aging: Evidence from the comparative approach. *Journal of Comparative Physiology. B*, *168*, 149–158.

Petronilli, V., Costantini, P., Scorrano, L., Colonna, R., Passamonti, S., & Bernardi, P. (1994). The voltage sensor of the mitochondrial permeability transition pore is tuned by the oxidation-reduction state of vicinal thiols. Increase of the gating potential by oxidants and its reversal by reducing agents. *Journal of Biological Chemistry*, *269*, 16638–16642.

Pinton, P., Rimessi, A., Marchi, S., Orsini, F., Migliaccio, E., Giorgio, M., et al. (2007). Protein kinase C beta and prolyl isomerase 1 regulate mitochondrial effects of the life-span determinant p66Shc. *Science*, *315*, 659–663.

Su, K. G., Savino, C., Marracci, G., Chaudhary, P., Yu, X., Morris, B., et al. (2012). Genetic inactivation of the p66 isoform of ShcA is neuroprotective in a murine model of multiple sclerosis. *The European Journal of Neuroscience*, *353*, 562–571.

Tomilov, A. A., Ramsey, J. J., Hagopian, K., Giorgio, M., Kim, K. M., Lam, A., et al. (2011). The Shc locus regulates insulin signaling and adiposity in mammals. *Aging Cell*, *10*, 55–65.

Trinei, M., Berniakovich, I., Beltrami, E., Migliaccio, E., Fassina, A., Pelicci, P., et al. (2009). P66Shc signals to age. *Aging*, *1*, 503–510.

Trinei, M., Giorgio, M., Cicalese, A., Barozzi, S., Ventura, A., Migliaccio, E., et al. (2002). A p53–p66Shc signalling pathway controls intracellular redox status, levels of oxidation-damaged DNA and oxidative stress-induced apoptosis. *Oncogene*, *21*(24), 3872–3878.

CHAPTER SEVEN

Detecting Disulfide-Bound Complexes and the Oxidative Regulation of Cyclic Nucleotide-Dependent Protein Kinases by H_2O_2

Joseph R. Burgoyne, Philip Eaton[1]

King's College London, Cardiovascular Division, The Rayne Institute, London, United Kingdom

[1]Corresponding author: e-mail address: philip.eaton@kcl.ac.uk

Contents

1. Introduction 112
   1.1 Intermolecular disulfide formation 112
   1.2 Oxidation increases PKARIs' affinity for its binding partners 116
   1.3 Oxidative PKG1α activation 119
2. Experimental Considerations and Procedures 121
   2.1 Preparation of tissue and cell lysates 121
   2.2 Diagonal SDS-PAGE to detect disulfide-bound complexes 121
   2.3 Nonreducing SDS-PAGE to detect PKG1α and PKARI disulfide dimers 123
   2.4 Data processing for determining percentage disulfide dimer formation 124
3. Summary 124
Acknowledgments 125
References 125

Abstract

Hydrogen peroxide regulates intracellular signaling by oxidatively converting susceptible cysteine thiols to a modified state, which includes the formation of intermolecular disulfides. This type of oxidative modification can occur within the cAMP- and cGMP-dependent protein kinases often referred to as PKA and PKG, which have important roles in regulating cardiac contractility and systemic blood pressure. Both kinases are stimulated through conical pathways that elevate their respective cyclic nucleotides leading to direct kinase stimulation. However, PKA and PKG can also be functionally modulated independently of cyclic nucleotide stimulation through direct cysteine thiol oxidation leading to intermolecular disulfide formation. In the case of PKG, the formation of an intermolecular disulfide between two parallel dimeric subunits leads to

Methods in Enzymology, Volume 528
ISSN 0076-6879
http://dx.doi.org/10.1016/B978-0-12-405881-1.00007-0

enhanced kinase affinity for substrate. For PKA, the formation of two intermolecular disulfides between antiparallel dimeric regulatory RI subunits increases the affinity of this kinase for its binding partners, the A-kinase anchoring proteins, leading to increased PKA localization to its substrates. In this chapter, we describe the methods for detecting intermolecular disulfide-bound proteins and monitoring PKA and PKG oxidation within biological samples.

1. INTRODUCTION

Protein posttranslational modifications govern the intracellular signaling events required for physiological signaling. In this chapter, we focus on the oxidative posttranslational modification of intermolecular disulfide bond formation, describing a technique for detecting proteins that can undergo this modification. 3′–5′-Cyclic adenosine monophosphate (cAMP)- and 3′,5′-cyclic guanosine monophosphate (cGMP)-dependent protein kinase, known, respectively, as PKA and PKG, are a particular focus as both of these kinases play fundamental roles in the cardiovascular system. Both kinases can be regulated by hydrogen peroxide (H_2O_2)-mediated formation of intermolecular disulfides, with this process being directly linked to physiological signaling as well as potential pathological dysfunction. Here, we also describe the experimental considerations that should be applied to the general study of protein oxidation.

1.1. Intermolecular disulfide formation

Cellular processes are largely governed by dynamic protein–protein interactions and posttranslational modifications that lead to an alteration in target function. The cellular proteome consists of numerous proteins that can undergo cysteine thiol oxidation—a process originally principally associated with dysfunction and disease but now widely accepted as a physiological process of cellular signaling (Burgoyne, Mongue-Din, Eaton, & Shah, 2012). Oxidants are generated as by-products of many cellular processes including the electron transport chain and catalysis by the intracellular oxidases but also in a regulated manner by being the primary signaling molecule of the NADPH oxidases. The NADPH oxidases generate superoxide (O_2^-) that is readily dismutated to the more stable H_2O_2 form either spontaneously or catalyzed by the enzyme superoxide dismutase (Cave et al., 2006; Leitch, Yick, & Culotta, 2009). The modification of cysteine thiols (SH) by oxidants such as H_2O_2 is a selective process that generally requires the target cysteine to be in the deprotonated thiolate anion form (S-). The reactivity of a cysteine is governed by its local environment with neighboring amino acids within the tertiary structure contributing to the cysteine's overall pK_a value. Neighboring basic amino acids such as histidines and lysines lower the pK_a of adjacent cysteine residues, increasing their propensity to undergo thiol oxidation. Reactive cysteine thiols can undergo a wide range of different types of oxidative modification, depending on the environment of the cysteine and type and quantity of oxidant present. Several reviews have comprehensively covered the biochemical and physiological relevance of various types of oxidative modification (Burgoyne, Oka, Ale-Agha, & Eaton, 2013; Hill & Bhatnagar, 2012; Reddie & Carroll, 2008). In this chapter, we focus on the study and detection of one specific type of oxidative modification, namely, intermolecular disulfide bond formation. For intermolecular disulfide formation to occur between proteins, certain criteria must be met; first, susceptible proteins must exist in very close proximity with an adjacent cysteine on each protein being close enough (i.e., vicinal thiols) to form a disulfide bond. In addition, at least one of the adjacent cysteines on each protein needs to be oxidant reactive (known as the peroxidatic cysteine), forming a reversible modification that can then be resolved by a proximal reduced cysteine (resolving cysteine) on the neighboring protein to form an intermolecular disulfide. Several different oxidants react with thiols to generate thiol-reducible intermediates that can undergo transition reactions that yield intermolecular disulfide bonds as outlined in Fig. 7.1. H_2O_2 gives rise to a sulfenic acid (Cys-SOH) intermediate, whereas oxidized glutathione disulfide leads to potential glutathiolation (Cys-S-SG) and S-nitrosothiol to potential S-nitrosylation (Cys-SNO) of the reactive sensor cysteine. All of these modifications can be resolved by a proximal cysteine thiol on an adjacent protein generating either a covalent homocomplex (disulfide between same protein molecules) or a covalent heterocomplex (disulfide between different protein molecules). This structural modification can directly alter the activity or interactions of the oxidatively modified proteins, providing a mode for redox regulation of protein function. Several targets of intermolecular disulfide bond formation have been identified including PKG1α (Burgoyne et al., 2007), cAMP-dependent protein kinase regulatory subunit 1 (PKARI; Brennan et al., 2006), NF-kappa-B essential modulator (Herscovitch et al., 2008), the C-terminal catalytic domain of receptor protein-tyrosine phosphatase alpha (van der Wijk, Overvoorde, & den Hertog, 2004), KEAP1, and adenosine-5′-triphosphate (ATP) synthase (Fourquet, Guerois, Biard, & Toledano,

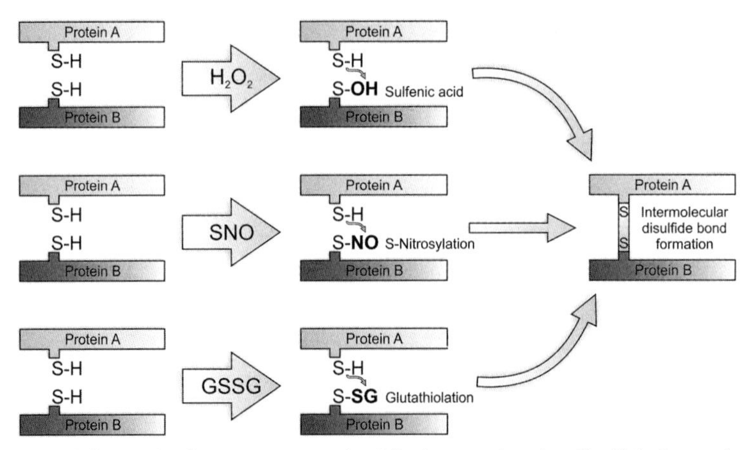

Figure 7.1 Schematic of processes required for intermolecular disulfide formation. For intermolecular disulfide formation, an adjacent cysteine must be present in each protein with close enough proximity for bond formation. In addition, one of the cysteines must contain a reactive thiol (shown here as the S-H in protein B) that can undergo oxidation. The reactive cysteine can be oxidized to a sulfenic acid by H_2O_2, S-nitrosylated by an S-nitrosothiol (SNO) or glutathiolated by oxidized disulfide glutathione (GSSG). Each of these intermediates can then be rapidly resolved by the adjacent cysteine (S-H in protein A) giving rise to an intermolecular disulfide bond. (See Color Insert.)

2010; Wang et al., 2011). Intermolecular disulfides can be removed from substrate proteins by thioredoxin (Trx). The reduced form of Trx directly reduces target protein disulfides by disulfide exchange leading to Trx oxidation. The disulfide oxidized form of Trx can then be catalyzed to the reduced form by Trx reductase in an NADPH- and FAD-dependent reaction. The reversibility of intermolecular disulfides by Trx allows a dynamic equilibrium to exist between the targets oxidized and reduced state, which can be shifted depending on the redox environment.

Here, we describe an adapted SDS-polyacrylamide gel electrophoresis (SDS-PAGE) technique for detecting and identifying targets of intermolecular disulfide formation. This method termed diagonal SDS-PAGE relies on the difference in migration of proteins on nonreducing SDS-PAGE compared with under reducing conditions where the disulfide-bound complexes are lost. The rational being proteins bound together by a disulfide will run at a higher combined molecular weight when resolved under non-reducing conditions, but when run in a second dimension on a reducing gel, they will separate at their natural individual molecular weights, as shown in Fig. 7.2. Therefore, any proteins that form part of a disulfide complex will migrate off the diagonal plane of the SDS-PAGE gel when separated under

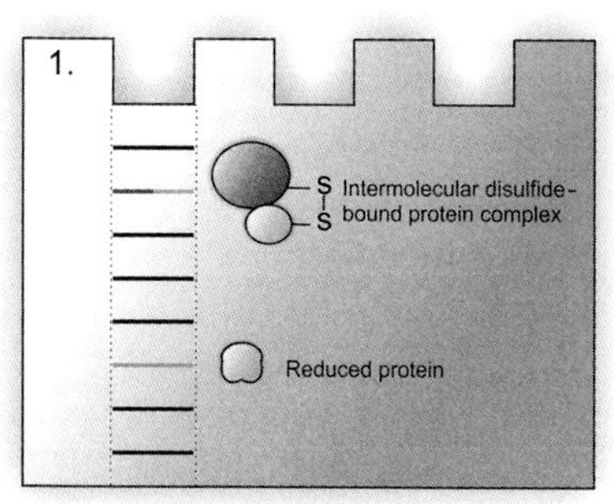

Resolve proteins on SDS-PAGE
under non reducing conditions

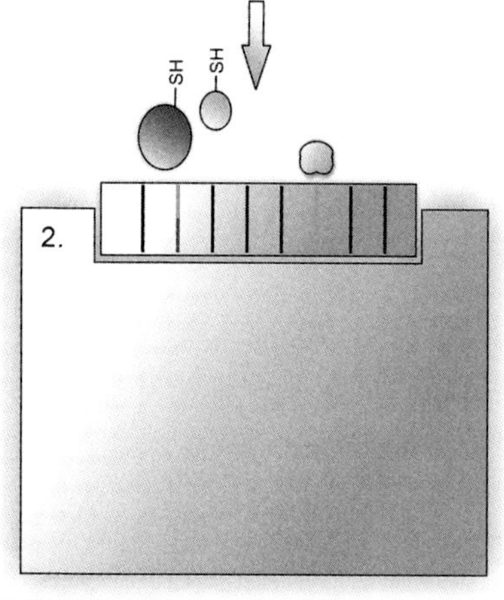

Excise resolved lane, soak in **2-Mercaptoethanol**,
and then place in large well of a fresh SDS-PAGE gel

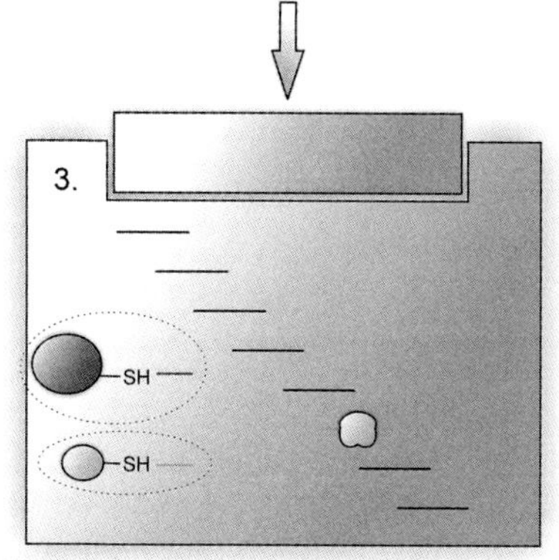

Resolve proteins under reducing conditions and then
stain proteins using colloidal Coomassie Blue

Excise stained proteins resolved below the diagonal
of the gel and identify using mass spectrometry

reducing conditions in the second dimension. Proteins that resolve from the diagonal can be easily identified following gel staining, by being excised from the gel and then analyzed by mass spectrometry.

1.2. Oxidation increases PKARIs' affinity for its binding partners

Cardiac contractility is controlled by dynamic changes in cardiac myocyte intracellular cytosolic Ca^{2+}, which is highly regulated by the cAMP-dependent PKA. When PKA is activated by elevated cAMP, it enhances cytosolic Ca^{2+} concentration as well as the Ca^{2+} in and out of the sarcoplasmic reticulum (SR). The principal activator of PKA is the cyclic nucleotide cAMP, which is synthesized in response to β-adrenergic receptor agonists (adrenaline or noradrenaline) and released into the circulatory system by sympathetic nerves. The β-adrenergic receptor agonists stimulate adenylate cyclase via the stimulatory G-protein-coupled signaling pathway, which converts ATP to cAMP (Rockman, Koch, Milano, & Lefkowitz, 1996). The activation of PKA by β-adrenergic receptor agonists is a principal mediator of the "flight-or-fight" response needed for rapid improvement in the body's ability to perform exercise involving elevated cardiac and skeletal muscle output.

PKA consists of a tetramer comprising of two homodimeric catalytic and two homodimeric regulatory subunits. Binding of cAMP to the regulatory subunits stimulates disassociation of the catalytic subunits, which exposes the active site allowing substrate phosphorylation. The ability of PKA to regulate cardiac contractility is through the coordinated phosphorylation of its targets located primarily at the plasma membrane, SR, and myofilaments. PKA potentiates intracellular cytosolic Ca^{2+} during systole by phosphorylating the L-type Ca^{2+} channel located in the T-tubular membrane (Bunemann, Gerhardstein, Gao, & Hosey, 1999). Phosphorylation of the α_{1C} subunit of

Figure 7.2 Diagonal gel SDS-PAGE. Proteins resolved under nonreducing conditions will retain their oxidation status, and therefore, intermolecular disulfide-bound proteins will run at a combined higher molecular weight. Once resolved, protein disulfides are removed from proteins by incubating the excised lane in the reducing agent 2-mercaptoethanol. By resolving a second time under reducing conditions, the proteins then resolve at their individual molecular weights. Once the gel has been stained with colloidal Coomassie Blue, proteins that were once disulfide bound can be identified as those that have migrated below the diagonal of the gel. The identity of these proteins can be determined by excising them from the gel followed by mass spectrometry analysis. (See Color Insert.)

the L-type Ca^{2+} at Ser-1928 and the β_2 subunit at Ser-478 and Ser-479 by PKA increases the channel open probability and therefore the influx of Ca^{2+}. In addition, PKA further enhances intracellular cytosolic Ca^{2+} by directly phosphorylating the ryanodine receptor on the SR at Ser-2808, which increases the open probability and enhances its sensitivity to Ca^{2+}-induced activation (Marx et al., 2000). Enhanced Ca^{2+} flux through the SR is also regulated by PKA-dependent SERCA activation. Phosphorylation of the SERCA accessory protein phospholamban at Ser-16 by PKA relieves its inhibitory effect on SERCA activity (MacLennan & Kranias, 2003). Increased SERCA activity enhances the reuptake of Ca^{2+} during diastole, thereby enhancing myofilament relaxation and also enhancing the SR Ca^{2+} store, allowing to a larger release of Ca^{2+} during subsequent systole.

Excessive levels of intracellular cytosolic Ca^{2+} can be proarrhythmogenic and therefore need to be dynamically regulated between systole and diastole. PKA prevents Ca^{2+} overload by regulating Na/K ATPase by phosphorylating its accessory protein phospholemman, which when phosphorylated at Ser-68 relieves its inhibitory effect on Na/K ATPase activity (Despa, Tucker, & Bers, 2008; Shattock, 2009). This limits the rise in intracellular Na^+, helping maintain flux through the Na^+/Ca^{2+} exchanger by maintaining it in its forward mode, whereby three Na^+ enter the cell per Ca^{2+} ion extruded. The control of cardiac contractility by PKA is also regulated at the myofilament level with phosphorylation of both myosin-binding protein-C (MBP-C) and troponin I. The phosphorylation of MBP-C by PKA accelerates myosin–actin cross-bridge cycling (Previs, Beck Previs, Gulick, Robbins, & Warshaw, 2012; Tong, Stelzer, Greaser, Powers, & Moss, 2008). Whereas phosphorylation of tropinin I decreases its interaction with cardiac troponin C, thereby lowering the affinity of the regulatory subunit of the troponin complex for Ca^{2+} (Feng, Chen, Weinstein, & Jin, 2008; Fink et al., 2001). This decreases force development, leading to accelerated myofilament relaxation, which enables the heart to beat faster.

As well as regulating contractility, PKA can also mediate blood vessel relaxation by phosphorylating many of the same targets in vascular smooth muscle cells as PKG, including HSP20 (Komalavilas et al., 2008), RhoA (Murthy, Zhou, Grider, & Makhlouf, 2003), PLN (Mundina-Weilenmann et al., 2000), with some evidence for myosin phosphatase target subunit 1 also being a substrate (Azam et al., 2007).

The regulatory I-alpha subunit of PKA (PKARI) was identified in a proteomic screen for proteins that form intermolecular disulfide complexes under oxidizing conditions (Brennan et al., 2004). This proteomic screen used diagonal SDS-PAGE comparing untreated myocytes and myocytes treated with the disulfide-inducing compound diamide. The presence of two disulfides in PKARI was first reported by Zick and Taylor (1982), with Cys16 on each subunit binding to Cys37 on the other to form an antiparallel disulfide-bound dimer. However, this structural modification was thought to be constitutive as the concentration of the reducing agent dithiothreitol required to reduce the disulfide bonds was exceptionally high at around 100 m*M*. This suggested that the disulfides were buried and not solvent accessible. The disulfides were, however, not required for PKARI dimerization as this complex is held together by virtue of an amphipathic leucine zipper. These disulfides in PKARI that were originally thought to be constitutive structural bonds were later discovered to be mostly absent in cardiac tissue under basal conditions, but instead formed during oxidative stress (Brennan et al., 2006). The oxidation state of PKARI was assessed in myocytes treated with increasing concentrations of H_2O_2 by lysing into sample buffer containing alkylating agent to prevent artificial air oxidation of reactive cysteine residues. The cell lysates were then resolved under non-reducing conditions (i.e., without dithiothreitol or mercaptoethanol) using SDS-PAGE to preserve protein oxidation and then analyzed using Western blotting with immunostaining for PKARI. Reduced PKARI monomer ran at 50 kDa, but in myocytes treated with H_2O_2, there was a concentration-dependent transition of PKARI from the monomer to the oxidized disulfide dimer at ~100 kDa. The high-molecular weight oxidized form of PKARI was reduced to the monomeric form by mercaptoethanol, demonstrating the complex was the result of intermolecular disulfide formation. The oxidized form of PKARI correlated with elevated PKA substrate phosphorylation, suggesting oxidation increased the activity of this kinase. In addition, oxidation of PKARI altered its localization leading to translocation from the cytosol to the myofilament, the nucleus, and, to a small extent, the membrane. The increase in PKA substrate phosphorylation and kinase relocalization was consistent with PKARI oxidation increasing the affinity for its binding partners, the A-kinase-anchoring proteins (AKAPs). This is supported by the intermolecular disulfides in PKARI being located within the AKAP interaction domain. Furthermore, studies have demonstrated that the binding of D-AKAP to PKARI is significantly attenuated by mutation of Cys16 or Cys37 to an alanine preventing disulfide formation (Sarma et al., 2010). The relevance of increased interaction of PKA for its AKAPs goes beyond just bringing PKA into closer proximity with its substrates as the presence of substrate itself was found to increase dissociation of the catalytic and regulatory subunits (Vigil, Blumenthal, Taylor, & Trewhella, 2005). The substrate-dependent activation of PKA appears to be confined to the type I isoform and perhaps explains how oxidation likely increases PKA substrate phosphorylation without elevation in cAMP.

In addition to H_2O_2, PKARI was also found to be oxidized by nitrosocysteine (CysNO), a transnitrosylating agent that induces vasorelaxation of aortic vessels independently of cAMP (Burgoyne & Eaton, 2009). Therefore, PKARI oxidation by thiol-reactive forms of nitric oxide may play a role in regulating blood pressure. The role of PKARI oxidation in regulating myocardial contractility is yet to be fully elucidated and is hampered by the simultaneous overriding negatively inotropic activation of PKG1α by exogenous oxidants.

1.3. Oxidative PKG1α activation

The conical activation of PKG1α by elevated cGMP is a key mediator of blood pressure regulation, which is critical for maintaining health, with hypertension increasing the risk of aortic aneurysms, peripheral artery disease, heart attacks, stroke, and kidney and heart failure (Germino, 2009). The lowering of blood pressure by increased vascular smooth muscle cell relaxation to generate vessel dilation is regulated by three major pathways. First, sheer stress or the binding of vasodilatory ligands (such as acetylcholine or bradykinin) to their respective ligands increases endothelial cell Ca^{2+} uptake (Yetik-Anacak & Catravas, 2006). The cationic metal ion Ca^{2+} directly stimulates nitric oxide synthase activity leading to elevated NO formation. Gaseous NO readily diffuses into smooth muscle cells where it binds to and directly stimulates the activity of guanylate cyclase, increasing the conversion of guanosine triphosphate to cGMP. The small cyclic nucleotide cGMP is a direct activator of the kinase PKG, which phosphorylates several targets, including the BKca channel (Fukao et al., 1999), IRAG, and RGS2 (Osei-Owusu, Sun, Drenan, Steinberg, & Blumer, 2007; Schlossmann et al., 2000), to lower smooth muscle cell intracellular Ca^{2+}. In addition, PKG also lowers smooth muscle Ca^{2+} sensitivity by phosphorylating the myosin-binding subunit (MYPT1) of myosin light chain phosphatase, preventing its inactivation by RhoA kinase. The dephosphorylation of myosin light chain inhibits ATPase activity, leading to a reduction in cross-bridge cycling (Wooldridge et al., 2004). The second signaling pathway involved in vascular smooth muscle cell relaxation is the activation of PKA by cAMP. This process is mediated by prostaglandins (such as PGE2) generated in endothelial cells by cyclooxygenase that then diffuse into smooth muscle cells where they stimulate adenylate cyclases ability to generate cAMP (Tanaka, Yamaki, Koike, & Toro, 2004). Once PKA is activated by cAMP, it can induce smooth muscle cell relaxation by phosphorylating many of the same substrates known for PKG. The third pathway of vessel relaxation involves the formation of a vasodilatory substance in endothelial cells known as endothelial-derived hyperpolarizing factor (EDHF) that diffuses into smooth muscle cells where it generates plasma membrane hyperpolarization, inhibiting import of Ca^{2+} required for constriction (Garland, Hiley, & Dora, 2011). The true identity of EDHF remains controversial with evidence for several potential candidates including epoxyeicosatrienoic acids, K^+, and H_2O_2. Although the identity of EDHF has not been conclusively determined, H_2O_2 at least constitutes part of this factor as the enzyme catalase, which decomposes this oxidant, attenuates EDHF-mediated vessel relaxation (Shimokawa, 2010). The ability of H_2O_2 to mediate the EDHF effect is through its ability to directly oxidize PKG1α, leading to increased catalytic activity. Oxidation of PKG1α generates a disulfide bond within the N-terminus between two homodimeric subunits that directly activate the kinase independently of cGMP by increasing its affinity for substrate (Burgoyne et al., 2007). This process of PKG1α oxidation predominantly occurs within small resistance vessels that control blood pressure rather than in large conduit vessels (Burgoyne, Prysyazhna, Rudyk, & Eaton, 2012). The ability of PKG1α to mediate the EDHF response is supported in human coronary arterioles where H_2O_2 dilates these vessels by oxidizing PKG1α leading to opening of smooth muscle BKCa channels (Zhang et al., 2012). In addition, mice that express PKG1α with cysteine 42 converted to a charge-conserved serine residue that cannot undergo disulfide activation have a deficit in their EDHF response (Prysyazhna, Rudyk, & Eaton, 2012). The oxidation of PKG1α explains at least in part how the EDHF response is mediated but also represents a potentially important process mediated by changes in cellular reducing/oxidant (redox) capacity.

In this chapter, we describe a technique for detecting PKG1α and PKARI intermolecular disulfide formation using nonreducing SDS-PAGE. The rational of this technique is that by lysing tissue or cells directly into an alkylating buffer, the redox state of the protein is preserved and stabilized by chemically blocking reduced cysteine thiols to prevent their artificial air oxidation during preparation. These samples can then be analyzed for PKG1α or PKARI oxidation by resolving on nonreducing SDS-PAGE, preventing loss of protein oxidation that would occur under reducing conditions.

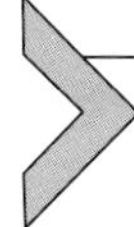

2. Experimental Considerations and Procedures

2.1. Preparation of tissue and cell lysates

Samples being assessed for interdisulfide bond redox status using nonreducing SDS-PAGE gels should be rapidly prepared in alkylating buffer to prevent artificial protein oxidation and stabilize oxidative modifications by preventing transfer reactions.

To analyze intermolecular disulfide formation in cultured cells, the cells should be first rinsed in phosphate-buffered saline (PBS) before being lysed directly into 1 × nonreducing alkylating sample buffer (50 m*M* Tris–HCl buffer pH 6.8, 2% SDS, 10% glycerol, 0.005% bromophenol blue, and 100 m*M* maleimide). Rinsing in PBS will remove any residual fetal calf serum that may interfere with subsequent immunostaining. The quantity of cells used in each experiment should be varied depending on the cell type and experimental protocol such that there is sufficient protein to be visualized on a diagonal SDS-PAGE gel or for immunodetection of PKARI or PKG1α oxidation.

For analyzing intermolecular disulfide formation in tissue, excised tissue or organs that are not *ex vivo* perfused should be rinsed briefly in Krebs solution (6.93 g of NaCl, 0.35 g of KCl, 0.16 g of KH_2PO_4, 2.1 g of $NaHCO_3$, 0.30 g of $MgSO_4$, and 2 g of glucose made up to 800 ml with deionized H_2O, bubbled with 5% CO_2/95% O_2 for 10–15 min before addition of 0.21 g of $CaCl_2$, and then made up to 1 l with deionized H_2O, followed by filtration through a 5 μ*M* filter membrane into a conical flask) to remove excess blood, before being snap frozen in liquid nitrogen to preserve samples until analysis.

For analysis, tissue/organs should be homogenized on ice using a Polytron-type hand-held homogenizer or an alternative rapid tissue homogenization buffer procedure (100 m*M* Tris–HCl pH 7.4, protease inhibitors, and 100 m*M* maleimide) to yield a 10% (weight/volume) homogenate. The required amount of homogenate can be added to an equal volume of 2 × nonreducing alkylating sample buffer for SDS-PAGE analysis with the remaining homogenate being snap frozen in liquid nitrogen and stored for future analysis.

2.2. Diagonal SDS-PAGE to detect disulfide-bound complexes

The detection of proteins that can form intermolecular covalently bound disulfide complexes can be identified using diagonal gel electrophoresis as shown in Fig. 7.2. This technique relies on the efficient removal of disulfides through reduction by reducing agents such as DTT or 2-mercaptoethanol. First, samples are run under nonreducing conditions so that disulfide complexes are maintained and resolved at their combined molecular weight. Taking these resolved proteins (by excising the entire lane they were resolved in, as shown in Fig. 7.2) and separating in a second dimension on a fresh SDS-PAGE gel under reducing conditions (breaking the disulfides) enables the proteins to run at their individual weights. Consequently, these proteins will run off the diagonal of the gel and can be visualized by total protein staining (e.g., Coomassie Blue) and then identified by excising the band and analysis by mass spectrometry. The treatment of tissues or cells with diamide (100–1000 μ*M* for 10 min) is an effective way to induce intermolecular disulfide formation and can serve as a useful way to identify susceptible proteins.

When comparing samples prepared, as stated above in Section 2.1, they should be resolved within separate lanes on an SDS-PAGE gel in the absence of any reducing agent, typically on a higher percentage (12.5–15%) or gradient gel to maximize protein coverage. Typically, our SDS-PAGE gels are cast within a $15 \times 10.6 \times 5.3$ cm gel cassette (Bio-Rad, Criterion gel cassettes) or an equivalent commercial system. Once the gel has fully resolved, the lanes containing each sample should be carefully excised using a scalpel bade and then submerged for 10 min in reducing sample buffer containing 2-mercaptoethanol (50 m*M* Tris–HCl buffer pH 6.8, 2% SDS, 10% glycerol, 0.005% bromophenol blue, and 5% mercaptoethanol). The mercaptoethanol will reduce protein disulfides, thus breaking disulfide-bound complexes into their monomeric units. Once soaked in reducing sample buffer, the gel strips can then be carefully placed into a large well (designed for immobilized pH gradient (IPG) strips) on individual $15 \times 10.6 \times 5.3$ cm IPG + 1-well SDS-PAGE gels (Bio-Rad). The IPG + 1-well SDS-PAGE gels should be assembled within the gel apparatus and submerged in running buffer after the gel strips have been firmly placed into the large wells. The surface of the gels should then be covered in a layer of reducing sample buffer. The single small well on each gel can be loaded with protein standards allowing the molecular weights of proteins to be determined after electrophoretic separation. Once the samples have been resolved, the gel can be stained in colloidal Coomassie Blue (100 g of ammonium sulfate, 200 ml ethanol, 30 ml ortho-phosphoric acid in 980 ml of H_2O_2 and then add 20 ml of 5% Coomassie Blue G-250). Directly before use, the colloidal Coomassie is shaken to generate a homogeneous solution. 40 ml of this then is mixed with 10 ml of methanol, using this to stain the gel overnight, with destaining using repeated rinsing with deionized water the following day. The gels should be destained until there is good contrast between the protein bands and background staining. Over-destaining can lead to loss of protein staining; however, if this occurs, gels can be restained again overnight. Novel differences in the presence or intensity of stained proteins resolved below the diagonal of each sample can be visualized using a light box. These proteins can then be excised using a sterile scalpel blade and analyzed by mass spectrometry to determine their identity. Proteins identified using the diagonal SDS-PAGE technique can have their ability to form a disulfide complex verified using the procedure outlined in the following section.

2.3. Nonreducing SDS-PAGE to detect PKG1α and PKARI disulfide dimers

Proteins identified using diagonal SDS-PAGE can be verified and analyzed using nonreducing SDS-PAGE. In this section, we describe how this method can be applied to the analysis of PKARI and PKG1α oxidation in cultured cells treated *in vitro* or in tissue from animals that have undergone an intervention. For example, cultured cells treated with either H_2O_2 (50–1000 μ*M* for 10 min) or diamide (50–1000 μ*M* for 10 min) will effectively induce intermolecular disulfide formation, and is a useful strategy to assess if a candidate protein is susceptible to forming a disulfide-bound complex. After a treatment or intervention, cells or animal tissue should be processed as described in Section 2.1 before being analyzed as described below.

Prepare standard SDS-PAGE gels (8% is suitable for analyzing PKARI and PKG1α oxidation) and fix into the gel apparatus and submerge in running buffer. Load prepared samples, resolve under nonreducing conditions, and then transfer proteins to polyvinylidene fluoride membranes using standard Western blotting techniques. The membrane should then be blocked in 5% milk in PBS+0.1% Tween (PBS-T) overnight at 4 °C or 10% milk in PBS-T for $\geq$1 h at room temperature. Once blocked, the membrane can be immunostained for PKARI, PKG1α, or other protein of interest using standard procedures. For example, when immunostaining blots for PKARI or PKG1α, incubation with primary antibody for 1 h at 1:1000 in 5% milk+PBS-T is sufficient. The incubation time with primary antibody should be increased if protein sample concentration is low, or optimized when probing for a newly identified candidate protein. After incubation with primary antibody, blots should be fully submerged and washed for 1 h in PBS-T by replacing the buffer every 15 min. This is then followed by 1-h incubation in the appropriate secondary antibody at 1:1000 in 5% milk + PBS-T and then another 1 h of washing as previously described. After washing (which can be increased if there is high background), blots can be analyzed by adding enhanced chemiluminescence (ECL), fixing into a film cassette between a plastic A4 sheet protector and then developing using ECL film and a film developer.

When immunostaining for PKARI, PKG1α, or protein of interest, it is important that the antibody is able to recognize both the reduced and oxidized disulfide-bound form of the protein. Often, this will require screening several antibodies for their suitability, as some have selective affinity for one redox state over the other. For PKARI, the antibody from BD Transduction labs (610165) and, for PKG1α, the antibody from Santa Cruz (sc-10338) both effectively recognize both the reduced and the oxidized forms.

2.4. Data processing for determining percentage disulfide dimer formation

The extent of protein disulfide dimer formation for PKAR1, PKG1α, or protein of interest can be determined by measuring the relative intensity of monomeric and disulfide-bound protein on immunostained blots as shown in Fig. 7.3. Once immunoblots have been developed, the ECL film can be digitized using a conventional scanner and then analyzed using commercial software designed for Western blot analysis. The % disulfide-bound dimer = (intensity of the high-molecular-weight dimeric protein band/the sum of the intensity of both the monomeric and dimeric protein bands) × 100.

3. Summary

Oxidants, such as H_2O_2, mediate physiological as well as pathological signaling by altering protein function via induction of posttranslational oxidative modifications, including intermolecular disulfide bonds. By using diagonal gel electrophoresis, proteins that form intermolecular disulfide-bound complexes can be identified within complex protein mixtures. Such proteins, which include PKARI and PKG1α, can subsequently be verified and quantitatively analyzed using nonreducing SDS-PAGE with Western immunoblotting, allowing the role of their oxidation to be determined during physiological and pathological signaling.

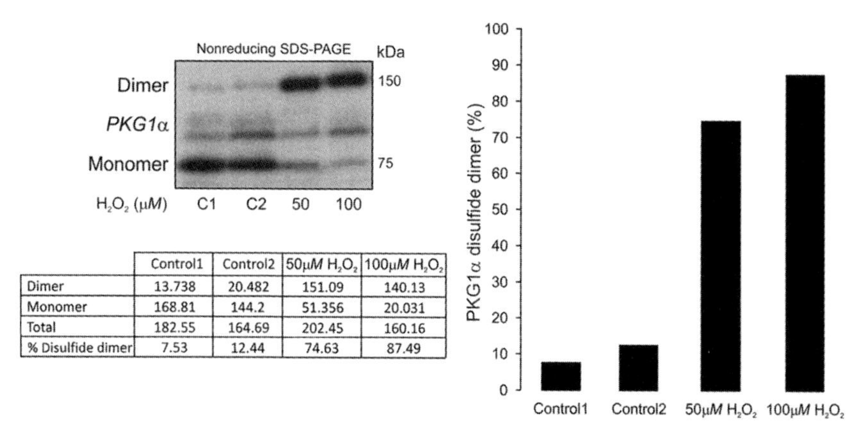

	Control1	Control2	50μM H_2O_2	100μM H_2O_2
Dimer	13.738	20.482	151.09	140.13
Monomer	168.81	144.2	51.356	20.031
Total	182.55	164.69	202.45	160.16
% Disulfide dimer	7.53	12.44	74.63	87.49

Figure 7.3 Analysis of PKG1α oxidation in H_2O_2 Langendorff-perfused rat hearts. Rat hearts perfused with Krebs buffer alone or Krebs buffer containing H_2O_2 for 10 min were snap frozen in liquid nitrogen and then analyzed for PKG1α oxidation using non-reducing SDS-PAGE. The % disulfide-bound dimeric PKG1α was determined after immunoblotting by measuring band intensities. The % disulfide bound = (intensity of dimeric PKG1α/the sum of the intensity of both monomeric and dimeric PKG1α) × 100. In hearts perfused with Krebs buffer alone, PKG1α is mostly reduced but becomes oxidized to the intermolecular disulfide dimeric form in those perfused with H_2O_2.

ACKNOWLEDGMENTS

We would like to acknowledge support from the Medical Research Council, the British Heart Foundation, the Leducq Foundation, and the Department of Health via the NIHR cBRC award to Guy's & St Thomas' NHS Foundation Trust. Also J. R. B. is supported by a Sir Henry Wellcome postdoctoral fellowship from The Wellcome Trust (sponsor reference 085483/Z/08/Z).

REFERENCES

Azam, M. A., Yoshioka, K., Ohkura, S., Takuwa, N., Sugimoto, N., Sato, K., et al. (2007). Ca^{2+}-independent, inhibitory effects of cyclic adenosine 5′-monophosphate on Ca^{2+} regulation of phosphoinositide 3-kinase C2alpha, Rho, and myosin phosphatase in vascular smooth muscle. *The Journal of Pharmacology and Experimental Therapeutics*, *320*(2), 907–916.

Brennan, J. P., Bardswell, S. C., Burgoyne, J. R., Fuller, W., Schroder, E., Wait, R., et al. (2006). Oxidant-induced activation of type I protein kinase A is mediated by RI subunit interprotein disulfide bond formation. *The Journal of Biological Chemistry*, *281*(31), 21827–21836.

Brennan, J. P., Wait, R., Begum, S., Bell, J. R., Dunn, M. J., & Eaton, P. (2004). Detection and mapping of widespread intermolecular protein disulfide formation during cardiac oxidative stress using proteomics with diagonal electrophoresis. *The Journal of Biological Chemistry*, *279*(40), 41352–41360.

Bunemann, M., Gerhardstein, B. L., Gao, T., & Hosey, M. M. (1999). Functional regulation of L-type calcium channels via protein kinase A-mediated phosphorylation of the beta(2) subunit. *The Journal of Biological Chemistry*, *274*(48), 33851–33854.

Burgoyne, J. R., & Eaton, P. (2009). Transnitrosylating nitric oxide species directly activate type I protein kinase A, providing a novel adenylate cyclase-independent cross-talk to beta-adrenergic-like signaling. *The Journal of Biological Chemistry*, *284*(43), 29260–29268.

Burgoyne, J. R., Madhani, M., Cuello, F., Charles, R. L., Brennan, J. P., Schroder, E., et al. (2007). Cysteine redox sensor in PKGIa enables oxidant-induced activation. *Science*, *317*(5843), 1393–1397.

Burgoyne, J. R., Mongue-Din, H., Eaton, P., & Shah, A. M. (2012). Redox signaling in cardiac physiology and pathology. *Circulation Research*, *111*(8), 1091–1106.

Burgoyne, J. R., Oka, S. I., Ale-Agha, N., & Eaton, P. (2013). Hydrogen peroxide sensing and signaling by protein kinases in the cardiovascular system. *Antioxidants & Redox Signaling*, *18*, 1042–1052.

Burgoyne, J. R., Prysyazhna, O., Rudyk, O., & Eaton, P. (2012). cGMP-dependent activation of protein kinase G precludes disulfide activation: Implications for blood pressure control. *Hypertension*, *60*(5), 1301–1308.

Cave, A. C., Brewer, A. C., Narayanapanicker, A., Ray, R., Grieve, D. J., Walker, S., et al. (2006). NADPH oxidases in cardiovascular health and disease. *Antioxidants & Redox Signaling*, *8*(5–6), 691–728.

Despa, S., Tucker, A. L., & Bers, D. M. (2008). Phospholemman-mediated activation of Na/K-ATPase limits [Na]i and inotropic state during beta-adrenergic stimulation in mouse ventricular myocytes. *Circulation*, *117*(14), 1849–1855.

Feng, H. Z., Chen, M., Weinstein, L. S., & Jin, J. P. (2008). Removal of the N-terminal extension of cardiac troponin I as a functional compensation for impaired myocardial beta-adrenergic signaling. *The Journal of Biological Chemistry*, *283*(48), 33384–33393.

Fink, M. A., Zakhary, D. R., Mackey, J. A., Desnoyer, R. W., Apperson-Hansen, C., Damron, D. S., et al. (2001). AKAP-mediated targeting of protein kinase a regulates contractility in cardiac myocytes. *Circulation Research*, *88*(3), 291–297.

Fourquet, S., Guerois, R., Biard, D., & Toledano, M. B. (2010). Activation of NRF2 by nitrosative agents and H_2O_2 involves KEAP1 disulfide formation. *The Journal of Biological Chemistry*, *285*(11), 8463–8471.

Fukao, M., Mason, H. S., Britton, F. C., Kenyon, J. L., Horowitz, B., & Keef, K. D. (1999). Cyclic GMP-dependent protein kinase activates cloned BKCa channels expressed in mammalian cells by direct phosphorylation at serine 1072. *The Journal of Biological Chemistry*, *274*(16), 10927–10935.

Garland, C. J., Hiley, C. R., & Dora, K. A. (2011). EDHF: Spreading the influence of the endothelium. *British Journal of Pharmacology*, *164*(3), 839–852.

Germino, F. W. (2009). The management and treatment of hypertension. *Clinical Cornerstone*, *9*(Suppl. 3), S27–S33.

Herscovitch, M., Comb, W., Ennis, T., Coleman, K., Yong, S., Armstead, B., et al. (2008). Intermolecular disulfide bond formation in the NEMO dimer requires Cys54 and Cys347. *Biochemical and Biophysical Research Communications*, *367*(1), 103–108.

Hill, B. G., & Bhatnagar, A. (2012). Protein S-glutathiolation: Redox-sensitive regulation of protein function. *Journal of Molecular and Cellular Cardiology*, *52*(3), 559–567.

Komalavilas, P., Penn, R. B., Flynn, C. R., Thresher, J., Lopes, L. B., Furnish, E. J., et al. (2008). The small heat shock-related protein, HSP20, is a cAMP-dependent protein kinase substrate that is involved in airway smooth muscle relaxation. *American Journal of Physiology. Lung Cellular and Molecular Physiology*, *294*(1), L69–L78.

Leitch, J. M., Yick, P. J., & Culotta, V. C. (2009). The right to choose: Multiple pathways for activating copper, zinc superoxide dismutase. *The Journal of Biological Chemistry*, *284*(37), 24679–24683.

MacLennan, D. H., & Kranias, E. G. (2003). Phospholamban: A crucial regulator of cardiac contractility. *Nature Reviews. Molecular Cell Biology*, *4*(7), 566–577.

Marx, S. O., Reiken, S., Hisamatsu, Y., Jayaraman, T., Burkhoff, D., Rosemblit, N., et al. (2000). PKA phosphorylation dissociates FKBP12.6 from the calcium release channel (ryanodine receptor): Defective regulation in failing hearts. *Cell, 101*(4), 365–376.

Mundina-Weilenmann, C., Vittone, L., Rinaldi, G., Said, M., de Cingolani, G. C., & Mattiazzi, A. (2000). Endoplasmic reticulum contribution to the relaxant effect of cGMP- and cAMP-elevating agents in feline aorta. *American Journal of Physiology. Heart and Circulatory Physiology, 278*(6), H1856–H1865.

Murthy, K. S., Zhou, H., Grider, J. R., & Makhlouf, G. M. (2003). Inhibition of sustained smooth muscle contraction by PKA and PKG preferentially mediated by phosphorylation of RhoA. *American Journal of Physiology. Gastrointestinal and Liver Physiology, 284*(6), G1006–G1016.

Osei-Owusu, P., Sun, X., Drenan, R. M., Steinberg, T. H., & Blumer, K. J. (2007). Regulation of RGS2 and second messenger signaling in vascular smooth muscle cells by cGMP-dependent protein kinase. *The Journal of Biological Chemistry, 282*(43), 31656–31665.

Previs, M. J., Beck Previs, S., Gulick, J., Robbins, J., & Warshaw, D. M. (2012). Molecular mechanics of cardiac myosin-binding protein C in native thick filaments. *Science, 337*(6099), 1215–1218.

Prysyazhna, O., Rudyk, O., & Eaton, P. (2012). Single atom substitution in mouse protein kinase G eliminates oxidant sensing to cause hypertension. *Nature Medicine, 18*(2), 286–290.

Reddie, K. G., & Carroll, K. S. (2008). Expanding the functional diversity of proteins through cysteine oxidation. *Current Opinion in Chemical Biology, 12*(6), 746–754.

Rockman, H. A., Koch, W. J., Milano, C. A., & Lefkowitz, R. J. (1996). Myocardial beta-adrenergic receptor signaling in vivo: Insights from transgenic mice. *Journal of Molecular Medicine (Berlin), 74*(9), 489–495.

Sarma, G. N., Kinderman, F. S., Kim, C., von Daake, S., Chen, L., Wang, B. C., et al. (2010). Structure of D-AKAP2:PKA RI complex: Insights into AKAP specificity and selectivity. *Structure, 18*(2), 155–166.

Schlossmann, J., Ammendola, A., Ashman, K., Zong, X., Huber, A., Neubauer, G., et al. (2000). Regulation of intracellular calcium by a signalling complex of IRAG, IP3 receptor and cGMP kinase Ibeta. *Nature, 404*(6774), 197–201.

Shattock, M. J. (2009). Phospholemman: Its role in normal cardiac physiology and potential as a druggable target in disease. *Current Opinion in Pharmacology, 9*(2), 160–166.

Shimokawa, H. (2010). Hydrogen peroxide as an endothelium-derived hyperpolarizing factor. *Pflugers Archiv: European Journal of Physiology, 459*(6), 915–922.

Tanaka, Y., Yamaki, F., Koike, K., & Toro, L. (2004). New insights into the intracellular mechanisms by which PGI2 analogues elicit vascular relaxation: Cyclic AMP-independent, Gs-protein mediated-activation of MaxiK channel. *Current Medicinal Chemistry. Cardiovascular and Hematological Agents, 2*(3), 257–265.

Tong, C. W., Stelzer, J. E., Greaser, M. L., Powers, P. A., & Moss, R. L. (2008). Acceleration of crossbridge kinetics by protein kinase A phosphorylation of cardiac myosin binding protein C modulates cardiac function. *Circulation Research, 103*(9), 974–982.

van der Wijk, T., Overvoorde, J., & den Hertog, J. (2004). H_2O_2-induced intermolecular disulfide bond formation between receptor protein-tyrosine phosphatases. *The Journal of Biological Chemistry, 279*(43), 44355–44361.

Vigil, D., Blumenthal, D. K., Taylor, S. S., & Trewhella, J. (2005). The conformationally dynamic C helix of the RIalpha subunit of protein kinase A mediates isoform-specific domain reorganization upon C subunit binding. *The Journal of Biological Chemistry, 280*(42), 35521–35527.

Wang, S. B., Foster, D. B., Rucker, J., O'Rourke, B., Kass, D. A., & Van Eyk, J. E. (2011). Redox regulation of mitochondrial ATP synthase: Implications for cardiac resynchronization therapy. *Circulation Research, 109*(7), 750–757.

Wooldridge, A. A., MacDonald, J. A., Erdodi, F., Ma, C., Borman, M. A., Hartshorne, D. J., et al. (2004). Smooth muscle phosphatase is regulated in vivo by exclusion of phosphorylation of threonine 696 of MYPT1 by phosphorylation of Serine 695 in response to cyclic nucleotides. *The Journal of Biological Chemistry*, *279*(33), 34496–34504.

Yetik-Anacak, G., & Catravas, J. D. (2006). Nitric oxide and the endothelium: History and impact on cardiovascular disease. *Vascular Pharmacology*, *45*(5), 268–276.

Zhang, D. X., Borbouse, L., Gebremedhin, D., Mendoza, S. A., Zinkevich, N. S., Li, R., et al. (2012). H_2O_2-induced dilation in human coronary arterioles: Role of protein kinase G dimerization and large-conductance Ca^{2+}-activated K^+ channel activation. *Circulation Research*, *110*(3), 471–480.

Zick, S. K., & Taylor, S. S. (1982). Interchain disulfide bonding in the regulatory subunit of cAMP-dependent protein kinase I. *The Journal of Biological Chemistry*, *257*(5), 2287–2293.

CHAPTER EIGHT

Redox Regulation of Protein Tyrosine Phosphatases: Methods for Kinetic Analysis of Covalent Enzyme Inactivation

Zachary D. Parsons*, **Kent S. Gates***,†,1

*Department of Chemistry, University of Missouri, Columbia, Missouri, USA
†Department of Biochemistry, University of Missouri, Columbia, Missouri, USA
[1]Corresponding author: e-mail address: gatesk@missouri.edu

Contents

1. Introduction 130
2. Rate Expressions Describing Covalent Enzyme Inactivation 133
3. Ensuring That the Enzyme Activity Assay Accurately Reflects the Amount of Active Enzyme 134
   3.1 Ensuring a linear response in Y as a function of $[E_{act}]$ 135
   3.2 Some causes of nonlinearity in instrument response and associated remedies 137
   3.3 Other potential sources of error in PTP assays 138
4. Assays for Time-Dependent Inactivation of PTPs 139
   4.1 General assay design considerations for a discontinuous "time-point" assay measuring time-dependent enzyme inactivation 139
   4.2 Inactivation of PTP1B by hydrogen peroxide 141
5. Analysis of the Kinetic Data 142
   5.1 "Traditional" linear analysis 143
   5.2 Nonlinear curve-fitting regression analysis 147
6. Obtaining an Inactivation Rate Constant from the Data 150
7. Summary 152
References 152

Abstract

Phosphorylation of tyrosine residues is an important posttranslational modification that modulates the function of proteins involved in many important cell signaling pathways. Protein tyrosine kinases and protein tyrosine phosphatases (PTPs) work in tandem to control the phosphorylation status of target proteins. Not surprisingly, the activity of some PTPs is regulated as part of the endogenous cellular mechanisms for controlling the intensity and duration of responses to various stimuli. One important mechanism for the regulation of PTPs involves endogenous production of hydrogen peroxide (H_2O_2) that inactivates enzymes via covalent modification of an active site cysteine

Methods in Enzymology, Volume 528
ISSN 0076-6879
http://dx.doi.org/10.1016/B978-0-12-405881-1.00008-2

thiolate group. Other endogenous metabolites and xenobiotics that inactivate PTPs via covalent mechanisms also have the potential to modulate signal transduction pathways and may possess either therapeutic or toxic properties. This chapter discusses methods for quantitative kinetic analysis of covalent inactivation of PTPs by small molecules.

1. INTRODUCTION

Phosphorylation of tyrosine residues is an important posttranslational modification that modulates the function of proteins involved in many important cell signaling pathways (Hunter, 2000; Lemmon & Schlessinger, 2010; Tarrant & Cole, 2009; Tonks, 2006). The phosphorylation status of proteins that are regulated in this manner is controlled by the balanced action of protein tyrosine kinases (PTKs) that add a phosphoryl group to the hydroxyl group of a target tyrosine side chain and protein tyrosine phosphatases (PTPs) that catalyze hydrolytic removal of the phosphoryl group (Hunter, 2000; Lemmon & Schlessinger, 2010; Tarrant & Cole, 2009; Tonks, 2006). It has long been known that the cellular activity of PTKs is tightly regulated (Lemmon & Schlessinger, 2010) and there is growing recognition that the activity of PTPs can also be regulated as part of the cellular mechanisms for controlling the intensity and duration of response to a given stimulus (den Hertog, Groen, & van der Wijk, 2005; Östman, Frijhoff, Sandin, & Böhmer, 2011; Tonks, 2006). One important mechanism for the regulation of PTP activity involves generation of endogenous hydrogen peroxide (H_2O_2) by the highly controlled activation of NADPH oxidase (Nox) enzymes (Lambeth, 2004; Lambeth, Kawahara, & Diebold, 2007; Ushio-Fukai, 2006) in response to growth factors, hormones, and cytokines such as platelet-derived growth factor, epidermal growth factor, VEGF, insulin, tumor necrosis factor-α, and interleukin-1β (Dickinson & Chang, 2011; Rhee, 2006; Tonks, 2006; Truong & Carroll, 2012). Inactivation of purified PTPs *in vitro* by hydrogen peroxide proceeds via oxidation of the catalytic cysteine thiolate residue (Scheme 8.1; Denu & Tanner, 1998; Hecht & Zick, 1992; Heffetz, Bushkin, Dror, & Zick, 1990; Lee, Kwon, Kim, & Rhee, 1998; Tanner, Parson, Cummings, Zhou, & Gates, 2011). Despite the high sequence and structural homology of the catalytic subunits of classical PTPs (Barr et al., 2009), the oxidized forms of different family members have the potential to adopt significantly different structures, with the active site cysteine residue existing as either a sulfenic acid, disulfide, or sulfenyl amide (Scheme 8.1; Tanner et al., 2011). Reaction of the oxidized enzymes with

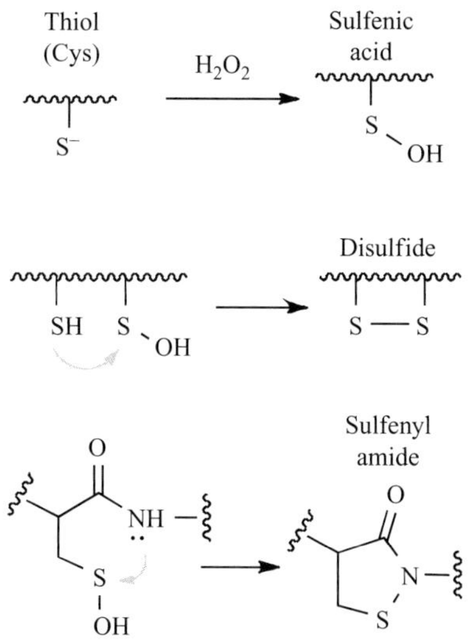

Thiol (Cys) S^- → H_2O_2 → Sulfenic acid S–OH

SH S–OH → Disulfide S—S

O NH S OH → Sulfenyl amide O N S

Scheme 8.1 Oxidative inactivation of PTPs.

low molecular weight or protein thiols leads to regeneration of the catalytically-active enzymes (Denu & Tanner, 1998; Parsons & Gates, 2013; Sivaramakrishnan, Cummings, & Gates, 2010; Sivaramakrishnan, Keerthi, & Gates, 2005; Tanner et al., 2011; Zhou et al., 2011).

The chemical and biochemical mechanisms underlying H_2O_2-dependent inactivation of PTPs in cells are not yet well understood. For example, the inactivation of intracellular PTPs during signaling events depends upon H_2O_2 and occurs rapidly (5–15 min; Lee et al., 1998; Mahedev, Zilbering, Zhu, & Goldstein, 2001; Meng, Buckley, Galic, Tiganis, & Tonks, 2004). However, the rate constants measured for inactivation of purified PTPs by H_2O_2 *in vitro* are modest (e.g., 10–40 M^{-1} s^{-1} for PTP1B; Denu & Tanner, 1998; Zhou et al., 2011). With rate constants in this range, the loss of PTP activity is anticipated to be rather sluggish at the low cellular concentrations of H_2O_2 expected to be present during signaling events ($t_{1/2}$=5–200 h at steady-state H_2O_2 levels of 0.1–1 μ*M*; Stone, 2006; Winterbourn, 2008). This kinetic discrepancy suggests that some unknown chemical or biochemical mechanism(s) in cells potentiates the ability of H_2O_2 to inactivate PTPs. Colocalization of PTP1B and Nox4 on the surface of the endoplasmic reticulum has been offered as a potential means for rapid and selective inactivation of this PTP during insulin signaling (Chen, Kirber, Yang, & Keaney, 2008). Others have provided evidence that localized inactivation of the peroxide-destroying enzyme, peroxiredoxin, during cell signaling events may yield regions of high peroxide concentration that can drive rapid inactivation of PTPs in these locales (Woo et al., 2010). Alternatively, or in addition, H_2O_2 may be spontaneously or enzymatically converted to a more reactive species that rapidly inactivates PTPs. For example, hydrogen peroxide has the potential to be converted to peroxymonophosphate or acyl peroxides capable of extremely rapid PTP inactivation (Bhattacharya, LaButti, Seiner, & Gates, 2008; LaButti, Chowdhury, Reilly, & Gates, 2007; LaButti & Gates, 2009). The conversion of H_2O_2 into peroxymonophosphate or acylperoxides would presumably require enzymatic assistance. The biological carbonate/bicarbonate buffer has been shown to potentiate the ability of H_2O_2 to inactivate PTPs (Zhou et al., 2011). This involves the spontaneous conversion of H_2O_2 to peroxymonocarbonate and seems likely to occur in the intracellular environment (Zhou et al., 2011). H_2O_2 can also react with lipids to generate lipid peroxides and, ultimately, decomposition products such as acrolein, *trans*-2-nonenal, and 4-hydroxynonenal (Conrad et al., 2010; Glascow et al., 1997; Seiner & Gates, 2007). Acrolein has been shown to inactivate purified PTP1B with an apparent second-order rate constant of (k_{inact}/K_I) of 87 M^{-1} s^{-1} (Seiner & Gates, 2007). Lipid peroxides, *trans*-2-nonenal, and 4-hydroxynonenal have been shown to inhibit PTP activity, although no rate or inhibition constants have been measured (Conrad et al., 2010; Glascow et al., 1997; Hernandez-Hernandez et al., 2005; Rinna & Forman, 2008). Along these lines, the only alkyl peroxide for which PTP inactivation rates have been measured is 2-hydroperoxytetrahydrofuran, shown to inactivate PTP1B with a rate constant of approximately 20 M^{-1} s^{-1} (Bhattacharya et al., 2008).

All of the potential mechanisms described above for the endogenous regulation of PTPs involve covalent enzyme modification. Furthermore, xenobiotics that inactivate PTPs via covalent mechanisms also have the potential to modulate signal transduction and may possess either therapeutic or toxic properties. Accurate measurement of the rate constants for chemical reactions leading to inactivation of PTPs by endogenous small molecules and xenobiotics is an important part of assessing their potential involvement in biological processes. For example, published rate constants allow the scientific community to estimate how quickly a given concentration of the agent will inactivate the protein. Covalent enzyme inactivation and reversible inhibition are fundamentally different processes, monitored and quantified by different means. The "on" rates for non-covalent binding of a reversible inhibitor to the enzyme are typically quite large (in the range of $1 \times 10^6\ M^{-1}\ s^{-1}$), and as a result, reversible enzyme inhibition usually can be observed almost immediately upon mixing of enzyme with inhibitor (within seconds or less, Pargellis et al., 1994; an exception to this rule is slow, tight-binding inhibitors, Merkler, Brenowitz, & Schramm, 1990). In contrast, the chemical reactions involved in the covalent modification of enzymes by irreversible enzyme inactivators are usually relatively slow. Thus, covalent enzyme inactivation is often described as a "time-dependent" process, and the assays used to quantitatively determine the kinetic constants associated with covalent enzyme modification involve measuring the loss of enzyme activity over the course of minutes or hours. Here, we describe the concepts and methods underlying quantitative kinetic analysis of the covalent inactivation of PTPs by small molecules.

2. RATE EXPRESSIONS DESCRIBING COVALENT ENZYME INACTIVATION

In the case where an agent inactivates an enzyme by covalent modification without prior noncovalent association, the process may be described by the rate expression:

$$\text{Rate of inactivation} = -\frac{\mathrm{d}[E_{\mathrm{act}}]}{\mathrm{d}t} = k_{\mathrm{obs}}[I][E_{\mathrm{act}}],$$

where $[I]$ and $[E_{\mathrm{act}}]$ are the concentrations of inactivator and active enzyme, respectively, and k_{obs} is the observed rate constant of interest (being second order, as shown). If the concentration of inactivator is in great excess over that of enzyme (10-fold or higher), $[I]$ is effectively constant during the course of the experiment, and the rate equation may be reduced to that of a first-order process:

$$\text{Rate} = -\frac{\mathrm{d}[E_{\mathrm{act}}]}{\mathrm{d}t} = k_{\psi}[E_{\mathrm{act}}],$$

where k_{ψ} is the pseudo-first-order rate constant for the reaction and is equal to $k_{\mathrm{obs}}*[I]$. Thus, the investigator will generally be concerned with experimentally determining k_{ψ} by first-order kinetic analysis and may determine k_{obs} by division of k_{ψ} by $[I]$.

Because loss of enzyme activity with respect to time follows first-order kinetics under pseudo-first-order conditions, such processes that "go to zero" may be described by the rate equation:

$$\frac{[E_{\text{act}}]_t}{[E_{\text{act}}]_0} = e^{-k_\psi * t},$$

where $[E_{\text{act}}]_0$ and $[E_{\text{act}}]_t$ are the concentrations of active enzyme at the start of reaction monitoring and at time t during the reaction, respectively. Note that if the reaction proceeds to some nonzero point at equilibrium, the relevant expression for generalized first-order processes is as follows:

$$\frac{[E_{\text{act}}]_t - [E_{\text{act}}]_\infty}{[E_{\text{act}}]_0 - [E_{\text{act}}]_\infty} = e^{-k_\psi * t},$$

where $[E_{\text{act}}]_\infty$ is the concentration of active enzyme at equilibrium (at $t = \infty$).

Because it is generally impractical to directly monitor the concentration of active enzyme in kinetics assays, it is common to rely on nonnatural substrates that release easy-to-measure products as an indirect readout of active enzyme concentration. A number of colorimetric and fluorometric substrates exist for the assay of PTPs (Montalibet, Skorey, & Kennedy, 2005). Thus, in the rate expressions above, we will henceforth substitute the term $[E_{\text{act}}]$ with Y, which denotes some *instrument reading*—detection of the enzyme's product—that reports indirectly upon $[E_{\text{act}}]$. Clearly, then, Y must faithfully report on the concentration of active enzyme to be of value. Therefore, we first discuss methods to ensure that the measurement of the product of enzymatic catalysis truly reflects the amount of remaining active enzyme in the PTP assay.

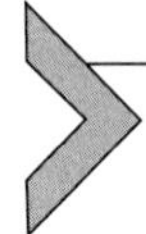

3. ENSURING THAT THE ENZYME ACTIVITY ASSAY ACCURATELY REFLECTS THE AMOUNT OF ACTIVE ENZYME

Before conducting enzyme inactivation assays, it is important to develop a set of assay conditions that allow accurate measurement of time-dependent losses of enzyme activity. Because *direct* mathematical substitution of Y for $[E_{\text{act}}]$ is to be made when solving the aforementioned rate equations, it is imperative that Y be linearly dependent on, and change solely as a function of, $[E_{\text{act}}]$. Note: this implicitly requires that the physical property of the molecule being measured reports linearly on its concentration – such as $\text{Abs}_{410\text{nm}}$ does for 4-nitrophenol, under our conditions (as determined in a separate experiment). Below, we discuss methods to test whether the assay conditions give a linear relationship between enzyme concentration and instrument response. We also discuss factors that can lead to nonlinear response and how to avoid these conditions.

3.1. Ensuring a linear response in *Y* as a function of [E_{act}]

3.1.1 Materials

PTP reaction buffer (*Buffer "R"*): 50 m*M* Tris, 50 m*M* Bis–Tris, 100 m*M* NaOAc, 10 m*M* DTPA, 0.5% Tween 80 (v/v), pH 7.0.

PTP activity assay buffer (*Buffer "A"*): 50 m*M* Bis–Tris, 100 m*M* NaCl, 10 m*M* DTPA, pH 6.0.

PTP substrate: 4-nitrophenyl phosphate (*p*NPP, disodium or di-Tris salt), made to 20 m*M* in PTP assay buffer.

Activity assay quench solution: 2 *M* NaOH in water.

Note: All buffers/aqueous solutions above are purely aqueous and should be made in highly purified water to minimize the concentration of redox-active transition metals which may oxidatively inactivate the PTP (air oxidation of thiolate in aqueous solution is a metal-dependent process; Misra, 1974).

Instruments: To maintain thermal equilibration in the reaction mixture and activity assay, a heating block or water bath will be required, and a quartz cuvette and UV–vis spectrophotometer required to measure the Abs_{410nm} of the product, 4-nitrophenolate (*p*NP).

4-Nitrophenyl phosphate provides a convenient method by which to monitor PTP activity, as PTP-mediated, catalytic hydrolysis of the substrate releases the chromophoric product 4-nitrophenol(ate), whose concentration can be measured at 410 nm and inorganic phosphate, which may also be measured if desired (Montalibet et al., 2005).

Note that, though acidification would also quench catalytic function of the PTP, it is the anion of *p*NP which absorbs strongly at 410 nm, and so a base (NaOH) quench is used to simultaneously halt the activity assay and increase signal.

3.1.2 Protocol

In a separate, preceding experiment, the concentration of the primary enzyme stock solution should be determined using standard methods (Bradford, 1976; Gilla & von Hippel, 1989; Pregel & Storer, 1997). Note that spectrophotometric determination of protein concentration requires precise knowledge of the protein sequence (Gilla & von Hippel, 1989). Additionally, these types of assays report on *total* concentration of protein in a sample; it is not necessarily the case that all protein present in the sample represents *catalytically active* enzyme. Thus, the protein concentration determined in these assays may be taken to represent the *maximum* amount of catalytically active enzyme possible in the sample. In the analyses below, exact knowledge of the fraction of total protein that is catalytically active is not required (it *is*, however, required for the determination of k_{cat} which is defined as v_{max}/[enzyme] in Michaelis–Menten kinetics). Here, because a calibrated instrument response, Y, is used in place of [E_{act}], we need not know its value precisely.

First, several dilutions from the concentrated primary enzyme stock are prepared and stored on ice until used. Then, to thermally-equilibrated samples of 490 μL of 20 m*M* *p*NPP in Buffer A at 30 °C are added 10 μL each of the varying-fold dilutions from the PTP stock. For the blank series, 10 μL of Buffer R alone are added to the activity assay mixture, and the sample treated identically to those containing PTP. Following addition of PTP, each sample is allowed to incubate for exactly 10 min prior to quenching of the activity assay by the addition of 500 μL of 2 *M* NaOH. The spectrophotometer is zeroed against the blank (no enzyme) assay, and the absorbance at 410 nm of each PTP-containing sample is recorded. The Abs_{410nm} readings are then plotted against the corresponding concentration of PTP in the sample (Fig. 8.1). Because each sample contains maximally active PTP, this calibration curve should represent maximal Y readings (Abs_{410nm}) that could be obtained if that particular concentration of enzyme were used in an inactivation experiment. Additionally, if a proper blank has been prepared, the

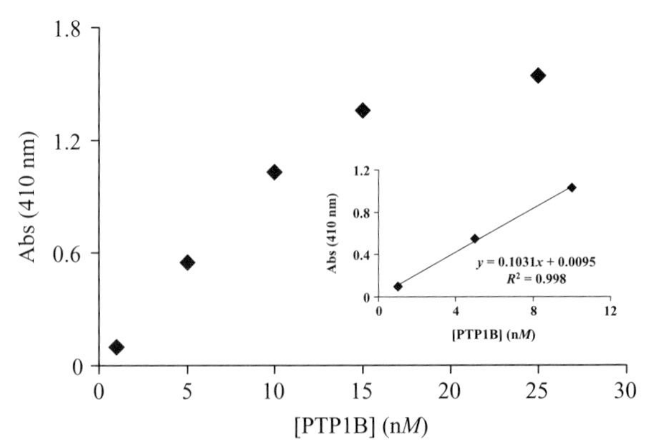

Figure 8.1 *Calibration curve for determining the range over which Y ($Abs_{410\ nm}$) is linearly dependent on concentration of PTP1B.* Several dilutions from a concentrated (20 μ*M*) stock of PTP1B were prepared and subjected to uniform activity assay conditions (20 m*M* *p*NPP, 30 °C, pH 6.0, 10 min; 500 μL total volume). Following quenching of the activity assay with 500 μL of 2 *M* NaOH, the absorbance at 410 nm was measured and plotted as a function of PTP concentration. The linear region (0–10 n*M* PTP1B) was determined to be the suitable concentration regime with which to conduct kinetics assays. (For color version of this figure, the reader is referred to the online version of this chapter.)

intercept of the regression line should be at the origin (x,y: $[\mathrm{PTP}_{\mathrm{active}}]=0$, $Y=0$). Enzyme concentrations that fall in the rising linear region of Fig. 8.1 will be suitable for enzyme inactivation experiments, as decreases in active enzyme concentration can be accurately detected in this range. On the other hand, use of enzyme concentrations in the plateau region of the plot will not be suitable for use in enzyme inactivation experiments. In this region, clearly, changes in concentration of active enzyme cannot be detected effectively. Below, we list some of the potential causes of this type of nonlinear instrument response in enzyme assays.

3.2. Some causes of nonlinearity in instrument response and associated remedies

As shown in Fig. 8.1, a plateau in Y is observed at high concentrations of PTP. In these assays, the human eye can detect that the solutions display differences in color density, yet the instrument did not detect these differences. Such lack of instrument response commonly arises when absorbance readings are above the usable range of the spectrophotometer. For most UV–vis spectrophotometers, the usable range is approximately 0–3 A.U. Note that, when absorbance readings are above the useable range of the instrument, a change in concentration of the species of interest (active PTP, reported on by [*p*NP]) does *not* necessarily afford a change in Y, thus rendering mathematical substitution of Y for $[E_{\mathrm{act}}]$ invalid. In cases where saturation of the instrument detector occurs, the investigator may decrease the concentration of enzyme or decrease the time for which the enzyme is incubated with the substrate during the activity assay.

Most UV–vis spectrophotometers do not issue a warning to the user when readings are above the usable range of the instrument. Telltale signs of detector saturation generally involve distortions in peak shape in the absorption spectrum: frequently, detector saturation is accompanied by a flattening-out of the peak around the $\lambda_{\max}$ and/or a jagged form to the top of the peak. New users should beware that, if the spectrophotometer has been blanked on a strongly absorbing solution, then although subsequent readings may be *numerically* well within the "acceptable 0–3 range," the readings may, in fact, be unreliable because the total absorbance of the solution is above the working range of the instrument.

Substrate depletion is another potential cause of a plateau such as that seen in Fig. 8.1. In the case of substrate depletion, the investigator may increase the concentration of substrate, decrease the concentration of enzyme, or decrease the incubation time of the enzyme with substrate during the activity assay. Last, we note that reaching the limit of enzyme solubility in the activity assay would also likely manifest as a plateau in Y versus $[E_{act}]$; however, this seems unlikely given that a large ($50\times$) dilution of the enzyme stock solution occurs when the PTP is introduced to the activity assay mixture.

3.3. Other potential sources of error in PTP assays

Because we equate Y to $[E_{act}]$, any process that diminishes or increases the instrument reading Y will contribute to errors in the $[E_{act}]$ values used in the kinetic analysis. Here, we consider just a few potential mechanisms that can lead to errors in the instrument readings Y.

Assay conditions. Because buffer pH and temperature, depletion of substrate, and time of reaction during signal amplification all contribute to the magnitude of Y, it is of paramount importance that these factors be held constant during the activity assay, such that the sole variable responsible for changes in the reading Y is $[E_{act}]$. In practice, this is easily accomplished by using buffers of appropriate strength, maintaining thermostated reaction mixtures, ensuring sufficient substrate concentration, and rigorously controlling reaction times. Similarly, the conditions (e.g., temperature, solvent, cuvette, instrument settings, wavelength) employed for the ultimate instrumental measurement of Y in all samples must be rigorously controlled to ensure all samples are treated in an identical manner.

Interference from substrate decomposition or another absorbing species. In some assays, a species other than that of interest may contribute to the instrument reading Y. For example, in a spectrophotometric assay, an enzyme inactivator may absorb at the same wavelength as the product of the substrate used to report on enzyme activity. Two approaches are useful to correct for the undesired contribution: (1) including the appropriate concentration of the interfering species in the blank for that series negates its contribution to Y, or (2) if the former is impractical, the contribution to Y by the species may be separately determined and its contribution corrected for in the terms Y_0 and Y_∞ (thus making changes in Y_t again solely a function of $[E_{act}]$). It is also possible that decomposition of an enzyme inactivator during the assay can lead to time-dependent changes in Y that would have to be accounted for to arrive at appropriate values.

In some cases, undesired contributions to Y may arise from decomposition of the substrate employed in the assay. For example, the PTP substrate *p*NPP (Scheme 8.2) undergoes slow, nonenzymatic hydrolysis to release the pNP chromophore. Thus, in kinetics assays involving extended incubation

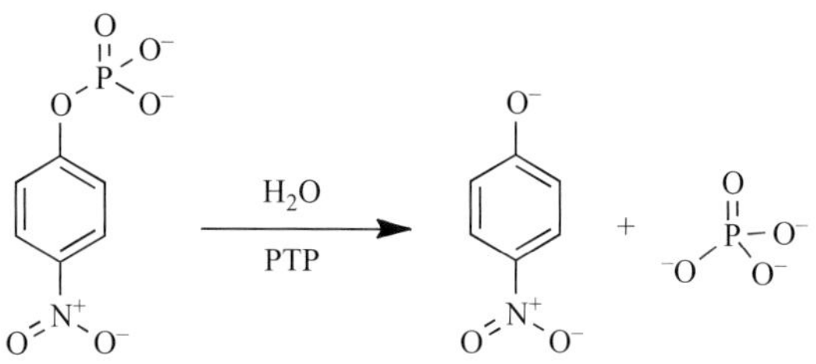

Scheme 8.2 Colorimetric PTP substrate *p*-nitrophenylphosphate (*p*NPP).

times (several hours or more), stock solutions of the enzyme substrate (*p*NPP) may begin to "yellow" and it may be necessary to rezero the spectrophotometer for each *time point* in the assay, against blanks that account for nonenzymatic hydrolysis. Doing so ensures that late reaction data do not "falsely report" an increase in concentration of active enzyme, due to the accumulation of nonenzymatically generated pNP.

Having addressed the means by which to ensure the instrument response Y faithfully reports on $[E_{act}]$, we may now turn our attention to assays that measure the rate of covalent enzyme inactivation.

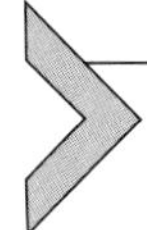

4. ASSAYS FOR TIME-DEPENDENT INACTIVATION OF PTPs

4.1. General assay design considerations for a discontinuous "time-point" assay measuring time-dependent enzyme inactivation

Using the methods and concepts discussed above, it should be possible to design assay conditions that allow accurate measurement of active enzyme concentrations in an inactivation assay mixture. Thus, we are prepared to discuss the design of a discontinuous "time-point" assay that measures the kinetics of time-dependent PTP inactivation. In this type of experiment, the enzyme and the time-dependent inactivator are mixed to create an *inactivation reaction*, and at various time points, aliquots of the mixture are removed and subjected to an *activity assay* that measures the amount of active enzyme remaining at that time. Ideally, each data point taken should represent a time-frozen "snapshot" of the enzyme inactivation process. In order to accomplish this, the enzyme inactivation reaction must be stopped and remaining enzyme activity assayed. In the following section, we list several methods for stopping the enzyme inactivation reaction prior to assessment of the amount of active enzyme remaining in the aliquot:

1. *pH perturbation*: In some cases, the enzyme inactivator may be effectively quenched by acid or base. For example, protonation of anionic groups may significantly decrease the affinity of some agents for the active site of PTPs, thus decreasing their ability to carry out inactivation of the enzyme. Some reactivity is likely to remain, and time between quench and the activity assay should remain short (and consistent for all samples).
2. *Decomposition of agent*: In some cases, convenient methods for rapid decomposition of an agent may be available. For example, catalase may be used to rapidly decompose the PTP inactivator hydrogen peroxide (Halliwell & Gutteridge, 1990). Control reactions should be carried out to confirm that neither the agent employed for decomposition of the inactivator nor the products of decomposition inactivate the enzyme.
3. *Dilution of agent:* Perhaps most commonly, the inactivation reaction is abruptly halted by dilution of an aliquot of the inactivation reaction mixture directly into the activity assay mixture used to measure remaining active enzyme. Like all aforementioned methods, this approach actually serves to dramatically decrease the *rate* of reaction, rather than to completely stop it, and the decrease in rate is directly proportional to the dilution factor. Clearly, assay designs employing large dilution factors at this step will give more accurate results. In cases where the enzyme inactivator is also capable of reversible noncovalent binding to the enzyme active site (*inhibition*), residual amounts of the inactivator present following dilution into the activity assay may be sufficient to inhibit the enzyme during the assay. In these cases, at time zero, when no covalent inactivation should have occurred, the enzyme activity may be significantly below that of a control enzyme sample that contains no agent. While this may alter the appearance of the data (inactivation time courses for higher inactivator concentrations in a plot such as Fig. 8.3 will be shifted downward), analysis of the *time-dependent* loss of enzyme activity should be largely unaffected.
4. *Physical removal of the agent*: Agents may be physically removed from inactivation mixtures by various means including extraction or gel filtration. If reaction times are short, it may be somewhat challenging to execute these methods at sharply defined time points.

After the aliquot is removed from the inactivation reaction mixture and the inactivation reaction stopped, the aliquot must be subjected to an *activity assay* to measure the amount of active enzyme remaining. For accurate and reproducible results, the activity assay must be carried out in an identical manner for each time point. This typically requires a rapid quench of enzymatic activity in the activity assay, which may be accomplished by "pH shock", addition of metal chelators (for metal-dependent enzymes), and rapid freezing.

4.2. Inactivation of PTP1B by hydrogen peroxide

Denu and Tanner (2002) provided an excellent discussion of the inactivation of PTPs by hydrogen peroxide as part of a broader review on the redox regulation of PTPs in a previous volume of this series. Our treatment of the subject is intended to provide additional technical detail. From the PTP concentration calibration curve shown in Fig. 8.1, it was determined that 8 nM PTP1B was suitable for use in the activity assay, under our conditions. Because this final concentration of PTP1B was achieved following a 50-fold dilution into the activity assay buffer, the requisite enzyme concentration in the inactivation reaction was calculated to be 400 nM. An example benchtop layout of standard laboratory equipment employed for conducting inactivation assays is presented in Fig. 8.2.

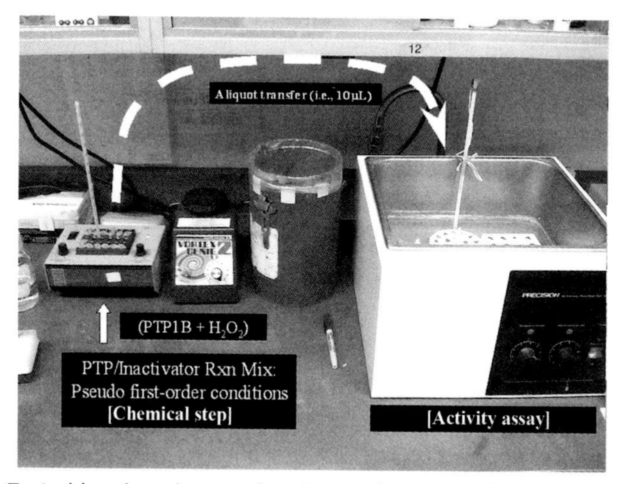

Figure 8.2 *Typical benchtop layout of equipment for enzyme kinetics assays.* A heating block (left), containing samples in which the "chemical step" (inactivation) is conducted, is maintained at the desired temperature under which the study is conducted (here, 25 °C). At predefined time intervals, aliquots are removed from the reaction mixture and diluted into ready-made activity assay mixtures (right). Here, remaining enzyme activity is assessed by permitting the enzyme to turn over substrate for a uniform period of time before quenching the reaction and measuring the concentration of product formed (e.g., $Abs_{410\ nm}$ for 4-nitrophenol). (For color version of this figure, the reader is referred to the online version of this chapter.)

4.2.1 Protocol

1. Free thiols were removed from a stock solution of concentrated PTP1B (20 μ*M*) by gel filtration/buffer exchange into Buffer R using a Zeba mini centrifugal desalting column (Pierce, catalog no. 89882). The stock was subsequently diluted to 0.8 μ*M* in Buffer R and stored on ice until used.
2. Prepared in Buffer R were 2×-concentrated solutions of H_2O_2: 500, 400, 300, 200, and 100 μ*M*, diluted from a 30% (wt/v) stock (Sigma). These were also stored on ice until used.
3. In 2 mL Eppendorf tubes, 490 μL (each) aliquots of 20 m*M* substrate (4-nitrophenyl phosphate, disodium hexahydrate; Sigma) in Buffer A were prepared, stored at room temperature until used (totaling 31 pre-prepared samples).
4. Just prior to starting the assay, microcentrifuge tubes containing 60 μL of 0.8 μ*M* PTP1B, and 60 μL of Buffer R containing hydrogen peroxide, were added to a heating block held at 25 °C and allowed to stand for 5 min. During that time, five activity assay samples were added to a water bath (30 °C) and allowed to thermally equilibrate.
5. The inactivation reaction was then initiated by combining 1:1 (v/v, 40 μL each) of 0.8 μ*M* PTP1B and 2×-H_2O_2 in buffer, and a timer started. At 1, 2, 4, 7, and 10 min time points during the reaction, 10 μL aliquots were removed from the inactivation mixture and diluted into the waiting 490 μL activity assay mixtures. The activity assay was allowed to proceed for 10 min before being quenched by addition of 500 μL of 2 *M* NaOH in ddH_2O (quench times of the activity assays occurred at 11, 12, 14, 17, and 20 min relative to the start of the inactivation reaction). This process was repeated for each concentration of H_2O_2 and for one control series lacking peroxide (Buffer R alone). A blank reaction sample was prepared in similar fashion: 10 μL of Buffer R alone was added to an activity assay sample and treated identically to the experimental series.
6. At the conclusion of the assay, the spectrophotometer was zeroed against the blank sample and the absorbance at 410 nm of the experimental series measured. The absorbance readings obtained in this manner were plotted as a function of time (Fig. 8.3).

5. ANALYSIS OF THE KINETIC DATA

There are many methods for analyzing enzyme inactivation data of this type. Here, we consider two broad approaches to analysis of kinetic data: (1) "traditional" linear regression analyses and (2) "modern" nonlinear curve-fitting analyses.

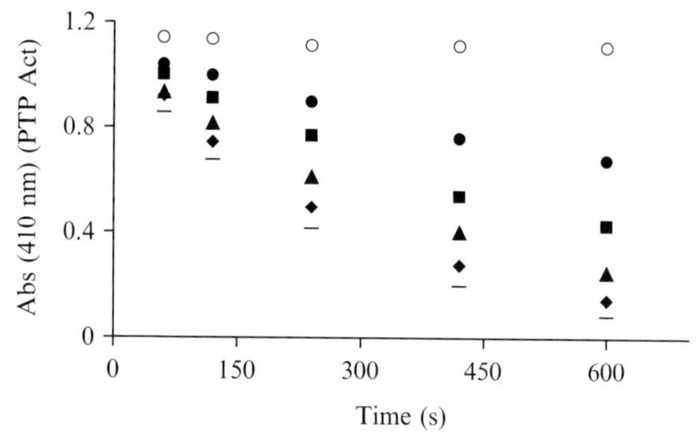

Figure 8.3 *Time-dependent inactivation of PTP1B by hydrogen peroxide under pseudo-first-order conditions.* 400 n*M* PTP1B was inactivated by various concentrations of H_2O_2 at 25 °C, pH 7.0: 250 μ*M* (dashes), 200 μ*M* (diamonds), 150 μ*M* (triangles), 100 μ*M* (squares), 50 μ*M* (closed circles), or no peroxide (open circles). Remaining enzyme activity was assayed at 1, 2, 4, 7, and 10 min intervals by dilution of 10 μL of the inactivation mixture into 490 μL of activity assay buffer. The activity assay was allowed to proceed for 10 min before quenching via addition of 500 μL of 2 *M* NaOH. The absorbance at 410 nm of each sample was then measured and plotted as a function of time, revealing exponential loss of enzyme activity versus time.

5.1. "Traditional" linear analysis

A plot of the absorbance data generated as described above yields a plot such as that seen in Fig. 8.3. Because inactivation of PTP1B by excess hydrogen peroxide affords *completely* inactive enzyme at $t=\infty$, and the blank used in the experiment accounted for any residual absorbance at 410 nm *not* from enzymatic hydrolysis, the Y_∞ value can be taken as zero. The initial measurement, Y_0, need not *actually* be that corresponding to the "real" $t=0$ (measurement of which is effectively impossible). Rather, Y_0 represents the first *monitoring* of the reaction, relative to which all Y_t values are spaced with respect to time. For our purposes here, the first collected data point in the control series (no H_2O_2) series has been defined as Y_0 for all experimental series.

5.1.1 Extraction of pseudo-first-order rate constants by linear analysis

Having defined all parameters (Y_0, Y_∞, and Y_t), a plot of the natural logarithm of percent remaining activity versus time may be generated; namely: ln $((Y_t/Y_0)*100)$ versus time. It is worth noting that plotting the natural logarithm of "simple" (not percent) remaining activity versus time affords the same numerical results, but a perhaps less aesthetically pleasing graphical form, as the natural logarithm of numbers less than 1 are negative.

Having generated the replot (Fig. 8.4), the resulting transformation should be carefully inspected for strict linearity. As will be described later, nonlinear behavior in this plot is indicative of either non-first-order kinetics or use of an inappropriate Y_∞ value (Fig. 8.5).

Once it has been determined that the data adhere to linearity in the "ln plot," linear regression analysis may be performed for each series (i.e., for each concentration of inactivator). Because the kinetics of a pseudo-first-order process is described by

$$\ln\frac{Y_t}{Y_o} = -k_\psi * t,$$

where k_ψ is the pseudo-first-order rate constant, a plot of $\ln(Y_t/Y_0)$ versus time affords lines with slopes $-k_\psi$. Thus, the negative of the slope of each line equals the pseudo-first-order rate constant for the corresponding concentration of inactivator (with units of reciprocal time corresponding to that used on the X-axis).

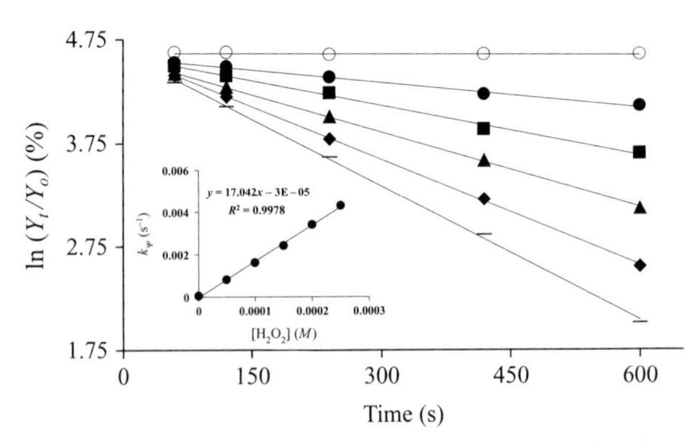

Figure 8.4 *Analysis of inactivation kinetic data via linear methods.* The $Abs_{410\ nm}$ readings from the inactivation assay were replotted as the natural logarithms of percent remaining activity (relative to the control) versus time. The replot afforded straight lines, in agreement with a (pseudo)-first-order process. The slope of the linear regression trendline from each series is equal to the negative of the pseudo-first-order rate constant, for inactivation of PTP1B by the corresponding concentration of H_2O_2. A replot of the pseudo-first-order rate constants versus concentration of inactivator affords a straight line which passes through the origin, in agreement with a simple bimolecular process in the rate-determining step (inset). The slope of this line is the apparent bimolecular rate constant for inactivation of PTP1B by H_2O_2 under our conditions, with units of $M^{-1}\ s^{-1}$.

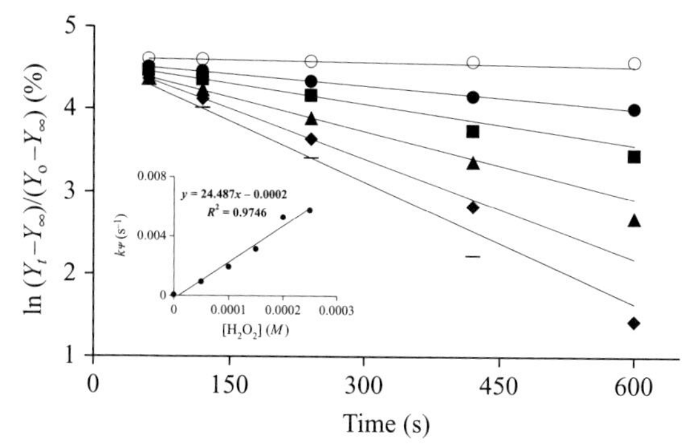

Figure 8.5 *Replot of natural logarithm of percent remaining activity with inappropriate choice of* Y_∞. Kinetic data from inactivation of PTP1B by H_2O_2 were analyzed by linear methods: here, with intentional choice of an inappropriate Y_∞ value (here, 0.1 A.U. instead of 0 A.U. at 410 nm). Appreciable curvature is noted in the data, especially as Y_t approaches Y_∞. This curvature is indicative of a poor determination of Y_∞. The calculated bimolecular rate constant of interest is "skewed" as a consequence of this selection of Y_∞ (inset). To better highlight curvature in the data, the regression trendlines shown only consider the early (first three) data points from each series.

Though, in principle, one could conduct a single assay under pseudo-first-order conditions and extract an apparent bimolecular rate constant by dividing k_ψ by the concentration of inactivator, this represents an incomplete and potentially erroneous approach to describing the kinetics of a process. The more rigorous and informative approach is to examine the dependence of the pseudo-first-order rate constants on concentration of inactivator. As shown in Fig. 8.4 (inset), the pseudo-first-order rate constant for inactivation of PTP1B by H_2O_2 is linearly dependent upon concentration of H_2O_2. Such linear dependence is not universal for all PTP inactivators, and the interpretation of these data is described at the end of this work.

5.1.2 Linear analysis methods: Strengths and weaknesses

Although contemporary computational abilities long ago dispensed with the *necessity* for linear (graphical) analyses of data, it is still a highly useful and straightforward approach. Linear analysis benefits from being "user-friendly," precise, and accurate when used in conjunction with good experimental data and requires little or no specialized software or training. Analysis of linearized data is typically very straightforward, and large deviations from linearity in a data set generally are easy to spot.

The chief drawbacks of linear methods for analysis involve treatment of "late data" and intercept evaluation. Regarding treatment of late data, as Y_t

approaches Y_∞, two issues arise: (1) because the dynamic changes in Y_t can become small with respect to the error inherent in making the measurement, the data sometimes become erratic and essentially useless in the ln replot; and (2) as the dynamic character of the measurement approaches zero (as Y_t approaches Y_∞, a constant), the data may again be unusable in linear analysis, as a consequence of there being no measurable change in activity versus time. If there is no measurable change in activity versus time, then there can be no change in ln (%) activity versus time, and the slope of the regression line in the ln replot will bend to zero. Additionally, in late data, small errors in the determination of Y_∞ become more pronounced, often resulting in an obvious curvature of the data in the ln replot, which *should be* linear (Fig. 8.5 depicts such curvature). As discussed in the next section, curve-fitting methods adapt well to, but are certainly not immune to these issues.

The second general weakness associated with linear regression analyses involves determination of values that are computed as intercepts, such as is done with the Kitz–Wilson plot (Fig. 8.6; Kitz & Wilson, 1962; Silverman, 2000). Values determined from intercepts along the *X*- or *Y*-axis may be subject to exaggerated error when the experimental data fall "far"

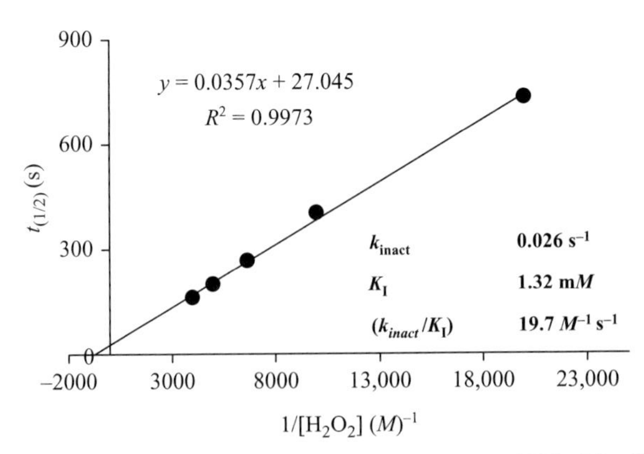

Figure 8.6 *Kitz–Wilson replot of kinetic data for inactivation of PTP1B by H_2O_2.* The Kitz–Wilson replot is commonly used in analysis of kinetic data of inactivators which have noncovalent affinities for their protein targets (e.g., affinity labeling agents). For inactivators which have no appreciable affinity, the linear regression trendline in the Kitz–Wilson plot should intercept at the origin. Here, small experimental errors afford a nonorigin intercept, as hydrogen peroxide is not known to possess any particular affinity for the active site of PTP1B. Thus, the physical interpretation of the intercept is ambiguous in this case.

from the axis at which the intercept occurs in the replot. Under these circumstances, small errors in the data may dramatically alter the values of the intercepts. For example, in the Kitz–Wilson plot, the y-intercept is equal to $\ln(2)/k_{inact}$, and the x-intercept to $(-)1/K_I$. Because the X-axis has units of reciprocal concentration, but the x-intercept is *negative*, the x-intercept *must* be calculated by projections made based on experimental data. It stands to reason, then, that the only way good projections can be made is by having collected excellent experimental data.

5.2. Nonlinear curve-fitting regression analysis

For this consideration of curve-fitting analysis, we will employ the same kinetic data used above in the linear regression analysis. The parameters defined above will be used in identical fashion here (those of Y_t, Y_0, and Y_∞). Conceptually, the approach is very similar: starting from the rearranged integrated rate law for a first-order process, we see that there is an exponential dependence of Y_t on both the (pseudo)-first-order rate constant and time:

$$Y_t = Y_o e^{-k_\psi * t}$$

Because Y_0 and Y_t are measured quantities, and the time intervals at which the process was monitored are known, the only parameter to be optimized to make the mathematical model fit the experimental data is the rate constant of interest, k_ψ.

The parameter optimization process (data fitting) may be accomplished with a variety of software suites, including Prism and Microsoft Excel. Independent of software package used, most commonly the essence of the fitting process involves computing the arithmetic differences between model and experimental data sets for each data point, squaring these differences, and finally minimizing the sum of these squared differences by modulating the parameter to be optimized (here, k_ψ). Finally, the optimized data set (from which k_ψ is determined) should be graphically compared against the empirical data set for visual inspection of agreement between the two (Fig. 8.7). Additionally, examining a plot of the residuals is a rigorous method by which to identify any systematic bias in the fitted data set. An excellent discussion of several ways by which this may be done is freely available on the NIST Web site (Anonymous). The optimized parameters that afford the best fit of the model to the experimental data set are taken to be the appropriate values of interest (here, k_ψ). Figure 8.7 shows an example

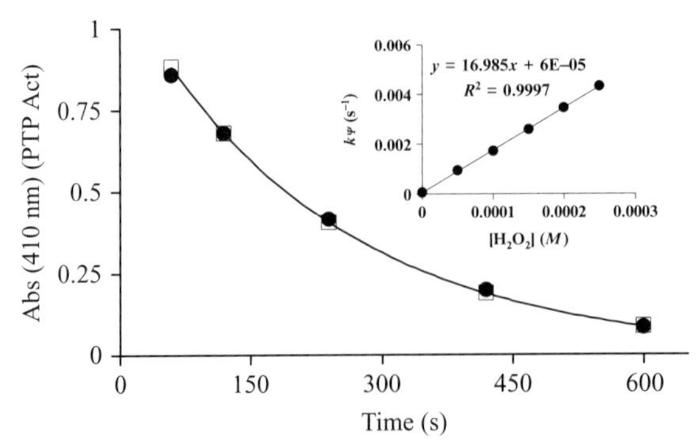

Figure 8.7 *Analysis of inactivation kinetic data via nonlinear regression analysis.* The untransformed experimental data ($Abs_{410\ nm}$ vs. time) were used to fit model first-order kinetic data, with k_ψ as the only adjustable parameter (Y_0 and Y_∞ were defined in identical fashion to that of the linear analysis method). No constraints were placed on k_ψ during the optimization process. The optimized, fitted model was graphically examined for goodness of fit. Shown are the experimental data for the inactivation of PTP1B by 250 μM H_2O_2 (closed circles) and the fitted model data (open squares and trendline) from which k_ψ was determined. This process was repeated for each concentration of H_2O_2, and the optimized k_ψ values were plotted as a function of corresponding concentration of H_2O_2. This afforded a straight line in the replot which passes through the origin, indicative of a simple bimolecular reaction. The calculated bimolecular rate constant is in excellent agreement with that determined by linear analysis methods (17 M^{-1} s^{-1}).

of a model set fitted to the experimental data, where the only parameter optimized was k_ψ. This process was repeated for each concentration of inactivator, and the resulting series of pseudo-first-order rate constants were plotted against corresponding concentrations of H_2O_2. A linear dependence of k_ψ on [H_2O_2] was again observed (Fig. 8.7, inset), and the apparent bimolecular rate constant was found to be in excellent agreement with that computed by linear graphical analysis methods (17 M^{-1} s^{-1} in both cases).

5.2.1 Curve-fitting analysis: Strengths and weaknesses

In order to speak meaningfully about the pitfalls of nonlinear regression analysis, we must first consider its strengths. Nonlinear regression analysis is an exceedingly powerful tool; it is very flexible, being well suited for the treatment of an entire data set, from "start to finish." Furthermore, if an accurate determination of Y_∞ was not experimentally made, the methods of nonlinear regression analysis allow Y_∞ itself to be an adjustable parameter, in addition to k_ψ. This facet is extremely useful for sluggish reactions, or for reactions which proceed to different Y_∞ values as a function of concentration of inactivating species. In these cases, collection of large amounts of "late-phase" reaction data actually facilitates accurate optimization of adjustable parameters, as the data are used to better determine appropriate values for the parameters being optimized (especially for Y_∞). Additionally, the investigator may define a numerical range into which the parameters must fall during/at the conclusion of the optimization process, such that values determined from empirically–collected data are used to bound upper and lower limits of any adjustable parameter. Taken together, nonlinear regression represents an exceedingly useful tool for kinetic analysis, allowing the investigator to decide which parameters to leave fixed or to make adjustable, and to define the window into which adjustable parameters must ultimately fall.

The reader may be struck by the last statement, understanding immediately some potentially crippling pitfalls of nonlinear regression analysis. If the investigator selects an inappropriate window into which any given parameter may fall, the kinetic analysis becomes invalid (the magnitude of invalidity being directly proportional to "how far off the mark" the prescribed window was set). It is no surprise, then, that parameters which are to be fit must be "bracketed" (bounded) with great care. Despite the obvious potential for mistakes in this regard, this aspect of data fitting is typically not of the greatest concern, as appropriate windows in which to bound optimized parameters are generally clear from the empirical data (e.g., following a reaction through three half-lives limits the maximum error in optimization of Y_∞ to roughly 10%).

Perhaps the greatest drawback to nonlinear regression analysis is the challenge of discerning when a good-*looking* fit of the model to the experimental data *does not* provide realistic kinetic values. This so-called chi-by-eye approach, referring to visual evaluation of goodness of fit by graphical methods, can be highly misleading. Furthermore, deviations from the expected graphical form may be more difficult to spot in curvilinear data than in linear data during visual analysis. It is the author's experience that the greatest potential for erroneous analysis of kinetic data by nonlinear regression methods arises when fitting "early reaction data" with a poorly defined Y_∞ value. It is a logical extension to suggest that such erroneous analysis would also apply when analyzing "late-phase" reaction data with a poorly defined Y_0 (though, for reasons stated earlier, this is probably less likely a concern). The fundamental rationale for these observations may be that, in many reactions, "early" and "late" data appear linear in form, and so do not adequately reflect the exponential nature of the data expected for the full time course (to which the model is fit). Unfortunately, when fitting data sets containing only "early data," the optimized model may look deceptively well fit to the experimental data, though the optimized parameters have very poor agreement with those of physical reality (the "true values"). For these reasons, it is crucial that the investigator adequately defines the window into which adjustable parameters are to fall and very critically evaluates the optimized parameters at the end of the fitting process, with respect to their empirical viability (i.e., do the values "make sense" with respect to what the investigator knows about the assay?).

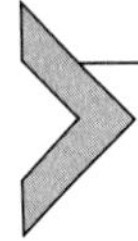

6. OBTAINING AN INACTIVATION RATE CONSTANT FROM THE DATA

Having collected and analyzed the kinetic data, and extracted pseudo-first-order rate constants by linear or curve-fitting methods, the investigator may now begin to characterize the kinetic properties of the process of interest. Here, we briefly cover two kinetic profiles observed for PTP inactivators and refer the reader elsewhere for a more complete discussion of analysis of pseudo-first-order kinetics (Espenson, 1995).

As shown in Figs. 8.4 and 8.7 (insets), a plot of the pseudo-first-order rate constants (in s^{-1}) versus molar concentration of the inactivator H_2O_2 affords a straight line that passes through the origin. The graphical form of this plot (linear, intercepting the origin) suggests a simple bimolecular process in the rate-determining step, that is first order in inactivator. The slope of the line is the observed bimolecular rate constant (in $M^{-1}\ s^{-1}$) for inactivation of PTP1B by H_2O_2, under these conditions. Indeed, this is perhaps the simplest kinetic profile for a bimolecular reaction.

Another common graphical form the investigator may encounter is the "saturation profile" in the plot of k_{ψ} versus concentration of inactivator (Fig. 8.8). In similar fashion to the principles underlying traditional Michaelis–Menten kinetics with *substrates*, this behavior is a result of reversible, noncovalent association of the inactivator with the enzyme prior to the chemical inactivation step (Scheme 8.3). Here, the enzyme first reversibly binds the inactivator, rapidly forming the $E\cdot I$ complex by the second-order rate constant k_{on}. The $E\cdot I$ complex may then collapse back to free enzyme and inhibitor by rate constant k_{off}, or covalent modification/inactivation of the enzyme may occur by rate constant k_{inact} (both rate constants being first order). Because the chemical step, represented by $k_{inact}[E\cdot I]$, is usually rate limiting, increasing the concentration of the $E\cdot I$ complex increases the observed rate of inactivation. At low concentrations of inhibitor, increasing

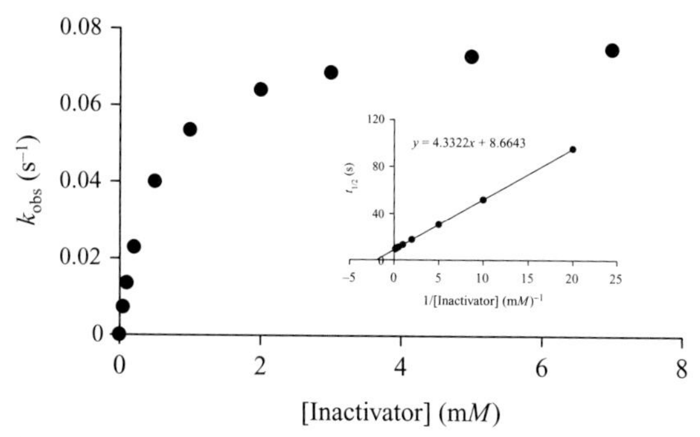

Figure 8.8 *Mock data: "saturation profile" for an inactivator which has affinity for its enzyme target.* For inactivating species which possess affinity for their targets (such as affinity labeling agents), saturation kinetics may be observed in the rate of inactivation at high concentrations of inactivator. The kinetic parameters k_{inact} and K_I may be extracted by linear methods (e.g., Kitz–Wilson replot, inset), or by curve-fitting methods. Here, the mock data were generated using $k_{inact}=0.08\ s^{-1}$ and $K_I=500\ \mu M$.

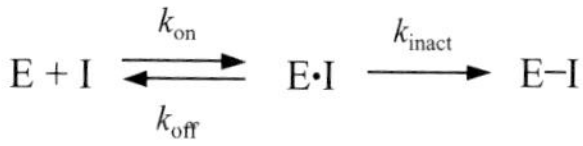

$$E + I \underset{k_{off}}{\overset{k_{on}}{\rightleftharpoons}} E{\cdot}I \xrightarrow{k_{inact}} E{-}I$$

Scheme 8.3 Inactivation of PTPs by affinity labeling agents.

the concentration of the inhibitor drives formation of the $E{\cdot}I$ complex, resulting in an increase in observed rate of inactivation. However, when essentially all enzyme is occupied in the $E{\cdot}I$ complex under saturating concentrations of inactivator, further increases in concentration of inactivator do *not* result in increased rates of inactivation. Consequently, the observed rate of inactivation becomes *independent* of inactivator concentration under saturating conditions, resulting in a plateau in the plot of k_{ψ} versus [inactivator]. Note that reaching the limit of solubility of the inactivator may also result in such a plateau, and the investigator should confirm that this is not the underlying cause of the plateau in observed rates of inactivation at high inactivator concentrations.

Analysis of this type of kinetic data may be accomplished by the Kitz–Wilson plot (Fig. 8.8, inset; Kitz & Wilson, 1962; Silverman, 2000). The Kitz–Wilson plot is a convenient tool with which the investigator may determine both K_I and k_{inact}. However, for reasons described earlier, the investigator should take care in data collection for and evaluation of intercept-based values. Alternatively, these data may be evaluated using curve-fitting methods. For example, Zhang's group identified inactivators of PTP1B which exhibit saturation behavior and determined the parameters K_I and k_{inact} by fitting to, essentially, the Michaelis–Menten equation, with K_I in place of K_m and k_{inact} in place of V_{max} (Liu et al., 2008).

7. SUMMARY

Here, we have described some fundamental considerations to be made when monitoring the activity of enzyme systems, the general conceptual bases underlying assay design when measuring the rates of covalent enzyme modification, and several methods by which the kinetic data may be analyzed. We have also reported a detailed method by which the rate of inactivation of PTP1B by hydrogen peroxide may be determined; this method may be extended to other inactivators, and to other PTPs. Finally, we have addressed the merits and drawbacks to numerous methods of data analysis, with an eye toward "choosing the right tool for the job."

REFERENCES

Anonymous. National Institutes of Standards and Technology website. http://wwwitlnistgov/div898/handbook/indexhtm.

Barr, A. J., Ugochukwu, E., Lee, W. H., King, O. N., Filippakopoulos, P., Alfano, I., et al. (2009). Large-scale structural analysis of the classical human protein tyrosine phosphatome. *Cell*, *136*, 352–363.

Bhattacharya, S., LaButti, J. N., Seiner, D. R., & Gates, K. S. (2008). Oxidative inactivation of PTP1B by organic peroxides. *Bioorganic & Medicinal Chemistry Letters*, *18*, 5856–5859.

Bradford, M. (1976). Rapid and sensitive method for the quantification of microgram quantities of protein utilizing the principle of protein-dye binding. *Analytical Biochemistry*, *72*, 248–254.

Chen, K., Kirber, M. T., Yang, Y., & Keaney, J. F. J. (2008). Regulation of ROS signal transduction by NADPH oxidase 4 localization. *The Journal of Cell Biology*, *181*, 1129–1139.

Conrad, M., Sandin, A., Förster, H., Seiler, A., Frijhoff, J., Dagnell, M., et al. (2010). 12/15-Lipoxygenase-derived lipid peroxides control receptor protein kinase signaling through oxidation of protein tyrosine phosphatases. *Proceedings of the National Academy of Sciences of the United States of America*, *107*, 15774–15779.

den Hertog, J., Groen, A., & van der Wijk, T. (2005). Redox regulation of protein-tyrosine phosphatases. *Archives of Biochemistry and Biophysics*, *434*, 11–15.

Denu, J. M., & Tanner, K. G. (1998). Specific and reversible inactivation of protein tyrosine phosphatases by hydrogen peroxide: Evidence for a sulfenic acid intermediate and implications for redox regulation. *Biochemistry*, *37*, 5633–5642.

Denu, J. M., & Tanner, K. G. (2002). Redox regulation of protein tyrosine phosphatases by hydrogen peroxide: Detecting sulfenic acid intermediates and examining reversible inactivation. *Methods in Enzymology*, *348*, 297–305.

Dickinson, B. C., & Chang, C. J. (2011). Chemistry and biology of reactive oxygen species in signaling and stress responses. *Nature Chemical Biology*, 7, 504–511.

Espenson, J. H. (1995). *Chemical kinetics and reaction mechanisms* (2nd ed.). New York: McGraw-Hill, Inc.

Gilla, S. C., & von Hippel, P. H. (1989). Calculation of protein extinction coefficients from amino acid sequence data. *Analytical Biochemistry, 182*, 319–326.
Glascow, W. C., Hui, R., Everhart, A. L., Jayawickreme, S. P., Angerman-Stewart, J., Han, B.-B., et al. (1997). The linoleic acid metabolite, (13S)-hydroperoxiyocta-decadienoic acid, augments the epidermal growth factor receptor signaling pathway by attenuation of receptor dephosphorylation. *The Journal of Biological Chemistry, 272*, 19269–19276.
Halliwell, B., & Gutteridge, J. M. C. (1990). Role of free radicals and catalytic metal ions in human disease: An overview. *Methods in Enzymology, 186*, 1–85.
Hecht, D., & Zick, Y. (1992). Selective inhibition of protein tyrosine phosphatase activities by H_2O_2 and vanadate in vitro. *Biochemical and Biophysical Research Communications, 188*, 773–779.
Heffetz, D., Bushkin, I., Dror, R., & Zick, Y. (1990). The insulinomimetic agents H_2O_2 and vanadate stimulate protein tyrosine phosphorylation in cells. *The Journal of Biological Chemistry, 265*, 2896–2902.
Hernandez-Hernandez, A., Garabatos, M. N., Rodriguez, M. C., Vidal, M. L., Lopez-Revuelta, A., Sanchez-Gallego, J. I., et al. (2005). Structural characteristics of a lipid peroxidation product, trans-2-nonenal, that favour inhibition of membrane-associated phosphotyrosine phosphatase activity. *Biochimica et Biophysica Acta, 1726*, 317–325.
Hunter, T. (2000). Signaling—2000 and beyond. *Cell, 100*, 113–127.
Kitz, R., & Wilson, I. B. (1962). Esters of methanesulfonic acid as irreversible inhibitors of acetylcholinesterase. *The Journal of Biological Chemistry, 237*, 3245–3249.
LaButti, J. N., Chowdhury, G., Reilly, T. J., & Gates, K. S. (2007). Redox regulation of protein tyrosine phosphatase 1B by peroxymonophosphate. *Journal of the American Chemical Society, 129*, 5320–5321.
LaButti, J. N., & Gates, K. S. (2009). Biologically relevant properties of peroxymonophosphate (=O3POOH). *Bioorganic & Medicinal Chemistry Letters, 19*, 218–221.
Lambeth, J. D. (2004). NOX enzymes and the biology of reactive oxygen. *Nature Reviews. Immunology, 4*, 181–189.
Lambeth, J. D., Kawahara, T., & Diebold, B. (2007). Regulation of Nox and Duox enzymatic activity and expression. *Free Radical Biology & Medicine, 43*, 319–331.
Lee, S. R., Kwon, K. S., Kim, S. R., & Rhee, S. G. (1998). Reversible inactivation of protein-tyrosine phosphatase 1B in A431 cells stimulated with epidermal growth factor. *The Journal of Biological Chemistry, 273*, 15366–15372.
Lemmon, M. A., & Schlessinger, J. (2010). Cell signaling by receptor tyrosine kinases. *Cell, 141*, 1117–1134.
Liu, S., Yang, H., He, Y., Jiang, Z.-H. , Kumar, S., Wu, L., et al. (2008). Aryl vinyl sulfonates and sulfones as active site-directed and mechanism-based probes for protein tyrosine phosphatases. *Journal of the American Chemical Society, 130*, 8251–8260.
Mahedev, K., Zilbering, A., Zhu, L., & Goldstein, B. J. (2001). Insulin-stimulated hydrogen peroxide reversibly inhibits protein-tyrosine phosphatase 1B in vivo and enhances the early insulin action cascade. *The Journal of Biological Chemistry, 276*, 21938–21942.
Meng, T.-C. , Buckley, D. A., Galic, S., Tiganis, T., & Tonks, N. K. (2004). Regulation of insulin signaling through reversible oxidation of the protein tyrosine phosphatases TC45 and PTP1B. *The Journal of Biological Chemistry, 279*, 37716–37725.
Merkler, D. J., Brenowitz, M., & Schramm, V. L. (1990). The rate constant describing slow-onset inhibition of yeast AMP deaminase by coformycin analogues is independent of inhibitor structure. *Biochemistry, 29*, 8358–8364.
Misra, H. P. (1974). Generation of superoxide free radical during the autooxidation of thiols. *The Journal of Biological Chemistry, 249*, 2151–2155.
Montalibet, J., Skorey, K. I., & Kennedy, B. P. (2005). Protein tyrosine phosphatase: Enzymatic assays. *Methods, 35*, 2–8.

Östman, A., Frijhoff, J., Sandin, A., & Böhmer, F.-D. (2011). Regulation of protein tyrosine phosphatases by reversible oxidation. *The Biochemical Journal, 150,* 345–356.
Pargellis, C. A., Morelock, M. M., Graham, E. T., Kinkade, P., Pav, S., Lubbe, K., et al. (1994). Determination of kinetic rate constants for the binding of inhibitors to HIV-1 protease and for the association and dissociation of active homodimer. *Biochemistry, 33,* 12527–12534.
Parsons, Z. D., & Gates, K. S. (2013). Thiol-dependent recovery of catalytic activity from oxidized protein tyrosine phosphatases. *Biochemistry,* Manuscript under revision.
Pregel, M. J., & Storer, A. C. (1997). Active site titration of tyrosine phosphatases SHP-1 and PTP1B using aromatic disulfides. *The Journal of Biological Chemistry, 272,* 23552–23558.
Rhee, S. G. (2006). H_2O_2, a necessary evil for cell signaling. *Science, 312,* 1882–1883.
Rinna, A., & Forman, H. J. (2008). SHP-1 inhibition by 4-hydroxynonenal activates Jun N-terminal kinase and glutamate cysteine ligase. *American Journal of Respiratory Cell and Molecular Biology, 39,* 97–104.
Seiner, D. R., & Gates, K. S. (2007). Kinetics and mechanism of protein tyrosine phosphatase 1B inactivation by acrolein. *Chemical Research in Toxicology, 20,* 1315–1320.
Silverman, R. B. (2000). *The organic chemistry of enzyme-catalyzed reactions.* San Diego: Academic Press.
Sivaramakrishnan, S., Cummings, A. H., & Gates, K. S. (2010). Protection of a single-cysteine redox switch from oxidative destruction: On the functional role of sulfenyl amide formation in the redox-regulated enzyme PTP1B. *Bioorganic & Medicinal Chemistry Letters, 20,* 444–447.
Sivaramakrishnan, S., Keerthi, K., & Gates, K. S. (2005). A chemical model for the redox regulation of protein tyrosine phosphatase 1B (PTP1B). *Journal of the American Chemical Society, 127,* 10830–10831.
Stone, J. R. (2006). Hydrogen peroxide: A signaling messenger. *Antioxidants & Redox Signaling, 8,* 243–270.
Tanner, J. J., Parson, Z. D., Cummings, A. H., Zhou, H., & Gates, K. S. (2011). Redox regulation of protein tyrosine phosphatases: Structural and chemical aspects. *Antioxidants & Redox Signaling, 15,* 77–97.
Tarrant, M. K., & Cole, P. A. (2009). The chemical biology of protein phosphorylation. *Annual Review of Biochemistry, 78,* 797–825.
Tonks, N. K. (2006). Protein tyrosine phosphatases: From genes, to function, to disease. *Nature Reviews. Molecular Cell Biology, 7,* 833–846.
Truong, T. H., & Carroll, K. S. (2012). Redox-regulation of EGFR signaling through cysteine oxidation. *Biochemistry, 51,* 9954–9965.
Ushio-Fukai, M. (2006). Localizing NADPH oxidase-derived ROS. *Science STKE, 349,* 1–6.
Winterbourn, C. C. (2008). Reconciling the chemistry and biology of reactive oxygen species. *Nature Chemical Biology, 4,* 278–286.
Woo, H. A., Yim, S. H., Shin, D. H., Kang, D., Yu, D.-Y. , & Rhee, S. G. (2010). Inactivation of peroxiredoxin I by phosphorylation allows localized hydrogen peroxide accumulation for cell signaling. *Cell, 140,* 517–528.
Zhou, H., Singh, H., Parsons, Z. D., Lewis, S. M., Bhattacharya, S., Seiner, D. R., et al. (2011). The biological buffer, bicarbonate/CO_2, potentiates H_2O_2-mediated inactivation of protein tyrosine phosphatases. *Journal of the American Chemical Society, 132,* 15803–15805.

SECTION II

H_2O_2 in the Redox Regulation of Transcription and Cell-Surface Receptors

CHAPTER NINE

Activation of Nrf2 by H_2O_2: *De Novo* Synthesis Versus Nuclear Translocation

Gonçalo Covas[1], H. Susana Marinho, Luísa Cyrne, Fernando Antunes[2]
Departamento de Química e Bioquímica and Centro de Química e Bioquímica, Faculdade de Ciências, Universidade de Lisboa, Lisboa, Portugal
[1]Present address: Bacterial Cell Surface and Pathogenesis, ITQB, Av. da República (EAN), Oeiras, Portugal
[2]Corresponding author: e-mail address: fantunes@fc.ul.pt

Contents

1. Introduction 158
2. Experimental Conditions and Considerations 159
   2.1 Cell culture 159
   2.2 Reagents 160
   2.3 H_2O_2 measurement 160
   2.4 Protein sample preparation 161
3. Pilot Experiments 163
   3.1 Glucose oxidase activity 163
   3.2 Kinetics of H_2O_2 consumption by HeLa cells 164
4. Experimental H_2O_2 Exposure 164
   4.1 Bolus addition 164
   4.2 Steady-state method 164
5. Data Handling and Analysis 166
6. Summary 170
Acknowledgments 170
References 170

Abstract

The most common mechanism described for the activation of the transcription factor Nrf2 is based on the inhibition of its degradation in the cytosol followed by its translocation to the nucleus. Recently, Nrf2 *de novo* synthesis was proposed as an additional mechanism for the rapid upregulation of Nrf2 by hydrogen peroxide (H_2O_2). Here, we describe a detailed protocol, including solutions, pilot experiments, and experimental setups, which allows exploring the role of H_2O_2, delivered either as a bolus or as a steady state, in endogenous Nrf2 translocation and synthesis. We also show experimental data, illustrating that H_2O_2 effects on Nrf2 activation in HeLa cells are strongly dependent both on the H_2O_2 concentration and on the method of H_2O_2 delivery. The *de novo* synthesis of Nrf2 is triggered within 5 min of exposure to low concentrations of H_2O_2,

Methods in Enzymology, Volume 528
ISSN 0076-6879
http://dx.doi.org/10.1016/B978-0-12-405881-1.00009-4

preceding Nrf2 translocation to the nucleus which is slower. Evidence of *de novo* synthesis of Nrf2 is observed only for low H_2O_2 steady-state concentrations, a condition that is prevalent *in vivo*. This study illustrates the applicability of the steady-state delivery of H_2O_2 to uncover subtle regulatory effects elicited by H_2O_2 in narrow concentration and time ranges.

1. INTRODUCTION

Nuclear factor erythroid-2-related factor 2 (Nrf2) is a transcription factor of the leucine zipper family, and Keap1 (Kelch-like ECH-associated protein 1) is its specific repressor, responsible for Nrf2 sequestration in the cytoplasm (Itoh et al., 1999; Xue & Cooley, 1993) as well as for its proteosomal degradation pathway (Kobayashi et al., 2006; Zhang & Hannink, 2003). These two proteins mediate cellular response to oxidative stress and to electrophilic xenobiotics (Osburn & Kensler, 2008). Included among the target genes regulated by Nrf2 are antioxidant enzymes, involved in electrophile conjugation, glutathione homeostasis, production of reducing equivalents, proteasome function, and other (Hayes & McMahon, 2009).

In the cell, regulation of Nrf2 levels and its activity occurs at several levels, including transcription, translation, degradation, translocation, and posttranslational modifications such as phosphorylation (Huang, Nguyen, & Pickett, 2000, 2002; Kong et al., 2001; Nioi & Hayes, 2004; Zhang & Hannink, 2003).

One of the most important mechanisms determining the increase of Nrf2 protein levels, involves a decreased rate of Nrf2 protein degradation. In the absence of any stress conditions, the normally low cellular concentrations of Nrf2 are maintained by proteasomal degradation, through a Keap1–Cullin 3–Roc1-dependent mechanism, in which Keap1 serves as the substrate adaptor subunit in the E3 holoenzyme. Activation of Nrf2 allows it to escape proteolysis and to rapidly accumulate in the nucleus inducing its target genes (Kobayashi et al., 2004, 2006; Zhang, Lo, Cross, Templeton, & Hannink, 2004). Keap1 is a cysteine-rich protein (Human- and murine Keap1 contain 27 and 25 cysteine residues, respectively) and so modifications in sulfhydryl-containing residues of this protein result in conformational changes (Itoh et al., 1999). In fact, oxidative stress conditions, and many exogenous chemicals, alter the redox status of Keap1 cysteine residues. As a consequence, there is a destabilization of the Keap1/Nrf2 complex, preventing Nrf2 degradation, which allows Nrf2 translocation to the nucleus.

Recent work indicates that Nrf2 *de novo* synthesis is an important mechanism for the rapid Nrf2 upregulation by oxidative stress (Purdom-Dickinson, Sheveleva, Sun, & Chen, 2007). Purdom-Dickinson, Lin, et al. (2007) and Purdom-Dickinson, Sheveleva, et al. (2007) found that in rat cardiomyocytes, treatment with low to mild doses of H_2O_2 caused a rapid increase in endogenous Nrf2 protein levels, by a process that is independent of Nrf2 protein stabilization. The authors suggested that H_2O_2 stress can cause selective protein translation, resulting in a rapid increase of Nrf2 protein.

Here, we describe a detailed protocol, which allows exploring the role of H_2O_2, in endogenous Nrf2 translocation and synthesis, by exposing HeLa cells to a wide range of H_2O_2 concentrations, delivered either as a bolus or as a steady state.

2. EXPERIMENTAL CONDITIONS AND CONSIDERATIONS

2.1. Cell culture

HeLa cells (American Type Culture Collection, Manassas, VA, USA) grown in RPMI 1640 media supplemented with 10% (v/v) fetal bovine serum (FBS), 100 U/mL of penicillin, 100 μg/mL of Streptomycin, and 2 m*M* of L-glutamine at 37 °C and 5% (v/v) CO_2 should be kept at exponential phase and in monolayer growth by periodic replanting every 2–3 days.

Cell culture conditions are of paramount importance for the reproducibility of the experiments. In the preparation of the biological material for the experimental procedures described here, the following precautions should be taken:

1. Cells must be seeded at a density of 0.5 million cells in a 100-mm dish. It is important to be accurate in the cell counting and to distribute homogenously cells in the dish.
2. Cells should be incubated for 46–48 h at 37 °C and 5% (v/v) CO_2. Reproducibility of the experiments is significantly affected if cells are incubated only overnight or for 24 h after seeding. At the day of the experiment, cells should show a confluence of about 60% (~1.5 million cells in a 100-mm dish).
3. Growth media must be renewed 1 h prior to the experimental procedures, using prewarmed and CO_2 preequilibrated media.

These conditions are to be used when delivering H_2O_2 either as bolus addition or as a steady-state addition.

2.2. Reagents

1. H_2O_2—Make fresh every day the solution using concentrated Perhydrol 30% (m/m) H_2O_2, density 1.11 g/mL, MW = 34.02, 9.79 *M*. To obtain the stock solution of H_2O_2 (~9–10 m*M*), dilute 1/1000 the concentrated H_2O_2 solution in water and confirm the concentration by reading the absorbance at 240 nm ($\varepsilon = 43.4\ M^{-1}\ cm^{-1}$). Keep on ice.
2. Catalase (bovine liver, Sigma C-1345, 2000–5000 units/mg protein) 1 mg/mL (in water). Can be stored for weeks.
3. Glucose oxidase from *Aspergillus niger*, Sigma G-0543, ≥200 units/mg protein, ≤0.1 units/mg catalase, buffered aqueous solution (in 100 m*M* sodium acetate, 40 m*M* KCl, with 0.004% thimerosal), pH 4.5, low catalase activity. Storage temperature 2–8 °C. A working diluted solution (1/100, 1/1000, or 1/10,000 dilution in water) should be made daily.
4. 0.1 *M* potassium phosphate buffer pH 6.5.
5. Phosphate buffered saline (PBS)—1.5 m*M* KH_2PO_4, pH 7.4, 137 m*M* NaCl, 3.0 m*M* KCl, and 8.0 m*M* Na_2HPO_4.
6. Cytosolic lysis buffer—50 m*M* HEPES, pH 7.2, 2 m*M* EDTA, 10 m*M* NaCl, 250 m*M* sucrose, 2 m*M* DTT, 10% (v/v) Nonidet P40, and protease inhibitors (Sigma-Aldrich, Inc., St. Louis, MO, USA): 1 m*M* PMSF, 1.5 mg/mL benzamidine, 10μg/mL leupeptin, and 1 μg/mL pepstatin, all freshly added.
7. Nuclear lysis buffer—Identical to the cytosolic proteins buffer except that 250 m*M* sucrose was replaced by 20% (v/v) glycerol and NaCl was 400 m*M*.
8. RIPA buffer—50 m*M* Tris–HCl pH 7.4, 400 m*M* NaCl, 1% (v/v) Nonidet P-40, 0.25% (w/v) Na-deoxycholate, and protease inhibitors (Sigma-Aldrich, Inc., Saint Louis, MO, USA): 1 m*M* PMSF, 1.5 mg/mL benzamidine, 10μg/mL leupeptin, and 1 μg/mL pepstatin, all freshly added.

2.3. H_2O_2 measurement

H_2O_2 is followed by the formation of O_2 after the addition of catalase (Eq. 9.1) using an oxygen electrode, as explained in Marinho, Cyrne, Cadenas, and Antunes (2013b):

$$2H_2O_2 \xrightarrow{\text{catalase}} O_2 + 2H_2O \qquad [9.1]$$

We use a chamber oxygen electrode (Oxygraph system, Hansatech Instruments, Ltd, Norfolk, UK), with a magnetic stirrer and temperature control. All measurements are performed at room temperature and with a final volume of 800 μL. The electrode should be giving a stable baseline, which is particularly important when measuring low concentrations of H_2O_2. For that, it is recommended to add 800 μL of distilled water and to connect the stirring a few hours before the measurements.

A typical measurement is as follows:

1. Take an 800 μL aliquot from the incubation media and add to the electrode chamber.
2. Start recording and, when a baseline is established, rapidly add 15 μL of catalase using a Hamilton syringe (being careful not to add air bubbles since they interfere in the measurement). After a new baseline is established, stop the recording. The value of the difference between the two baselines is converted to H_2O_2 concentration with the help of a calibration curve.
3. Remove the content of the oxygen electrode chamber and clean thoroughly with distilled water (fill the chamber up until the middle at least four times and then up until the top four times also) in order to be sure to remove all the catalase before the next H_2O_2 assay.

A H_2O_2 calibration curve should be made daily as described in Marinho et al. (2013b) briefly:

1. Make H_2O_2 solutions in water with known concentrations. Keep at room temperature.
2. Add 400 μL of H_2O_2 from one of the test tubes, starting with the lowest concentration, to 400 μL of 0.1 *M* potassium phosphate buffer pH 6.5 already in the electrode chamber. Readings can also be done without using the buffer, but we found that with the buffer the oxygen electrode has a more stable output.
3. Measure H_2O_2 by adding catalase as explained earlier.

2.4. Protein sample preparation

In order to prevent protein denaturation and/or degradation all following procedures must be done on ice and with precooled reagents.

2.4.1 Cytosol/nucleus differential protein extraction

The differential protein extraction from the cytosol and nucleus is performed according to the method described by Roebuck, Rahman, Lakshminarayanan, Janakidevi, and Malik (1995):

1. After exposing cells to H_2O_2, the incubation media in the 100-mm plates is removed, and cells are washed twice with 1 mL of PBS.
2. Afterward lysis is promoted by addition of 300 μL of cytosolic lysis buffer and scrapping. The lysate is transferred to a cooled *Eppendorf* tube, and the cells remaining in the dish are collected with additional 100 μL of cytosolic lysis buffer. Note that in order to prevent variation on the total time cells are exposed to H_2O_2, the wash and breaking step must be as brief and reproducible as possible. It is not advisable to process a large number of samples simultaneously.
3. The cytosolic protein samples are obtained as the supernatant after centrifugation at 3000 × *g* for 4 min at 4 °C.
4. The pellet is used for extraction of nuclear proteins. It is washed by resuspending it in 300 μL of cytosolic lysis buffer and centrifuged again at 3000 × *g* for 4 min at 4 °C. Next, the pellets are resuspended in 30 μL of nuclear lysis buffer and allowed to incubate 20 min on ice, with vortexing every 5 min. After that, the nuclear protein samples are obtained by centrifugation at 10,000 × *g* for 15 min at 4°C.
5. Samples are stored at −80 °C.

2.4.2 Total protein extraction

Total protein extracts are obtained according to the method described by Luo et al. (2004):

1. Washing and lysis are as described in the previous section, except that RIPA buffer is used instead of lysis buffer.
2. After incubating for 10 min at 4 °C with RIPA buffer, the total protein extracts are obtained from the supernatant of a centrifugation at 10,000 × *g* at 4 °C for 10 min.

2.4.3 Detection and protein quantification by Western blot

1. Protein samples are quantified using the Bradford method (Bradford, 1976).
2. 50 μg of sample is resolved in a 12.5% (w/v) SDS-PAGE gel and transferred to a nitrocellulose membrane by semidry electroblotting.
3. Membranes are stained with Ponceau S red, in order to confirm protein loading, before being blocked by incubation for 1 h with a solution of 5% (w/v) lyophilized fat-free cow milk.
4. Antibody incubations are carried at room temperature. The primary antibody against Nrf2 (1/600; clone 383727 from R&D Systems,

Inc., USA) is incubated for 2 h, and the secondary antibody (1/2000; Sc2005 from Santa Cruz Biotechnology Inc., USA) is incubated for 1 h. Membranes are washed three times with a PBS solution of 0.1% (v/v) Tween-20 for 15 min after incubation with the primary antibody and three times with PBS for 15 min after incubation with the secondary antibody.

5. Immunoreactivity is detected using the ECL kit (GE Healthcare Life Sciences, USA) according to supplier's instructions.

3. PILOT EXPERIMENTS

To implement the delivery of H_2O_2 as a steady state, two pilot experiments are needed:

1. The determination of the kinetics of H_2O_2 consumption by cells.
2. The rate of formation of H_2O_2 catalyzed by glucose oxidase under the experimental conditions to be used for the steady-state incubation.

As described in detail in this volume (Marinho, Cyrne, Cadenas, & Antunes, 2013a), to set up the steady state, the concentration of H_2O_2 added initially is maintained constant by adding a source of H_2O_2 that matches the cellular consumption of this oxidant. Glucose oxidase is an excellent candidate because it uses the glucose present in the growth media to produce H_2O_2.

3.1. Glucose oxidase activity

The determination of the rate of H_2O_2 production by glucose oxidase is done in growth media in the cell incubator:

1. Add 10 μL of glucose oxidase to 990 μL of water (1/100 dilution of the original stock solution).
2. Add 10 μL of the previous solution to a 100-mm cell culture dish with 8 mL of prewarmed and CO_2 preequilibrated medium.
3. Put dish in the cell incubator.
4. Take 800 μL aliquots at different times to measure H_2O_2.

A plot of H_2O_2 concentration versus time should be linear. From the slope calculate glucose oxidase activity as nanomolar of H_2O_2 produced per minute per microliter of the 1/100 glucose oxidase solution. The activity of the vial of glucose oxidase used in the experiments described here was 4.28 nmol/(min μL).

3.2. Kinetics of H_2O_2 consumption by HeLa cells

Cells are seeded as described in Section **2.1**.

1. To start the experiment add 100 μM H_2O_2 (bolus addition) to the medium.
2. Take 800 μL aliquots at different times to measure H_2O_2 in the range 10–90 μM. Initially, when consumption is faster, take aliquots every 5 min then, when consumption is slower, 10 min between aliquots is appropriate.

At least six time points should be recorded. Use the calibration curve to calculate H_2O_2 concentrations and make a plot of $\ln[H_2O_2]_{corrected}$ versus time. A correction of the H_2O_2 concentrations is needed to take into account the decrease in the incubation volume caused by the removal of the aliquots to measure H_2O_2. The correction is described in detail in Marinho et al. (2013a) and is calculated as the ratio between the volume in which H_2O_2 consumption is measured ($Vol_{measurement}$) over the initial reaction volume ($Vol_{initial}$) as shown in Eq. (9.2):

$$[H_2O_2]_{corrected} = [H_2O_2]_{experimental} \times Vol_{measurement}/Vol_{initial} \quad [9.2]$$

The rate constant obtained from the slope of the plot of $\ln[H_2O_2]_{corrected}$ versus time was 0.50 min^{-1} $mL/10^6$ cells.

4. EXPERIMENTAL H_2O_2 EXPOSURE

After determining the kinetics of H_2O_2 consumption by cells and the glucose oxidase activity, a full set of H_2O_2 experiments can be carried out.

4.1. Bolus addition

When delivering H_2O_2 as a bolus addition, no pilot experiments are needed. The H_2O_2 profiles observed upon a bolus addition are strongly dependent on experimental conditions, such as the number of cells and the volume of incubation media. For all experiments, the conditions described in Section 2.1 are used for the cell culture. Indicated initial concentrations of H_2O_2 (50, 100, and 200 μM) are added to cells, and no further measurements to monitor H_2O_2 concentrations are done.

4.2. Steady-state method

A H_2O_2 steady state—$[H_2O_2]ss$—can be obtained by adding the desired concentration of H_2O_2 simultaneously with a quantity of glucose oxidase that counters the cellular H_2O_2 consumption by producing H_2O_2, and thereby, the concentration of this compound is kept constant throughout the test.

The rate of H_2O_2 production needed is calculated by applying Eq. (9.3), where the desired steady state, the rate constant k obtained in Section **3.2**, and the conditions of the assay are used:

$$V_{production} = k \times [H_2O_2]ss \times \text{number of cells/reaction volume} \quad [9.3]$$

1. The $V_{production}$ obtained has units of concentration × time^{-1}. For example, to establish a 12.5 μ*M* steady state in an assay with 1.5 million HeLa cells in a 100-mm dish with 10 mL of reaction volume, $V_{production}$ is: (0.50 min^{-1} mL/10^6 cells) × 12.5 μ*M* × 1.5 × 10^6 cells/ 10 mL = 0.938 μ*M*/min.
2. The number of molecules of H_2O_2 produced is obtained by multiplying the value obtained in step 1 by the reaction volume: (0.938 μ*M*/min) × 10 mL = 9.38 nmol/min.
3. By dividing the number obtained in the previous step by the activity calculated in Section 3.1, we obtain the volume in microliter of the 1/100 glucose dilution needed to obtain the desired steady state: (9.38 nmol/ min)/(4.28 nmol/(min μL)) = 2.2 μL.

We need 2.2 μL of a 1/100 dilution of the original glucose oxidation. To minimize pipetting error, a 22.0 μL of a 1/1000 diluted solution would be advisable.

4.2.1 Monitoring cell exposure to H_2O_2

For long exposure times (longer than 1 h), the H_2O_2 concentration must be monitored and corrected. To this end, the H_2O_2 concentration is measured every hour by removing an 800 μL aliquot of the medium. Only assays in which the H_2O_2 concentration variation to the desired value is not higher than 20% should be accepted.

Table 9.1 depicts an example of the monitoring and corrections made for a 12.5 μ*M* H_2O_2 steady state lasting for 6 h. After 1 h, a H_2O_2 concentration of 11.8 μ*M* was measured. The difference from the desired concentration (12.5 μ*M*) is (11.8–12.5)/12.5 × 100 = −6%. In order to correct the H_2O_2 steady state to the desired value, glucose oxidase and H_2O_2 must be added. For that, we consider that the difference between the H_2O_2 concentration measured and the target one is directly proportional to the

Table 9.1 Monitoring H_2O_2 concentration during incubation with a steady state of 12.5 μ*M*

Time (h)	$[H_2O_2]$ (μ*M*)	Difference to desired concentration (%)	H_2O_2 or GO removed in aliquot (%)	GO and H_2O_2 correction (%)
1	11.8	−6	+8	+14
2	15.0	+20	+8	−12
3	15.0	+20	+8	−12
4	11.8	−6	+8	+14
5	10.1	−19	+8	+27
6	15.0	+20	+8	−12

The steady state is initiated by delivering 22 μL of a 1/1000 glucose oxidase (GO) dilution and 13.9 μL of 9 m*M* H_2O_2 to a 100-mm dish containing 1.5 million HeLa cells and 10 mL incubation media. Correction is calculated as–(Difference to desired concentration (%) − % of H_2O_2 or GO removed).

correction needed. However, an additional correction is needed. Because the steady-state concentration of H_2O_2 is maintained from a balance between H_2O_2 consumption, by the cells, and H_2O_2 formation by glucose oxidase, when an aliquot is taken from the media, given that the cells are attached to the bottom of the culture dish, an unbalance is introduced by the removal of the glucose oxidase present in the aliquot. Therefore, since the H_2O_2 concentration was 6% lower than the desired one and that 8% of the initial glucose oxidase and H_2O_2 were removed with the 800 μL aliquot (i.e., 800 μL represent 8% of the original 10 mL incubation volume), we need to add 14% of the original glucose oxidase addition, that is, 3.1 μL of glucose oxidase 1/1000 dilution, plus 14% of the original H_2O_2 addition, that is, 1.9 μL of 9 m*M* H_2O_2, in 800 μL of incubation media to restore original volume. For practical reasons, we would add 31 μL of a 1/10,000 glucose oxidase solution and 19 μL of 900 μ*M* H_2O_2.

For incubations of 1 h or shorter, corrections are not introduced, although H_2O_2 is monitored. Figure 9.1 shows typical H_2O_2 concentration profiles obtained either for short or long steady-state incubations.

5. DATA HANDLING AND ANALYSIS

Following cell exposure to H_2O_2, Western blot data concerning activation of Nrf2 are processed:

1. The films are scanned, and the intensity of bands obtained is determined relative to the control band (protein extract from cells that were not exposed to any treatment), using the *ImageJ* software (Rasband, 1997).

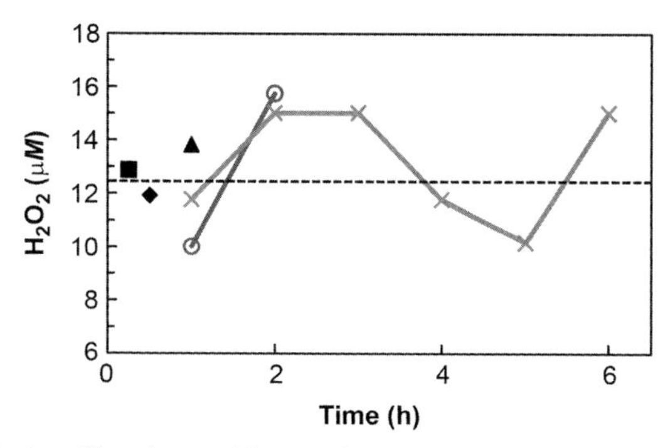

Figure 9.1 Typical profiles obtained for steady-state H_2O_2 incubations with HeLa cells. For up to 1-h incubations (■, ◆, ▲), the steady state is monitored at the end of the experiment. For longer incubation times (O, X), H_2O_2 is monitored every hour and corrections are made when necessary (see corrections for a 6-h incubation in Table 9.1). (For color version of this figure, the reader is referred to the online version of this chapter.)

2. In order to correct for differences in the total amount of protein applied to each well of the gel, we use as correction factor the relative intensity of staining with Ponceau S red area corresponding to each well.
3. Data are plotted according to Figs. 9.2–9.4.

When delivered as a bolus addition, H_2O_2 caused, in a concentration-dependent way, an increase in the nuclear, but not in the cytosolic levels of Nrf2 within 30 min of incubation (Fig. 9.2). This is consistent with the model in which Keap1 senses H_2O_2, causing the termination of both Nrf2 ubiquitination and degradation, allowing accumulation of Nrf2 in the nucleus. The absence of variation in the cytosolic levels of Nrf2 can be explained by a balance resulting from its decreased degradation and its increased nuclear translocation. This scenario changed when H_2O_2 was delivered as a steady state. A fast significant accumulation of cytosolic Nrf2 was observed in the first 15 min, while nuclear levels only changed significantly after 2 h (Fig. 9.3). The increase in cytosolic levels was mirrored by total Nrf2 cellular levels (Fig. 9.3). These observations are consistent with the hypothesis that the exposure of HeLa cells to a 12.5 μM H_2O_2 steady state led to a triggering of Nrf2 *de novo* synthesis, as described by Purdom-Dickinson, Lin, et al. (2007) and Purdom-Dickinson, Sheveleva, et al. (2007).

These results are confirmed if a concentration study is done (Fig. 9.4). At 15-min time, a bolus addition within a wide range of H_2O_2 concentrations—12.5–400 μM—did not increase Nrf2 cytosolic concentration, while a nuclear accumulation of Nrf2 was observed for 50 and 100 μM H_2O_2. The increase in cytosolic Nrf2 levels was only observed for 12.5 μM H_2O_2

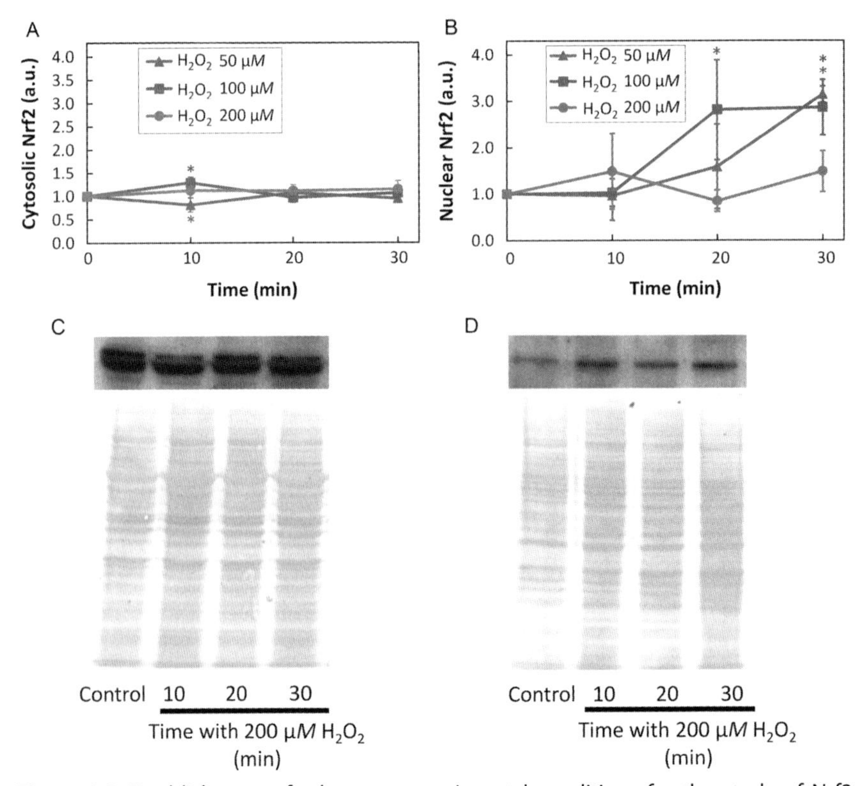

Figure 9.2 Establishment of adequate experimental conditions for the study of Nrf2 activation by H_2O_2 using a bolus addition. The indicated H_2O_2 concentrations were added as a bolus to HeLa cells and Nrf2 levels in (A) cytosol and (B) nucleus were determined by Western blot. Representative Western blot for cytosolic (C) and nuclear (D) extracts are shown. Quantification of Nrf2 protein levels was performed by signal intensity analysis using the ImageJ software and is shown in arbitrary units relative to control. Control of protein loading was performed by analysis of the membrane stained with Ponceau S red. Results are mean ± standard deviation of two to nine independent experiments. $^*P < 0.01$ versus control using ANOVA followed by a Holm–Sidak post test. (For color version of this figure, the reader is referred to the online version of this chapter.)

steady state, while higher H_2O_2 steady-state concentrations did not trigger this accumulation. Conversely, the fast nuclear accumulation of Nrf2 was only observed for higher H_2O_2 steady-state concentrations (25 and 50 μ*M*). It is important to note that a 25 μ*M* steady-state concentration of H_2O_2 already elicits cell toxicity after 6 h (Oliveira-Marques, Cyrne, Marinho, & Antunes, 2007).

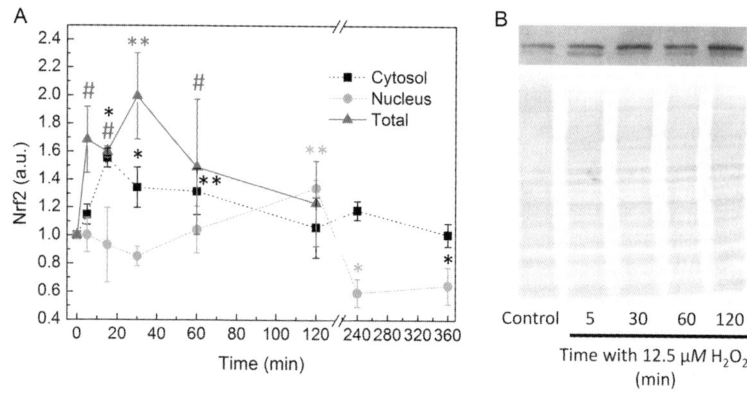

Figure 9.3 The *de novo* synthesis of Nrf2 is triggered very rapidly by H_2O_2, while Nrf2 translocation to the nucleus is slower. HeLa cells were treated with 12.5 μ*M* steady-state H_2O_2 for the indicated times, and Nrf2 levels were determined in cytosolic, nuclear, and total cellular extracts by Western blot. (A) Quantification of Nrf2 protein levels performed as described in Fig. 9.2. (B) Representative Western blot for total cellular extracts are shown. Results are mean ± standard deviation of 2–14 independent experiments. $^{*}P<0.01$, $^{**}P=0.01$, $^{\#}P<0.05$ versus control using ANOVA followed by a Holm–Sidak post test. (For color version of this figure, the reader is referred to the online version of this chapter.)

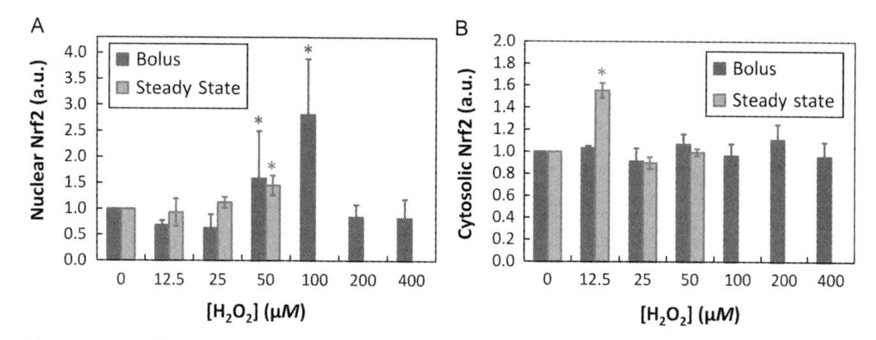

Figure 9.4 Nrf2 activation by H_2O_2 is strongly dependent both on the concentrations of H_2O_2 and on the method of H_2O_2 delivery. Titration of the effect of exposing HeLa cells for 15 min to increasing H_2O_2 concentrations added either as a bolus or as a steady state on nuclear (A) and cytosolic (B) Nrf2 levels. Nrf2 levels in protein extracts were determined as described in Fig. 9.2. Results are mean ± standard deviation of two to nine independent experiments. $^{*}P<0.01$ versus control using ANOVA followed by a Holm–Sidak post test. (For color version of this figure, the reader is referred to the online version of this chapter.)

Overall, these results suggest that low sustained H_2O_2 concentrations trigger preferentially Nrf2 *de novo* synthesis without nuclear translocation. For Nrf2 nuclear translocation to occur, an higher H_2O_2 concentration or, alternatively, cell exposure to a sustained low H_2O_2 concentration for a long period of time is needed. Fast nuclear accumulation of Nrf2 may only be triggered by H_2O_2 at high potentially toxic levels that cannot be coped by the cell for prolonged periods of time. Whether Nrf2 *de novo* synthesis is blocked by high concentrations of H_2O_2 cannot be evaluated from the data presented here because an absence of Nrf2 accumulation in the cytosol at high H_2O_2 concentrations may be due to its fast translocation to the nucleus.

6. SUMMARY

This chapter describes the application of H_2O_2 steady-state incubations to the study of Nrf2 activation. H_2O_2 effects on Nrf2 activation are strongly dependent both on the H_2O_2 concentration and on the method of H_2O_2 delivery. The *de novo* synthesis of Nrf2 is triggered within 5 min of exposure to low concentrations of H_2O_2, while Nrf2 translocation to the nucleus is slower. Evidence of *de novo* synthesis of Nrf2 is observed only for low H_2O_2 steady-state concentrations, a condition that is prevalent *in vivo*.

ACKNOWLEDGMENTS

Funding from Fundação para a Ciência e a Tecnologia (FCT), Portugal (Grants PTDC/QUI/69466/2006 and PEst-OE/QUI/UI0612/2013) is acknowledged.

REFERENCES

Bradford, M. M. (1976). A rapid and sensitive method for the quantitation of microgram quantities of protein utilizing the principle of protein-dye binding. *Analytical Biochemistry*, *72*, 248–254.

Hayes, J. D., & McMahon, M. (2009). NRF2 and KEAP1 mutations: Permanent activation of an adaptive response in cancer. *Trends in Biochemical Sciences*, *34*, 176–188.

Huang, H. C., Nguyen, T., & Pickett, C. B. (2000). Regulation of the antioxidant response element by protein kinase C-mediated phosphorylation of NF-E2-related factor 2. *Proceedings of the National Academy of Sciences of the United States of America*, *97*, 12475–12480.

Huang, H. C., Nguyen, T., & Pickett, C. B. (2002). Phosphorylation of Nrf2 at Ser-40 by protein kinase C regulates antioxidant response element-mediated transcription. *The Journal of Biological Chemistry*, *277*, 42769–42774.

Itoh, K., Wakabayashi, N., Katoh, Y., Ishii, T., Igarashi, K., Engel, J. D., et al. (1999). Keap1 represses nuclear activation of antioxidant responsive elements by Nrf2 through binding to the amino-terminal Neh2 domain. *Genes & Development*, *13*, 76–86.

Kobayashi, A., Kang, M. I., Okawa, H., Ohtsuji, M., Zenke, Y., Chiba, T., et al. (2004). Oxidative stress sensor Keap1 functions as an adaptor for Cul3-based E3 ligase to regulate proteasomal degradation of Nrf2. *Molecular and Cellular Biology*, *24*, 7130–7139.

Kobayashi, A., Kang, M. I., Watai, Y., Tong, K. I., Shibata, T., Uchida, K., et al. (2006). Oxidative and electrophilic stresses activate Nrf2 through inhibition of ubiquitination activity of Keap1. *Molecular and Cellular Biology*, *26*, 221–229.

Kong, A. N., Owuor, E., Yu, R., Hebbar, V., Chen, C., Hu, R., et al. (2001). Induction of xenobiotic enzymes by the MAP kinase pathway and the antioxidant or electrophile response element (ARE/EpRE). *Drug Metabolism Reviews*, *33*, 255–271.

Luo, L., Li, D. Q., Doshi, A., Farley, W., Corrales, R. M., & Pflugfelder, S. C. (2004). Experimental dry eye stimulates production of inflammatory cytokines and MMP-9 and activates MAPK signaling pathways on the ocular surface. *Investigative Ophthalmology & Visual Science*, *45*, 4293–4301.

Marinho, H. S., Cyrne, L., Cadenas, E., & Antunes, F. (2013a). H_2O_2 delivery to cells: Steady-state versus bolus addition. *Methods in Enzymology*, *526*, 159–173.

Marinho, H. S., Cyrne, L., Cadenas, E., & Antunes, F. (2013b). The cellular steady-state of H_2O_2: Latency concepts and gradients. *Methods in Enzymology*, *527*, 3–19.

Nioi, P., & Hayes, J. D. (2004). Contribution of NAD(P)H:quinone oxidoreductase 1 to protection against carcinogenesis, and regulation of its gene by the Nrf2 basic-region leucine zipper and the arylhydrocarbon receptor basic helix-loop-helix transcription factors. *Mutation Research*, *555*, 149–171.

Oliveira-Marques, V., Cyrne, L., Marinho, H. S., & Antunes, F. (2007). A quantitative study of NF-kB activation by H2O2: Relevance in inflammation and synergy with TNF-alpha. *Journal of Immunology*, *178*, 3893–3902.

Osburn, W. O., & Kensler, T. W. (2008). Nrf2 signaling: An adaptive response pathway for protection against environmental toxic insults. *Mutation Research*, *659*, 31–39.

Purdom-Dickinson, S. E., Lin, Y., Dedek, M., Morrissy, S., Johnson, J., & Chen, Q. M. (2007). Induction of antioxidant and detoxification response by oxidants in cardiomyocytes: Evidence from gene expression profiling and activation of Nrf2 transcription factor. *Journal of Molecular and Cellular Cardiology*, *42*, 159–176.

Purdom-Dickinson, S. E., Sheveleva, E. V., Sun, H., & Chen, Q. M. (2007). Translational control of nrf2 protein in activation of antioxidant response by oxidants. *Molecular Pharmacology*, *72*, 1074–1081.

Rasband, W. S. (1997). *ImageJ*. Bethesda, Maryland, USA: U. S. National Institutes of Health [Computer software].

Roebuck, K. A., Rahman, A., Lakshminarayanan, V., Janakidevi, K., & Malik, A. B. (1995). H2O2 and tumor necrosis factor-alpha activate intercellular adhesion molecule 1 (ICAM-1) gene transcription through distinct cis-regulatory elements within the ICAM-1 promoter. *The Journal of Biological Chemistry*, *270*, 18966–18974.

Xue, F., & Cooley, L. (1993). kelch encodes a component of intercellular bridges in Drosophila egg chambers. *Cell*, *72*, 681–693.

Zhang, D. D., & Hannink, M. (2003). Distinct cysteine residues in Keap1 are required for Keap1-dependent ubiquitination of Nrf2 and for stabilization of Nrf2 by chemopreventive agents and oxidative stress. *Molecular and Cellular Biology*, *23*, 8137–8151.

Zhang, D. D., Lo, S. C., Cross, J. V., Templeton, D. J., & Hannink, M. (2004). Keap1 is a redox-regulated substrate adaptor protein for a Cul3-dependent ubiquitin ligase complex. *Molecular and Cellular Biology*, *24*, 10941–10953.

CHAPTER TEN

H_2O_2 in the Induction of NF-κB-Dependent Selective Gene Expression

Luísa Cyrne*, Virgínia Oliveira-Marques†, H. Susana Marinho*, Fernando Antunes*,[1]

*Departamento de Química e Bioquímica and Centro de Química e Bioquímica, Faculdade de Ciências, Universidade de Lisboa, Lisboa, Portugal

†Thelial Technologies S.A., Cantanhede, Portugal

[1]Corresponding author: e-mail address: fantunes@fc.ul.pt

Contents

1. Introduction 174
2. Experimental Components and Considerations 175
    2.1 Reagents 175
    2.2 Cell culture preparation 176
    2.3 Methodological considerations 177
    2.4 H_2O_2 measurement 178
3. Pilot Experiments 179
    3.1 Calibrating the system: Cellular H_2O_2 consumption 179
    3.2 Calibrating the system: Glucose oxidase activity 181
4. Steady-State Titration Experiments 181
5. NF-κB Family Protein Levels 182
    5.1 Protein extraction 183
    5.2 Western blot 183
6. NF-κB-Dependent Gene Expression 184
    6.1 Plasmid constructs 184
    6.2 Cellular transfection and reporter gene assays 184
7. Summary 187
Acknowledgments 187
References 187

Abstract

NF-κB is a transcription factor that plays key roles in health and disease. Learning how this transcription factor is regulated by hydrogen peroxide (H_2O_2) has been slowed down by the lack of methodologies suitable to obtain quantitative data. Literature is abundant with apparently contradictory information on whether H_2O_2 activates or inhibits NF-κB. There is increasing evidence that H_2O_2 is not just a generic modulator of transcription factors and signaling molecules but becomes a specific regulator of

Methods in Enzymology, Volume 528
ISSN 0076-6879
http://dx.doi.org/10.1016/B978-0-12-405881-1.00010-0

individual genes. Here, we describe a detailed protocol to obtain rigorous quantitative data on the effect of H_2O_2 on members of the NF-κB/Rel and IκB families, in which H_2O_2 is delivered as a steady-state addition instead of the usual bolus addition. Solutions, pilot experiments, and experimental set-ups are fully described. In addition, we outline a protocol to measure the impact of alterations in the promoter κB regions on the H_2O_2 regulation of the expression of individual genes. As important as evaluating the effects of H_2O_2 alone is the evaluation of the modulation elicited by this oxidant on cytokine regulation of NF-κB. We illustrate this for the cytokine tumor necrosis factor alpha.

1. INTRODUCTION

The NF-κB/Rel family of transcription factors consists of homo- and heterodimers of five distinct proteins p65 (RelA), RelB, c-Rel, p50 (and its precursor p105), and p52 (and its precursor p100). NF-κB has key regulatory roles in inflammation, innate and adaptive immune response, proliferation, and apoptosis (Chen & Greene, 2004; Ghosh, May, & Kopp, 1998). NF-κB activation leads to its translocation from the cytosol to the nucleus and has as an outcome an inflammatory response characterized by an increased expression of: proinflammatory cytokines, for example, tumor necrosis factor alpha (TNF-α), interleukin-1 (IL-1), and interleukin-6 (IL-6); chemokines, for example, monocyte chemotactic protein-1 (MCP-1) and interleukin-8 (IL-8); adhesion molecules, for example, intercellular Adhesion Molecule-1 (ICAM-1), vascular cell adhesion protein-1 (VCAM-1) and E-selectin; growth factors; and, enzymes that produce secondary inflammatory mediators such as cyclooxygenase-2 (COX-2) and inducible NO synthase (iNOS) (Brigelius-Flohé & Flohé, 2011). The role of H_2O_2 in NF-κB activation *in vivo* is highly controversial. Studies involving H_2O_2 either alone or in conjunction with cytokines do not show a consistent activation of NF-κB, and H_2O_2 can activate, inhibit, or have no effect on NF-κB (reviewed in Oliveira-Marques, Marinho, Cyrne, & Antunes, 2009b). Partly this is due to the fact that most studies on NF-κB do not apply calibrated and controlled methods of H_2O_2 delivery to cells, such as the steady-state titration (Antunes & Cadenas, 2001; Antunes, Cadenas, & Brunk, 2001). Recent studies using delivery of H_2O_2 in a steady state to cells (steady-state titration) have allowed the emergence of a new paradigm where H_2O_2 acts not as an inducer of NF-κB but as a modulator of the activation of the NF-κB pathway by other agents (Oliveira-Marques, Cyrne, Marinho, & Antunes, 2007; Oliveira-Marques et al., 2009b). In fact, by using the steady-state titration, we previously showed using MCF-7 and HeLa cells that H_2O_2, at concentrations close to those occurring during an inflammatory situation, that is, 5–15 μ*M* (Liu & Zweier, 2001; Test & Weiss, 1984), has a synergistic effect on TNF-α-dependent translocation of p65 from the cytosol to the nucleus. This increased nuclear translocation of p65 in the presence of H_2O_2 and TNF-α has as outcome the enhanced gene expression of a subset of NF-κB-dependent genes, including proinflammatory genes, for example, IL-8; MCP-1; TLR2; and TNF-α, and anti-inflammatory genes, for example, heme oxygenase-1 (Oliveira-Marques et al., 2007). NF-κB, once inside the nucleus, binds to the promoter/enhancer regions of target genes (Moynagh, 2005), the κB sites, which have the general consensus sequence GGGR NNYYCC (R is purine, Y is pyrimidine, and N is any base). The differential gene regulation caused by H_2O_2 depends, among other factors, on the apparent affinity of κB sites in the gene-promoter regions toward NF-κB and, the lower the affinity, the higher the range of TNF-α concentrations where H_2O_2 upregulates gene expression (Oliveira-Marques, Marinho, Cyrne, & Antunes, 2009a). This selective gene expression has potential implications in personalized medicine because many single-nucleotide polymorphisms are found in the κB sites of the human genome (Kasowski et al., 2010).

Here, we present a detailed protocol that allows exploring the role of H_2O_2 as a modulator of NF-κB-dependent gene expression by exposing cells to H_2O_2 and cytokines in inflammatory-like conditions.

2. EXPERIMENTAL COMPONENTS AND CONSIDERATIONS

2.1. Reagents

1. H_2O_2—Make fresh every day the solution using concentrated Perhydrol 30% (m/m) H_2O_2, density 1.11 g/ml, MW = 34.02, 9.79 *M*. To obtain the stock solution of H_2O_2 (~9–10 m*M*), dilute 1/1000 the concentrated H_2O_2 in water and confirm the concentration by reading the absorbance at 240 nm ($\varepsilon = 43.4\ M^{-1}\text{cm}^{-1}$). Keep on ice.
2. Catalase (bovine liver, Sigma C-1345, 2000–5000 units/mg protein) 1 mg/mL (in water). Can be stored for weeks.
3. Glucose oxidase from *Aspergillus niger*, Sigma G-0543, ≥200 units/mg protein, ≤0.1 units/mg catalase, buffered aqueous solution (in 100 m*M* sodium acetate, 40 m*M* KCl, with 0.004% thimerosal), pH 4.5, low catalase activity. Storage temperature 2–8 °C. A working

diluted solution (1/100, 1/1000, or 1/10,000 dilution in water) should be made daily.

4. 0.1 *M* Potassium phosphate buffer pH 6.5. Optionally needed for the H_2O_2 calibration curve.
5. Phosphate-buffered saline (PBS)—1.5 m*M* KH_2PO_4 pH 7.4, 137 m*M* NaCl, 3.0 m*M* KCl, and 8.0 m*M* Na_2HPO_4.
6. Cytosolic proteins buffer—50 m*M* HEPES, pH 7.2, 2 m*M* EDTA, 10 m*M* NaCl, 250 m*M* sucrose, with freshly added protease inhibitor cocktail protease inhibitors (1 m*M* PMSF, 1.5 μg/mL benzamidine, 10 μg/mL leupeptin, and 1 μg/mL pepstatin), 2 m*M* dithiothreitol, and the detergent IGEPAL CA-630 0.1% (v/v), all from Sigma, Saint Louis, MO, USA.
7. Nuclear proteins buffer—Identical to the cytosolic proteins buffer except that 250 m*M* sucrose was replaced by 20% (v/v) glycerol and NaCl was 400 m*M*.

2.2. Cell culture preparation

To ensure that a reproducible and stable H_2O_2 steady state is obtained, it is fundamental to standardize cell culture preparation, especially when using adherent cells. When the experiment is performed, adherent cells should have completely recovered from splitting, for example, recovered their normal shape with no signs of toxicity, with a uniform distribution throughout the plate/flask and with a confluence between 60% and 80%. This means that the characteristics of the adherent cell lines being used should be known in advance and that conditions should be rigorously maintained for all experiments to have reproducible results. Too many cells in a plate/flask imply a higher H_2O_2 consumption and decrease in the steady state while lower cell numbers or unrecovered cells will be more sensitive to H_2O_2, causing an increase in H_2O_2 concentration and possibly inducing oxidative stress. Therefore, we recommend counting and plating the cells 46 h before the experiment. The number of cells plated is entirely dependent on the cells doubling time to achieve 60–80% confluence at the day of the experiment. As an example, to achieve the desired confluence in a 100-mm Petri dish, 0.9×10^6 MCF-7 (European Collection of Cell Cultures, Salisbury Wiltshire, UK) and 0.5×10^6 HeLa (American Type Culture Collection, Manassas, VA, USA) cells are plated to get 1.8×10^6 and 1.5×10^6 cells, respectively, after 46 h.

For an experiment of protein or mRNA extraction, we recommend using 100-mm Petri dishes to have enough extracted material per condition and at the same time to have enough cell culture volume to not disturb the system when taking aliquots for steady-state confirmation (Section 4). For other types of experiments, such as viability assays that do not require so many cells, it is possible to use multi-well plates (e.g., 96-well, 24-well plates), and in such situations, the number of cells has to be adjusted to achieve a 60–80% confluence at the day of the experiment.

2.3. Methodological considerations

When working with H_2O_2, it is important to understand that cells are equipped with mechanisms to rapidly consume this molecule. Although H_2O_2 is a mild oxidant, high H_2O_2 levels can cause oxidative stress and cytotoxicity. Only at the low/moderate levels occurring *in vivo* does H_2O_2 fulfill its potential as a signaling molecule. The steady-state titration arises exactly from this need of studying signaling pathways and works with doses close to the physiological ones. The addition of a single low H_2O_2 initial dose—bolus addition—might not be enough to switch-on signaling pathways because of the rapid H_2O_2 consumption within cells. For example, in our experiments using MCF-7 cells, a single bolus addition of 100 μ*M* H_2O_2 is consumed in less than 30 min of incubation with a confluence of 60% in a 100-mm Petri dish with 10 mL of incubation media. To overcome this issue, many have opted to increase the initial H_2O_2 dose up to the millimolar range that is far from the physiological relevance required. Again, in our hands, more than 90% of H_2O_2 is consumed during the first hour after a 1-m*M* bolus addition to 1.8×10^6 MCF-7 cells in 10 mL incubation media (Oliveira-Marques et al., 2007).

When using the steady-state titration, H_2O_2 consumption is balanced by its production via glucose oxidase, an enzyme that catalyzes the formation of H_2O_2 through the oxidation of glucose present in cell culture medium. When combining the addition of glucose oxidase to the addition of H_2O_2 at the desired concentration, the assay starts and keeps running at the same concentration of H_2O_2. This method allows using low/moderate concentrations of H_2O_2, which is fundamental to study signaling pathways, for example, NF-κB (Marinho, Cyrne, Cadenas, & Antunes, 2013a). When MCF-7 cells are exposed to a typical bolus addition of 1 m*M* H_2O_2 or to a steady state of 25 μ*M* H_2O_2 for 2 h, there is a translocation of NF-κB into the nucleus (Oliveira-Marques et al., 2007). However, as shown in

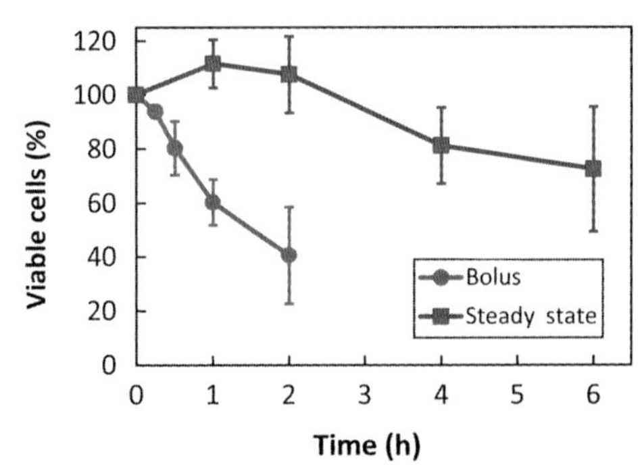

Figure 10.1 Comparison of cell viability of MCF-7 cells upon incubation with 1 m*M* H_2O_2 bolus dose (•) or a 25 μ*M* H_2O_2 steady state (■). Cell viability was assessed by following MTT reduction (McGahon et al., 1995). (For color version of this figure, the reader is referred to the online version of this chapter.)

Fig. 10.1, there is a significant difference in cell viability, and so translocation of NF-κB into the nucleus under bolus addition conditions occurs when cell viability is already lost. This clearly illustrates the advantages of using a H_2O_2 steady state, where it is possible to adjust the concentration of H_2O_2, even for long incubations, with no loss in cell viability.

2.4. H_2O_2 measurement

To have the steady-state method working properly, it is necessary to measure the actual concentration of H_2O_2 present in the cell culture medium so that the H_2O_2 concentration is not just an assumed value based on the amounts of glucose oxidase and H_2O_2 added. With either an oxygen (O_2) or a H_2O_2 electrode, it is possible to measure the actual concentration of H_2O_2 present in the cell culture medium at a certain incubation point. An O_2 electrode allows time point measurements by adding few microliters of 1 mg/mL catalase (10–15 μL) to convert H_2O_2 present in the medium to O_2. We use the Oxygraph system (Hansatech Instruments Ltd., Norfolk, UK), directly linked to a PC for registration. The O_2 permeable membrane is replaced every week and tested for quality before an experiment. When not in use, the chamber is always filled with distilled H_2O so that the membrane does not dry.

The following generic method describes the typical H_2O_2 measurements in cell culture medium of adherent cell lines using an O_2 electrode (see also Marinho, Cyrne, Cadenas, & Antunes, 2013b):

1. Approximately, 12 h before starting the experiment, leave the O_2 electrode on, with H_2O in the chamber, so that the system is fully stabilized in the day of the experiment;
2. Measure H_2O_2 present in the medium the following way:
   a. Aspirate H_2O from the electrode chamber;
   b. Add the medium/solution with H_2O_2 and cover with the electrode chamber cap;
   c. Start recording a baseline that should be horizontal and when stable add 10–15 μL of catalase to convert H_2O_2 into O_2. A steep change of the slope should be observed, if any H_2O_2 is present in the medium;
   d. When H_2O_2 is fully converted, the slope should return to baseline type and recording can be stopped;
   e. Aspirate medium from the chamber and clean thoroughly by filling the chamber up until the middle at least four times and up until the top four times also with distilled H_2O to ensure that all catalase is removed before the next assay.

The difference between both baselines represents the quantity of H_2O_2 present in the medium and can be converted in concentration by making a calibration curve of H_2O_2. We prepare daily a 9-m*M* stock solution of H_2O_2 to assess the real concentration in a spectrophotometer at 240 nm ($\varepsilon_{240nm} = 43.4\ M^{-1}cm^{-1}$) and make a calibration curve of H_2O_2 solutions (in H_2O) ranging from 9 to 90 μ*M* as described in Marinho et al. (2013b). Each measurement is performed as explained earlier.

3. PILOT EXPERIMENTS

Pilot experiments are required to calibrate the system every time there is a change in components of the steady-state method, such as the cell line to be used, the cell culture medium, or the batch of glucose oxidase.

3.1. Calibrating the system: Cellular H_2O_2 consumption

To have a continuous H_2O_2 source working properly, preliminary studies to calibrate the system are required. Removal of added H_2O_2 depends on several factors, such as the cellular capacity to remove H_2O_2, cell density, and consumption of H_2O_2 by the growth medium (Oliveira-Marques et al., 2009b). Therefore, in order to know the amount of glucose oxidase to add to the medium studies to estimate H_2O_2 consumption by cells need to be performed. Every time the conditions change, such as new medium or the cell number in the experiment, we recommend recalibrating the system. We typically use in our cell culture experiments the following incubation medium: RPMI 1640 medium supplemented with 10% (v/v) of fetal bovine serum, 100 U/mL penicillin, 100 μg/mL streptomycin, and 2 m*M* L-glutamine (Lonza, Basel, Switzerland), which does not interfere with the H_2O_2 steady-state method. Unless otherwise explained, throughout this chapter, we use as medium the described supplemented RPMI 1640 medium.

The following generic method describes the typical calibration for cellular H_2O_2 consumption (see also Marinho et al., 2013a):

1. Seed cells approximately 46 h before the experiment at two/three different final numbers. We recommend using 100-mm Petri dishes because they are easier to handle than flasks;
2. Approximately 12 h before the experiment, leave the oxygen electrode on, with H_2O in the chamber so that the system is fully stabilized in the day of the experiment;
3. On the day of the experiment, replace cells medium with prewarmed new medium (9 mL) and wait for 1 h. For experiments that start in the morning and total volumes above 50 mL, medium can be left inside the incubator overnight; otherwise, it will take several hours to attain 37 °C;
4. During this time, prepare H_2O_2 stock solution and make the calibration curve as described in Section 2.4;
5. Add 100 μ*M* of H_2O_2 to cells culture medium, swirl gently, and start counting time (cells are kept inside the incubator);
6. After 5 min, take the first aliquot of medium (800 μL minimum) and measure H_2O_2 concentration with the O_2 electrode as described in Section 2.4;
7. Continue the process every 5/7 min until no H_2O_2 is detectable;
8. Calculate the rate of H_2O_2 consumption for the different conditions and use this value to calibrate the steady-state titration experiments. Take into consideration that when using attached cells a correction for the medium volume is needed. The correction is calculated as the ratio between the volume in which H_2O_2 consumption was measured ($Vol_{measurement}$) over the initial reaction volume ($Vol_{initial}$) as shown in Equation 10.1

$$[H_2O_2]_{corrected} = [H_2O_2]_{experimental} \times Vol_{measurement}/Vol_{initial}; \qquad [10.1]$$

9. The rate constant is obtained from the slope of the plot of ln $[H_2O_2]_{corrected}$ versus time. For HeLa and MCF-7 cells, rate constants obtained were 0.50 and 0.43 min^{-1} mL/10^6 cells, respectively.

3.2. Calibrating the system: Glucose oxidase activity

In the presence of O_2, glucose oxidase uses glucose present in the culture medium to produce H_2O_2 and D-glucono-1,5-lactone. Glucose oxidase (*A. niger*) activity should be tested when using a new commercial flask since it is a fundamental parameter to establish accurate H_2O_2 steady states. And the activity will depend on the assay conditions. So the correct way to calculate glucose oxidase activity is keeping constant the exact experimental conditions such as the cells culture medium, the incubation temperature, and the oxygen electrode for H_2O_2 measurements (see Marinho et al., 2013a).

1. Warm up 10 mL of incubation medium for 1 h in the cells incubator;
2. Dilute 1/100 glucose oxidase in H_2O_2 and add to the medium 10 μL;
3. Measure the actual H_2O_2 concentration produced every 5–7 min with the O_2 electrode as described in Section 2.4.

A plot of H_2O_2 concentration versus time should be linear. From the slope, calculate glucose oxidase activity as nanomole of H_2O_2 produced per minute per microliter of the 1/100 glucose oxidase solution.

4. STEADY-STATE TITRATION EXPERIMENTS

With the system fully calibrated (Section 3) and cell cultures prepared as explained in Section 2.1, everything is set up for a steady-state titration experiment. Prepare the number of plates for each condition to be used in the experiment plus an extra plate for a steady-state pretest of the day. Although all parameters are previously calculated, day-by-day errors will be corrected with this pretesting to be sure that the H_2O_2 concentration is the desired one during the assay.

Repeat points 1–4 from Section 3.1, with the exception that 7 mL of medium is used, and continue through the following steps:

1. Add to the pretest dish the calculated volumes for H_2O_2 and glucose oxidase to achieve the desired steady-state concentration; see (Covas, Marinho, Cyrne, & Antunes, 2013; Marinho et al., 2013a) for examples of calculations;
2. After 1 h of incubation, measure the external H_2O_2 concentration in a 800-μL aliquot. If the H_2O_2 concentration is not the desired one,

recalculate the amount of glucose oxidase to be added to the assays, assuming that the difference between the H_2O_2 concentration measured and the wanted one is directly proportional to the correction needed. For example, if the calculated amount of glucose oxidase is 50 μL, and the steady state measured in the pretest dish is 10% higher than the desired one, add 90% of 50 μL to the assays;

3. Start the experiments by adding the glucose oxidase units taking into account the result obtained in the previous step;
4. Monitor H_2O_2 concentration every hour and at the end of the experiment;
5. For experiments up to 1 h, no corrections are introduced. For longer experiments, correct steady states taking into account the deviations from the target steady state assuming a direct proportionality between the deviation and the correction. Take also into account the removal of H_2O_2 and glucose oxidase in the aliquots. A detailed calculation is shown in Covas et al. (2013);
6. Importantly, do not make additional measurements than the needed ones to avoid disturbing unnecessarily the system.

It is worth mentioning that if using 96-well plates it is important to have several replicates to be able to pool the medium from several wells, for example, four wells with 200 μL each, and measure an average for H_2O_2 steady-state concentration.

5. NF-κB FAMILY PROTEIN LEVELS

Cells treated with H_2O_2 in steady state can be processed to analyze signaling pathways. For the NF-κB/Rel and IκB families, we typically perform the analysis of protein levels by immunoblot. We set up the H_2O_2 steady-state conditions as explained in previous sections in the range of 5–25 μM H_2O_2. Petri dishes of 100-mm usually give enough protein material for good immunoblot signals and are adequate and easy to handle for H_2O_2 steady-state methodology.

We recommend organizing your samples in groups of treatment that should be directly compared and keep in mind having time to measure all H_2O_2 concentrations in the O_2 electrode before protein extraction. Have an untreated control plate for each time point. For the NF-κB family members' protein levels, we use four conditions per time: control; steady-state H_2O_2; TNF-α; steady-state H_2O_2 plus TNF-α added simultaneously.

TNF-α (Human Recombinant, Sigma, Saint Louis, MO, USA) is used at 0.37 ng/mL and does not interfere with the steady-state H_2O_2 level.

We next describe the steps for cellular sub-fractioning to collect both cytosolic and nuclear proteins from the same samples.

5.1. Protein extraction

Several protein extraction protocols might be used. Here, we detail a fractionated protocol we have been using to analyze NF-κB/Rel- and IκB family members levels separately in cytosolic and nuclear compartments. The following procedure should be made in cold (Oliveira-Marques et al., 2007).

1. At the end of the incubation time, check the H_2O_2 concentration as explained before and follow to extraction procedure. If the H_2O_2 concentration differs more than 20% from the desired one, the experiment is discarded;
2. Aspirate the medium and wash the 100-mm plates with cold PBS;
3. Add 500 μL of cytosolic proteins buffer, scrape cells, and transfer the mixture to a sterile 1.5-mL tube;
4. Repeat the procedure with 100 μL of cytosolic proteins buffer and pool mixtures;
5. Centrifuge the tubes at 3000 *g* for 4 min at 4 °C;
6. Collect the supernatant that contains the cytosolic proteins to new tubes;
7. Wash the pellet with 300 μL of cytosolic buffer and centrifuge as in step 4;
8. Discard supernatant and resuspend the pellet with 30 μL of nuclear proteins buffer;
9. Keep tubes on ice for 20–25 min and extract proteins by vortexing three times during that period;
10. Centrifuge samples at 10,000 g for 10 min at 4 °C;
11. Collect to new tubes the enriched supernatant with nuclear proteins;
12. Quantify protein concentration by the Bradford method, which has low interferences.

5.2. Western blot

The following conditions have been used by us for studying the effect of H_2O_2 on NF-κB/Rel and IκB families. All proteins are analyzed on either 8% or 12.5% polyacrylamide gels. LMW-SDS protein markers from GE Healthcare Life Sciences (Uppsala, Sweden) or LMW protein markers from

NZYTech (Lisboa, Portugal) are used. Antibodies sc-372 (1:1000), sc-70 (1:300), sc-371 (1:800), sc-945 (1:400), and sc-7156 (1:800) are incubated for 2 h and used to identify p65, c-Rel, IκB-α, IκB-β, and IκB-ε, respectively (all from Santa Cruz Biotechnology, Santa Cruz, California, USA). The bands corresponding to each protein are then quantified by signal intensity analysis, with normalization to the protein loading (membrane stained with Ponceau S). We use the *ImageJ* software for band intensity quantification (Rasband, 1997).

6. NF-κB-DEPENDENT GENE EXPRESSION

The H_2O_2 steady-state method can also be applied to study gene expression regulated by NF-κB. All current methods to assess gene expression, such as real-time PCR and gene expression microarrays, can be adapted for cell culture exposure to H_2O_2 in steady state by following the steps extensively described in previous sections. One of the first decisions to make is the quantity of material that will be needed to measure gene expression and set up the system with the appropriate number of cells for the Petri dishes or multi-well plates chosen.

To exemplify, we describe the study of NF-κB regulation of gene expression using HeLa cells transiently transfected with a reporter plasmid containing different κB regions. As before, we use TNF-α as a classical NF-κB inducer and analyze the modulatory effects elicited by a H_2O_2 steady state.

6.1. Plasmid constructs

Experimental κB reporter plasmids are generated using common molecular biology techniques by inserting a minimal promoter in the pGL3-basic vector (Promega, Madison, WI, USA) with *Bgl*II (5′end) and *Hin*dIII (3′end) restriction enzymes (New England Biolabs, Ipswich, England): 5′-GATCTGGGTATATAATGGATCCCCGGGTACGCAGCTCA-3′. The κB sequences (Udalova, Mott, Field, & Kwiatkowski, 2002) are inserted upstream the minimal promoter, between the KpnI/SacI restriction site, with the following general sequence: 5′-GCT-κB-CTGGCTCCT-κB-CTCAGCT-3′. We tested three different κB sequences: κB1-GGGGACTTCC; κB2-GGGGATTCCC and κB3-GGGAATTTCC.

6.2. Cellular transfection and reporter gene assays

For reporter gene assays, we use the Dual-Luciferase Reporter Assay System (Promega, Madison, WI, USA) where cells are cotransfected with the experimental plasmid that has the firefly luciferase and a second control plasmid pRL-SV40 (Promega, Madison, WI, USA) that bears the renilla luciferase, important to normalize luminescence. Setting up the right amount of DNA to transfect is important for transfection efficiency, taking also in consideration the ratio between the plasmids to avoid interferences between promoters. An excess of cytotoxicity that might be introduced by transfection methods should be avoided, even if at the cost of a lower transfection efficiency, since H_2O_2 steady state could become toxic if cells are not healthy when starting the experiment. The first step is to choose the appropriate type of transfection reagent. Lipofectamine (Invitrogen, Carlsbad, California, USA) is widely used, but here we tested fugeneHD (Roche, Mannheim, Germany), which gave a good balance between transfection efficiency and toxicity. As recommended by the manufacturer, we do several tests to choose the appropriate transfection reagent quantity to use, but also test for the time to let cells recover from transfection, before starting the steady state. We recommend at this stage to run viability assays, such as MTT (McGahon et al., 1995), or alamar blue (O'Brien, Wilson, Orton, & Pognan, 2000), before initiating with luminescence experiments. For example, we observed that transfection *per se* was leading to approximately 50% loss in cell viability. Importantly, we had to reduce glucose oxidase volume to 50% of the usual volume to maintain the desired H_2O_2 steady state (Oliveira-Marques et al., 2009a).

The following procedure exemplifies a typical transfection experiment and H_2O_2 steady-state treatment to assess NF-κB-dependent gene expression.

1. Plate HeLa cells onto 24-well plates at a density of 4.5×10^4 cells/well in 500 μL of medium;
2. Let cells recover for 24 h;
3. Prepare the transfection mixture of fugeneHD: DNA 5:2 (v/m) in Opti-MEM medium (Invitrogen, Carlsbad, California, USA) and incubate for 20 min at room temperature;
4. Replace cells medium with 500 μL of fresh medium without antibiotics;
5. Add in a drop-wise manner 18 μL of the transfection mixture. This mixture contains 180 ng of kB experimental plasmid, 9 ng of pRL-SV40 control plasmid, and 171 ng of pGL3-basic plasmid and 0.9 μL of fugeneHD;
6. Swirl the wells to ensure distribution over the entire plate surface. Perform the assay within 24 h;
7. Expose cells to steady-state H_2O_2 as explained in Section 4 and TNF-α concentrations ranging from 0.18 to 50 ng/mL. Use 800 μL of medium

per well to allow an accurate measurement of H_2O_2 with the oxygen electrode;

8. Incubate cells for 4 h. Exceptionally, intermediate corrections of H_2O_2 should not be made, because they would imply the use of several replicate wells, as all medium from one well is needed to measure H_2O_2. This is a cost/quality balance choice. For a system calibrated to achieve 25 μM H_2O_2 steady state, we normally measured 21 μM of H_2O_2 after a 4 h incubation;

9. Lysis and luciferase analysis were assayed accordingly to the manufacturer instructions. Luminescence is read with the luminometer Zenyth 3100 with 1 s of integration time, one sample at a time because of the rapid decreased of the renilla signal. Each sample is read in triplicate.

Figure 10.2 shows the effect of H_2O_2 on NF-κB-dependent gene expression as a function of the κB site. The effect of H_2O_2 is dependent on the affinity of the κB site toward NF-κB, with genes with high-affinity sites (κB2) being modulated by H_2O_2 at lower levels of TNF-α, while genes with low-affinity sites (κB1) are modulated by H_2O_2 at higher levels of TNF-α (Oliveira-Marques et al., 2009a). The medium-affinity site (κB3) gives an intermediate response.

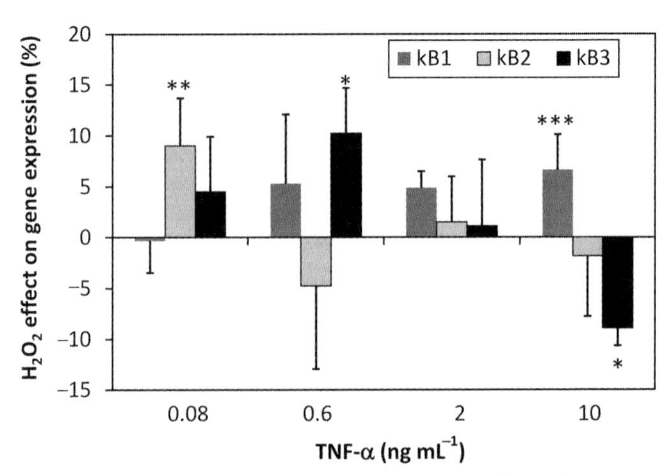

Figure 10.2 Effect of H_2O_2 on gene expression of NF-κB-dependent-reporter genes. The effect shown is the change elicited by H_2O_2 on TNF-α-dependent expression. $^{*}p<0.001$; $^{**}p=0.012$; $^{***}p=0.002$. Data are replotted from Oliveira-Marques et al. (2009a). (For color version of this figure, the reader is referred to the online version of this chapter.)

7. SUMMARY

This chapter presents an description of experimental components necessary to study NF-κB activation by H_2O_2 in a rigorous quantitative way. For that, cells are exposed to H_2O_2 steady states, by balancing the cellular H_2O_2 consumption with the production of H_2O_2 with glucose oxidase, which catalyzes the oxidation of glucose present in the growth media. H_2O_2 is monitored during the assays, and so the experimental H_2O_2 profiles are independent of the experimental conditions, facilitating the acquisition of reproducible data. Under these conditions, the variation of subtle biological responses as a function of H_2O_2 concentration can be obtained. This contrasts with experiments where H_2O_2 is delivered as a single initial dose—bolus addition—where H_2O_2 profiles and the amount of H_2O_2 delivered per cell are strongly dependent on the experimental conditions. We illustrated the advantage of the steady-state delivery methodology by showing that selective gene regulation by H_2O_2 occurs for genes that have κB sites in the promoter region with different affinity toward NF-κB. Thus H_2O_2 regulation may play an important role in the design of personal medicine drugs that target NF-κB, because single-nucleotide polymorphisms present in the κB sites are responsible for different gene expression patterns in humans.

ACKNOWLEDGMENTS

Funding from Fundação para a Ciência e a Tecnologia (FCT), Portugal (Grants PTDC/QUI/69466/2006 and PEst-OE/QUI/UI0612/2013) is acknowledged.

Conflicts of interest

VOM is a full-time employee of Thelial Technologies S.A. This present report precedes her current employment and there is no overlap in interests.

REFERENCES

Antunes, F., & Cadenas, E. (2001). Cellular titration of apoptosis with steady state concentrations of H(2)O(2): Submicromolar levels of H(2)O(2) induce apoptosis through Fenton chemistry independent of the cellular thiol state. *Free Radical Biology & Medicine, 30*, 1008–1018.

Antunes, F., Cadenas, E., & Brunk, U. T. (2001). Apoptosis induced by exposure to a low steady-state concentration of H2O2 is a consequence of lysosomal rupture. *The Biochemical Journal, 356*, 549–555.

Brigelius-Flohé, R., & Flohé, L. (2011). Basic principles and emerging concepts in the redox control of transcription factors. *Antioxidants & Redox Signaling, 15*, 2335–2381.

Chen, L. F., & Greene, W. C. (2004). Shaping the nuclear action of NF-kappaB. *Nature Reviews. Molecular Cell Biology, 5*, 392–401.

Covas, G., Marinho, H. S., Cyrne, L., & Antunes, F. (2013). Activation of Nrf2 by H_2O_2: De novo synthesis versus nuclear translocation. *Methods in Enzymology, 528*, 157–171.
Ghosh, S., May, M. J., & Kopp, E. B. (1998). NF-kappa B and Rel proteins: Evolutionarily conserved mediators of immune responses. *Annual Review of Immunology, 16*, 225–260.
Kasowski, M., Grubert, F., Heffelfinger, C., Hariharan, M., Asabere, A., Waszak, S. M., et al. (2010). Variation in transcription factor binding among humans. *Science, 328*, 232–235.
Liu, X., & Zweier, J. L. (2001). A real-time electrochemical technique for measurement of cellular hydrogen peroxide generation and consumption: Evaluation in human polymorphonuclear leukocytes. *Free Radical Biology & Medicine, 31*, 894–901.
Marinho, H. S., Cyrne, L., Cadenas, E., & Antunes, F. (2013a). H_2O_2 delivery to cells: Steady-state versus bolus addition. *Methods in Enzymology, 526*, 159–173.
Marinho, H. S., Cyrne, L., Cadenas, E., & Antunes, F. (2013b). The cellular steady-state of H_2O_2: Latency concepts and gradients. *Methods in Enzymology, 527*, 3–19.
McGahon, A. J., Martin, S. J., Bissonnette, R. P., Mahboubi, A., Shi, Y., Mogil, R. J., et al. (1995). The end of the (cell) line: Methods for the study of apoptosis in vitro. *Methods in Cell Biology, 46*, 153–185.
Moynagh, P. N. (2005). The NF-kappaB pathway. *Journal of Cell Science, 118*, 4589–4592.
O'Brien, J., Wilson, I., Orton, T., & Pognan, F. (2000). Investigation of the Alamar Blue (resazurin) fluorescent dye for the assessment of mammalian cell cytotoxicity. *European Journal of Biochemistry, 267*, 5421–5426.
Oliveira-Marques, V., Cyrne, L., Marinho, H. S., & Antunes, F. (2007). A quantitative study of NF-kB activation by H2O2: Relevance in inflammation and synergy with TNF-alpha. *Journal of Immunology, 178*, 3893–3902.
Oliveira-Marques, V., Marinho, H. S., Cyrne, L., & Antunes, F. (2009a). Modulation of NF-kB-dependent gene expression by H2O2: A major role for a simple chemical process in a complex biological response. *Antioxidants & Redox Signaling, 11*, 2043–2053.
Oliveira-Marques, V., Marinho, H. S., Cyrne, L., & Antunes, F. (2009b). Role of hydrogen peroxide in NF-kappaB activation: From inducer to modulator. *Antioxidants & Redox Signaling, 11*, 2223–2243.
Rasband, W. S. (1997). *ImageJ [Computer software]*. Bethesda, Maryland, USA: U.S. National Institutes of Health.
Test, S. T., & Weiss, S. J. (1984). Quantitative and temporal characterization of the extracellular H2O2 pool generated by human neutrophils. *The Journal of Biological Chemistry, 259*, 399–405.
Udalova, I. A., Mott, R., Field, D., & Kwiatkowski, D. (2002). Quantitative prediction of NF-kappa B DNA-protein interactions. *Proceedings of the National Academy of Sciences of the United States of America, 99*, 8167–8172.

CHAPTER ELEVEN

Detection of H_2O_2-Mediated Phosphorylation of Kinase-Inactive PDGFRα

Hetian Lei, Andrius Kazlauskas[1]

The Schepens Eye Research Institute, Massachusetts Eye and Ear Infirmary, Department of Ophthalmology, Harvard Medical School, Boston, Massachusetts, USA

[1]Corresponding author: e-mail address: andrius_kazlauskas@meei.harvard.edu

Contents

Abstract

Platelet-derived growth factor (PDGF) receptor α (PDGFRα) belongs to the 58-member family of receptor tyrosine kinases and contributes to a variety of physiological and pathological settings. Activation of PDGFRα proceeds by at least two mechanisms. The traditional route involves PDGF-dependent dimerization and activation of the receptor's intrinsic kinase activity. The second mechanism proceeds intracellularly and involves reactive oxygen species and Src family kinases, which activate monomeric PDGFRα. Herein we describe an assay to investigate reactive oxygen species-mediated phosphorylation of PDGFRα that is independent of the receptor's intrinsic kinase activity.

Platelet-derived growth factor (PDGF) receptor α (PDGFRα) is one member of the receptor tyrosine kinases family that contains 58 members in humans (Robinson, Wu, & Lin, 2000). PDGFRα consists of 1067 amino acids with 5 extracellular immunoglobulin loops, a single hydrophobic transmembrane-spanning domain and a split intracellular tyrosine kinase domain. PDGFRα knockout mice die during embryonic development; it appears that PDGFRα is required for neural crest cell development and for normal patterning of the somites (Olson & Soriano, 2009; Soriano, 1997). Growing evidence suggests a pivotal role for PDGFRα signaling in

Methods in Enzymology, Volume 528
ISSN 0076-6879
http://dx.doi.org/10.1016/B978-0-12-405881-1.00011-2

various types of mesenchymal cell/fibroblast-driven pathologies such as gastrointestinal stromal tumor and atherosclerosis (Andrae, Gallini, & Betsholtz, 2008; Olson & Soriano, 2009).

In the PDGF family there are five members: PDGF-A, -B, -AB, -C, and -D (Lei, Rheaume, & Kazlauskas, 2010). PDGF-A is specific for PDGFRα, that is, it assembles only PDGFRα homodimers. PDGF-AB, -C, and -B not only assemble PDGFRα homodimers but also induce the formation of PDGFRα/β heterodimers. In addition, PDGF-B is capable of inducing PDGFRβ homodimers. Ligand-driven dimerization of PDGFR activates its intrinsic kinase activity and triggers intracellular signaling cascades driven by enzymes such as Ras/mitogen-activated protein kinases, phosphoinositide 3-kinase/Akt, and phospholipase Cγ/protein kinase C. These signaling cascades mediate PDGF-induced cellular responses such as proliferation, survival, and migration (Andrae et al., 2008; Olson & Soriano, 2009).

Activation of PDGFRα via PDGFs is not the only way that this receptor can be activated. Growth factors outside of the PDGF family indirectly activate PDGFRα by triggering an intracellular mechanism involving reactive oxygen species and Src family kinases (Lei & Kazlauskas, 2009). Because this mechanism of activating PDGFRα does not induce dimerization and subsequent internalization and degradation (Lei, Rheaume, Velez, Mukai, & Kazlauskas, 2011), the receptor's output is prolonged and triggers signaling events that culminate in suppression of p53 (Lei et al., 2011). Reducing the level of p53 increases cell viability (Baker et al., 1989); hence indirectly activating PDGFRα may be a strategy to survive a variety of stressful conditions. The key point of the preceding section is that PDGFRα can be activated by more than one mechanism, which triggers nonidentical signaling events and cellular responses.

1. CONSTRUCTION OF KINASE-DEAD PDGFRα

The increase in tyrosine phosphorylation of PDGFRα that results from exogenously added H_2O_2 is likely to arise from either inactivation of tyrosine phosphatases and/or activation of kinases (Lei & Kazlauskas, 2009). These kinases include the receptor itself, which is capable of autophosphorylating, and/or other tyrosine kinases (such as Src family kinases) that phosphorylate PDGFRα (Lei & Kazlauskas, 2009). To investigate if autophosphorylation was required for H_2O_2-mediated tyrosine phosphorylation of PDGFRα, we tested if a kinase-inactive PDGFRα mutant (which is incapable of autophosphorylating) was tyrosine phosphorylated when cells were exposed to H_2O_2.

A kinase-inactive mutant of PDGFRα (K627R, called R627 herein) was constructed in *pLHDCX³-PDGFRα* (human wild type (WT)) using the following set of mutagenic oligonucleotide primers: (gi:61699224: 2014–2041) 5′-*GAAAGTTGCAGTGAGGATGCTAAAACCC*-3′ and its complimentary oligonucleotide by following instructions provided with QuickChange XL Site-Directed Mutagenesis Kit (Stratagene, La Jolla, CA). Briefly, the first step was to synthesize the mutant strands in the reaction (reaction buffer, *pLHDCX³-PDGFRα* (WT), the oligonucleotide primers, dNTP, and pfu turbo DNA polymerase) by thermal cycling (95 °C for 1 min, 95 °C for 50 s, 60 °C for 50 s, 68 °C for 8 min for 18 cycles, and 68 °C for 7 min). Then the *Dpn*I restriction enzyme was used to digest the parental DNA (WT) by incubating the enzyme with the reaction mixture at 37 °C for 1 h. Two microliters of the *Dpn*I treated DNA was mixed with the ultracompetent cells and incubated on ice for 30 min. The tubes containing the mixture were heat-pulsed in a 42 °C water bath for 90 s and then incubated on ice for 5 min. The mixture was then plated on an agar plate supplemented with ampicillin overnight. Three clones were picked, and one of them was characterized as outlined below.

The sequence of the mutated PDGFRα in the plasmid was determined by DNA sequencing. These two constructs (WT and mutant) were then transfected into 293GPG cells and the resulting retroviruses were used to infect F cells, which are immortalized fibroblasts from mice with both knockout of *pdgfrα* and *pdgfrβ*. The successfully infected cells named Fα and R627 were selected in histidine-free Dulbecco's Modified Eagle Medium (DMEM) supplemented with histidinol (5 m*M*) and 10% FBS for 2 weeks.

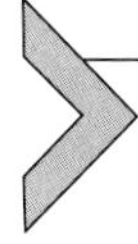

2. CHARACTERIZATION OF THE KINASE-INACTIVE RECEPTOR

The following experiment was to test the kinase activity of mutant PDGFRα. Fα and R627 cells were grown to near confluence in DMEM supplemented with 10% FBS in 10-cm cell-culture dishes, and then serum starved for 24 h. These cells were treated with or without PDGF-A (50 ng/ml) for 10 min, and the cell lysates were prepared in ice-cold extraction buffer (EB) [10 m*M* Tris–HCl (pH 7.4), 5 m*M* ethylenediaminetetraacetic acid (EDTA), 50 m*M* NaCl, 50 m*M* NaF, 20 μg/ml aprotinin, 1 m*M* phenylmethanesulfonyl fluoride, 2 m*M* Na_3VO_4, and 1% Triton X-100]. The lysates were quantitated using a Bicinchoninic Acid Protein Assay Kit.

The following steps were taken to immunoprecipitate PDGFRα from the cell lysates. Protein A conjugated agarose was transferred into microtubes (50 μl each tube) and then washed once with 1 ml EB buffer. Anti-PDGFRα rabbit serum (1 μl, 27P) was then incubated with the agarose in EB (500 μl) at 4 °C with gently rocking for 1 h. After a 15 s spin at 13,000 × *g*, the supernatant was removed, lysates (1 mg) were transferred into each tube (1 ml per tube) and gently rocked for 4 h at 4 °C. The agarose beads were washed five times with EB, sample buffer (EDTA 5 m*M*, SDS 2%, dithiothreitol 100 m*M*, sucrose 10%, Tris–HCl 100 m*M*, and bromophenol blue 0.1%) was added and boiled for 5 min.

The proteins recovered by immunoprecipitation were resolved on a 10% sodium dodecyl sulfate polyacrylamide gel electrophoresis (SDS-PAGE) gel and transferred onto a polyvinylidene difluoride (PVDF) membrane, which was subjected to Western blotting: the membrane was blocked for 1 h at room temperature in a blocking buffer with gently rocking; an antiphospho-tyrosine (pY20) antibody was used as a primary antibody (1–1000 dilution in blocking buffer (2% BSA, 0.05% Tween 20, 0.005% NaN3, 2 m*M* Trizma Base, 8 m*M* Trizma HCl, 154 m*M* NaCl), overnight incubation at 4 °C with gently rocking); the membrane was washed with washing buffer + 0.05% Tween three times, 10 min each time; an antimouse horseradish peroxidase conjugated antibody was used as a secondary antibody (1–5000 dilution in a Blotto buffer (2.5 % Non-fat dry milk, 0.05% Tween 20, 2 m*M* Trizma Base, 8 m*M* Trizma HCl, 154 m*M* NaCl) for 30 min at room temperature with gently rocking). The Western blot was developed using chemiluminescence (SuperSignal West Pico, Thermo Scientific Inc.). The membrane was stripped and then reprobed with the anti-PDGFRα 27P antibody (Lei & Kazlauskas, 2009) (Fig. 11.1).

The results showed that PDGF-A stimulated phosphorylation of WT but not the mutant PDGFRα. These findings support the idea that autophosphorylation of PDGFRα is essential to ligand-induced tyrosine phosphorylation of PDGFRα, and that the R627 mutant does not own this function.

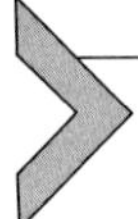

3. DETECTION OF H_2O_2-MEDIATED PHOSPHORYLATION OF KINASE-INACTIVE PDGFRα

Like PDGF, H_2O_2 can stimulate phosphorylation of PDGFRα (Sundaresan, Yu, Ferrans, Irani, & Finkel, 1995). To investigate if autophosphorylation was required for this H_2O_2-driven event, we

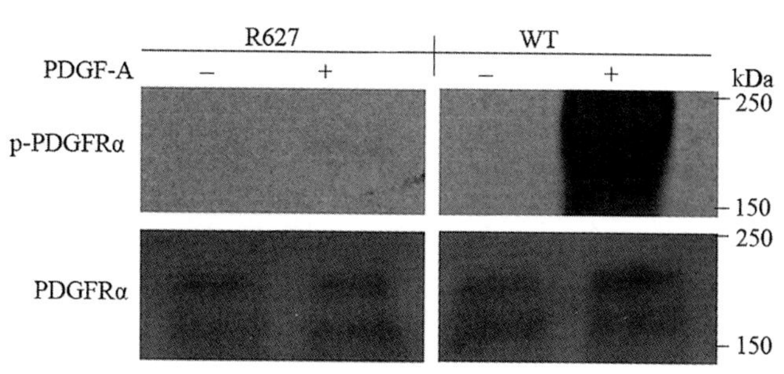

Figure 11.1 PDGF-A failed to stimulate phosphorylation of the mutant PDGFRα. PDGFRα was immunoprecipitated from lysates that were prepared from resting (−) or PDGF-stimulated (+) mouse fibroblasts (pdgfrα−/− and pdgfrβ−/−) expressing wild type (WT) or mutant K627R (R627) PDGFRα. The resulting samples were subjected to Western blot analysis using an antiphospho-tyrosine antibody (top panel). The membrane was stripped and reprobed with the anti-PDGFRα antibody (bottom panel).

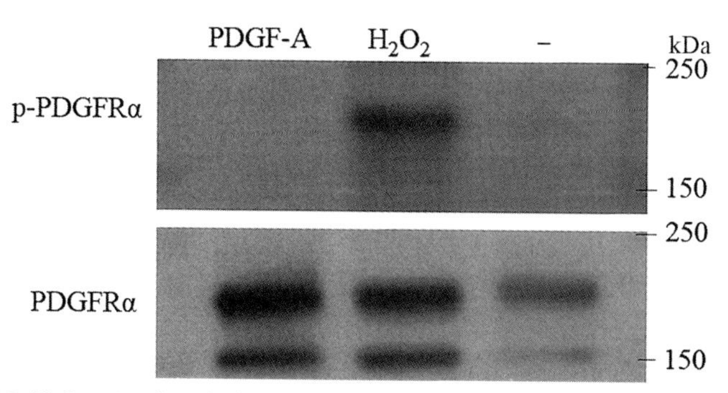

Figure 11.2 H_2O_2 stimulated phosphorylation of kinase-inactive PDGFRα. R627 cells were serum starved for 24 h and treated with PDGF-A or H_2O_2 for 10 min. The clarified lysates were immunoprecipitated with an anti-PDGFRα antibody, and the resulting samples were subjected to Western blot with an antiphospho-tyrosine antibody (top panel). The membrane was stripped and reprobed with the anti-PDGFRα antibody (bottom panel).

performed experiments as shown in Fig. 11.2. R627 cells were treated with PDGF-A, H_2O_2, or vehicle for 10 min. The lysates were subjected to immunoprecipitation with the PDGFRα 27P antibody, and the recovered proteins were separated in 10% SDS-PAGE and then transferred onto a PVDF membrane, which was subjected to Western blot analysis using an antiphospho-tyrosine antibody (PY20). The membrane was stripped and reprobed with the anti-PDGFRα 27P antibody (Lei & Kazlauskas, 2009) (Fig. 11.2).

In contrast to PDGF, H_2O_2 promoted tyrosine phosphorylation of PDGFRα (Fig. 11.2). This observation revealed that the two agonists engaged different mechanism to increase tyrosine phosphorylation of PDGFRα. While the PDGF-mediated mechanism was absolutely dependent on the kinase activity of PDGFRα, the H_2O_2-mediated route appeared to involve kinases other the PDGFRα itself. This observation suggests that there are tyrosine kinases capable of phosphorylating PDGFRα, and that their ability to do so is enhanced by H_2O_2.

4. IMPLICATION

The studies presented in this review article highlight that there are H_2O_2-activateable tyrosine kinases that are capable of phosphorylating PDGFRα. Src family kinases can be both activated by H_2O_2 and phosphorylate PDGFRα (Lei & Kazlauskas, 2009; Lei et al., 2011). Thus H_2O_2/Src family kinases-mediated activation of PDGFRα may be a previously unappreciated mechanism contributing to physiological and/or pathological settings in which H_2O_2 is elevated.

ACKNOWLEDGMENT

Funding for this work was provided by NIH Grant EY012509 to A. K.

REFERENCES

Andrae, J., Gallini, R., & Betsholtz, C. (2008). Role of platelet-derived growth factors in physiology and medicine. *Genes & Development*, *22*, 1276–1312.

Baker, S. J., Fearon, E. R., Nigro, J. M., Hamilton, S. R., Preisinger, A. C., Jessup, J. M., et al. (1989). Chromosome 17 deletions and p53 gene mutations in colorectal carcinomas. *Science*, *244*, 217–221.

Lei, H., & Kazlauskas, A. (2009). Growth factors outside of the PDGF family employ ROS/SFKs to activate PDGF receptor alpha and thereby promote proliferation and survival of cells. *The Journal of Biological Chemistry*, *284*, 6329–6336.

Lei, H., Rheaume, M. A., & Kazlauskas, A. (2010). Recent developments in our understanding of how platelet-derived growth factor (PDGF) and its receptors contribute to proliferative vitreoretinopathy. *Experimental Eye Research*, *90*, 376–381.

Lei, H., Rheaume, M. A., Velez, G., Mukai, S., & Kazlauskas, A. (2011). Expression of PDGFR{alpha} is a determinant of the PVR potential of ARPE19 cells. *Investigative Ophthalmology & Visual Science*, *52*, 5016–5021.

Olson, L. E., & Soriano, P. (2009). Increased PDGFRalpha activation disrupts connective tissue development and drives systemic fibrosis. *Developmental Cell*, *16*, 303–313.

Robinson, D. R., Wu, Y. M., & Lin, S. F. (2000). The protein tyrosine kinase family of the human genome. *Oncogene*, *19*, 5548–5557.

Soriano, P. (1997). The PDGF alpha receptor is required for neural crest cell development and for normal patterning of the somites. *Development*, *124*, 2691–2700.

Sundaresan, M., Yu, Z. X., Ferrans, V. J., Irani, K., & Finkel, T. (1995). Requirement for generation of H_2O_2 for platelet-derived growth factor signal transduction. *Science*, *270*, 296–299.

SECTION III

H_2O_2 and Regulation of Cellular Processes

CHAPTER TWELVE

Genetic Modifier Screens to Identify Components of a Redox-Regulated Cell Adhesion and Migration Pathway

Thomas Ryan Hurd*[,1], Michelle Gail Leblanc*, Leonard Nathaniel Jones*, Matthew DeGennaro†, Ruth Lehmann*

*Department of Cell Biology, HHMI and Kimmel Center for Biology and Medicine of the Skirball Institute, New York University School of Medicine, New York, USA

†Laboratory of Neurogenetics and Behavior, The Rockefeller University, New York, USA

[1]Corresponding author: e-mail address: thomas.hurd@med.nyu.edu

Contents

1. Introduction 198
2. Mutations in a *D. melanogaster* Gene Encoding a Peroxiredoxin Cause Germ Cell Adhesion and Migration Defects 199
3. Dominant Modifier Screens 200
4. Conducting a Dominant Modifier Screen to Identify Missing Components of a Redox-Regulated Germ Cell Migration Pathway 201
   4.1 *D. melanogaster* strains 201
   4.2 *D. melanogaster* culture and husbandry 202
   4.3 Choosing a suitable genetic background 202
   4.4 Inducing second-site mutations 204
   4.5 Crossing schemes 206
   4.6 Assaying germ cell migration 207
   4.7 Identifying enhancers 208
5. Limitations to Dominant Modifiers Screens 211
6. Concluding Remarks 212
Acknowledgments 213
References 213

Abstract

Under normal physiological conditions, cells use oxidants, particularly H_2O_2, for signal transduction during processes such as proliferation and migration. Though recent progress has been made in determining the precise role H_2O_2 plays in these processes, many gaps still remain. To further understand this, we describe the use of a dominant enhancer screen to identify novel components of a redox-regulated cell migration and adhesion pathway in *Drosophila melanogaster*. Here, we discuss our methodology and

Methods in Enzymology, Volume 528
ISSN 0076-6879
http://dx.doi.org/10.1016/B978-0-12-405881-1.00012-4

progress as well as the benefits and limitations of applying such an approach to study redox-regulated pathways. Depending on the nature of these pathways, unbiased genetic modifier screens may prove a productive way to identify novel redox-regulated signaling components.

1. INTRODUCTION

Oxidants are generated within organisms as unwanted by-products of aerobic respiration, but they also function as signaling molecules in physiological processes such as proliferation, cell migration, and adhesion (Finkel, 2011; Flohe, 2010; Janssen-Heininger et al., 2008; Rhee, Bae, Lee, & Kwon, 2000). These physiological functions are carried out by mild oxidants, such as H_2O_2, in a compartmentalized and coordinated fashion through the reversible oxidation of target proteins (Finkel, 2011; Flohe, 2010; Janssen-Heininger et al., 2008; Rhee et al., 2000). Though much progress has been made identifying such protein targets and the regulatory components of redox signaling pathways, significant gaps in our knowledge still remain. Most efforts have focused on proteomic methods to identify redox sensitive proteins that may be pathway components (Chouchani, James, Fearnley, Lilley, & Murphy, 2011; Thamsen & Jakob, 2011). Here, we describe an alternative approach, in which a specialized type of genetic screen, a dominant modifier screen (Simon, Bowtell, Dodson, Laverty, & Rubin, 1991; St Johnston, 2002), is used to elucidate redox-regulated pathway components. Dominant modifier screens have been used extensively to identify new components of phosphorylation-based signaling pathways (Jorgensen & Mango, 2002; St Johnston, 2002) and could potentially yield novel and hitherto unsuspected components when applied to H_2O_2-regulated processes.

To illustrate how such an approach would work in practice, we discuss an *in vivo* modifier screen currently being conducted to identify unrecognized components of a redox-regulated cell adhesion and migration pathway. For this screen, we use a *Drosophila melanogaster* fly strain harboring a mutation in a gene encoding a H_2O_2-degrading peroxiredoxin called *jafrac1*. Mutations in this peroxiredoxin gene cause defects in the adhesion of a particular cell type, germ cells, during their migration into *Drosophila* embryos. We then introduce second-site mutations and test their ability to modify the germ cell migration phenotype. We expect the genes identified from this screen to encode proteins that impact on novel components of this pathway. Before discussing modifier screens further, we briefly describe germ cell migration, and the mutant phenotype we are trying to better understand.

2. MUTATIONS IN A *D. MELANOGASTER* GENE ENCODING A PEROXIREDOXIN CAUSE GERM CELL ADHESION AND MIGRATION DEFECTS

Regulated adhesion between cells and their environment is critical for normal cell migration. To study this process *in vivo*, we use *Drosophila* germ cell migration as a model (Richardson & Lehmann, 2010). Germ cells initially form at the posterior pole of the embryo (Fig. 12.1A), and as the embryo develops, they are carried along into the embryo proper (Fig. 12.1B). Maintaining the appropriate contacts between germ cells and the rest of the embryo is critical during this stage of migration (DeGennaro et al., 2011).

Through a screen, we previously identified mutations in *jafrac1* that caused germ cells to fail to be internalized into the embryo during their migration, leaving some germ cells trailing on the outside of the embryo (Fig. 12.1C and D; DeGennaro et al., 2011). *jafrac1* encodes a homolog of the human cytosolic typical 2-cysteine peroxiredoxin II, which acts as a peroxidase catalyzing the reduction of H_2O_2 and alkyl hydroperoxides (Kang et al., 1998; Lee et al., 2009; Rodriguez, Agudo, Van Damme, Vandekerckhove, & Santaren, 2000).

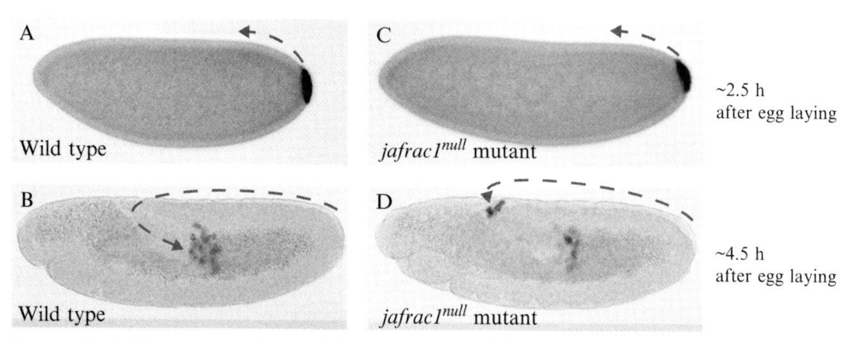

Figure 12.1 The peroxiredoxin, Jafrac1 is required for germ cell internalization during embryogenesis. (A and B) Images of representative wild-type embryos (w^{1118}) approximately 2.5 h (A) and 4.5 h (B) after egg laying. (C and D) Images of representative embryos from *jafrac1*^{null} homozygous mutant mothers approximately 2.5 h (C) and 4.5 h (D) AEL. Embryos were stained for Vasa protein to mark germ cells (brown). Embryos are oriented with the posterior to the right, anterior to the left, dorsal up, and ventral down. (See Color Insert.)

Further experimentation showed that in embryos from *jafrac1* mutant mothers, germ cells lose adherence to the embryo during their internalization due to a disruption in the cell–cell adhesion molecule DE-cadherin (DeGennaro et al., 2011; Harris & Tepass, 2010). Indeed, mutations in the gene *shotgun (shg)*, which encodes DE-cadherin, cause similar germ cell internalization defects (DeGennaro et al., 2011). As Jafrac1 is a peroxidase, we wanted to know whether increased H_2O_2 levels were responsible for the germ cell adhesion defect. Treating embryos with H_2O_2 caused a decrease in the protein levels of DE-cadherin suggesting that Jafrac1 might regulate germ cell adhesion by controlling H_2O_2 levels (DeGennaro et al., 2011). Given that *jafrac1* mutants have a distinct cell migration phenotype that is caused at least in part by redox changes such as those brought about by H_2O_2, we decided to harness the power of *Drosophila* genetics and conduct an unbiased modifier screen.

3. DOMINANT MODIFIER SCREENS

Modifier screens have been successfully used to identify missing components of genetic pathways (Jorgensen & Mango, 2002; Rubin et al., 1997; St Johnston, 2002). Such a screen normally begins with a strain of defined genetic composition and robust phenotypic defect. Second-site mutations are then introduced into this strain and assayed to determine if they enhance (worsen) or suppress (ameliorate) the starting phenotype. Second-site mutations that modify the phenotype may be in genes that fall within the same pathway that is perturbed in the starting strain. Secondary screens must then be conducted to confirm whether the modifying gene acts in the same pathway, a parallel one, or nonspecifically influences the process (Jorgensen & Mango, 2002; Karim et al., 1996; Rubin et al., 1997; St Johnston, 2002).

A particularly useful type of modifier screen is a sensitized screen for dominant modifiers. This screen is based on the premise that while a 50% reduction in the wild-type levels of a protein is most often sufficient for normal function (there are very few loci that have observable haploid phenotypes (Deutschbauer et al., 2005; Lindsley et al., 1972)) when a particular process is already partially disrupted by another mutation, this 50% reduction might no longer suffice (Rubin et al., 1997; St Johnston, 2002). It has been shown that for most, but not all, genes, the level of expression of a locus is proportional to the gene copy number (Malone et al., 2012; Springer, Weissman, & Kirschner, 2010). Therefore, if mutating one copy of a gene modifies a phenotype in a genetic background where a pathway of interest is already partially perturbed (sensitized), then this gene is likely to be involved, directly or otherwise, in the pathway. Rubin et al. carried out some of the first and most successful screens of this type, identifying several signal transduction components downstream of the receptor tyrosine kinase, Sevenless (Rubin et al., 1997; Simon et al., 1991).

This approach has two important advantages. First, recessive lethal mutations can be isolated since they are screened in a heterozygous state. Second, unlike most screens which require at least two generations (F2 and F3 screens) to isolate homozygous mutants, the progeny of the mutagenized flies can be screened directly (F1 generation) because they need not be made homozygous. This increases the throughput of the screen by an order of magnitude (St Johnston, 2002). Given these advantages, we decided to use this approach to search for new components of the *jafrac1* pathway.

4. CONDUCTING A DOMINANT MODIFIER SCREEN TO IDENTIFY MISSING COMPONENTS OF A REDOX-REGULATED GERM CELL MIGRATION PATHWAY

In the following sections, we discuss the methods and experiments conducted to identify components of the *jafrac1*-regulated germ cell migration pathway. We also discuss alternative ways of screening where applicable.

4.1. *D. melanogaster* strains

The following strains were used in this study:

- y^1 *P{SUPor-P}Jafrac1*^{KG05372} (referred to as *jafrac1*^{null})
- *jafrac1*^{null}; *P{hs-hid, w*$^+$*} wg*$^{Sp-1}$*/CyO* (*P{hs-hid, w*$^+$*} wg*$^{Sp-1}$ is referred to as *Sp hs-hid*)
- *PBac{PB}Jafrac1*^{f08066} (is referred to as *jafrac1*^{hypo})
- *jafrac1*^{hypo}; *hs-hid/CyO*
- w^{1118};*Df(2R)BSC814/SM6a* (*Df(2R)BSC814* is referred to as *shg*^{null})
- Deficiencies spanning 2L and 2R from the Bloomington deficiency kit (Cook et al., 2012; Parks et al., 2004)
- *pr*^{1} *cad*^{2} *P{neoFRT}40A/CyO* (*pr*^{1} *cad*^{2} *P{neoFRT}40A* is referred to as *cad*^{2})
- y^1 w^{67c23}; *P{EPgy2}Pomp*^{EY06518}*/CyO* (*P{EPgy2}Pomp*^{EY06518} is referred to as *pomp*^{EY})

- w^*; *Df(2L)vari*^{48EP} *Pomp*^{48EP} *vari*^{48EP}/*CyO* (*Df(2L)vari*^{48EP} *Pomp*^{48EP} *vari*^{48EP} is referred to as *vari*^{48EP})
- y^1 w^*; *Mi{MIC}CG9328*^{MI02953}/*SM6a* (*Mi{MIC}CG9328*^{MI02953} is referred to as *CG9328*^{MI})
- y^1; *P{SUPor-P}CG9328*^{KG09432}; ry^{506} (*P{SUPor-P}CG9328*^{KG09432} is referred to as *CG9328*^{KG09})
- y^1; *P{SUPor-P}CG9328*^{KG02184}/*CyO*; ry^{506} (*P{SUPor-P}* *CG9328*^{KG02184} is referred to as *CG9328*^{KG02})
- w^{1118}; *Mi{ET1}CG33322*^{MB07355}(*Mi{ET1}CG33322*^{MB07355} is referred to as *CG33322*^{MB}).

All strains used in this study were obtained from the Bloomington *Drosophila* Stock Center at Indiana University except for the *jafrac1*^{hypo} strain, which was from Exelixis at Harvard Medical School (Thibault et al., 2004).

4.2. *D. melanogaster* culture and husbandry

Flies were maintained at 25 °C and 60% humidity on media (3.6% (v/v) molasses (LabScientific Inc.), 0.532% (w/v) agar (MoorAgar Inc.), 3.6% (w/v) cornmeal (LabScientific Inc.), 1.488% (w/v) yeast (LabScientific Inc.), 0.112% (w/v) Tegosept® (Sigma), 1.12% (v/v) ethanol (Fisher Scientific), 0.38% (v/v) propionic acid (Fisher Scientific)) in polystyrene vials. Virgin females were used for all crosses conducted for this study. Since female flies do not mate within the first 8 h of adulthood, virgin females can be obtained by collecting flies within 8 h of their eclosion (emergence from their pupal case). For each cross, approximately 20 virgin females were mated with 10 males in polystyrene vials containing 20–24 ml of culture media. Files were transferred to fresh vials every 1 or 2 days for up to 10 days.

4.3. Choosing a suitable genetic background

In order to attempt a dominant modifier screen, an appropriate genetic background must first be identified. The most important consideration before starting such a screen is that the phenotype and the methods used to measure it are robust and consistent.

If one is interested in finding genes that enhance a phenotype, then a sensitized background must be identified, where the expression of the gene or the activity of its product is substantially altered, but a weak or absent phenotype is exhibited. For example, we identified a loss-of-function allele of *jafrac1* (called *jafrac1*^{hypo}) that has greatly reduced RNA expression (Fig. 12.2A) but exhibits largely normal germ cell migration when

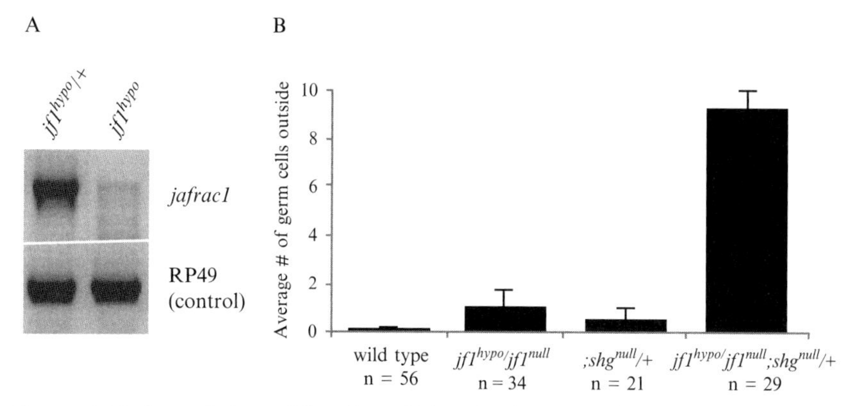

Figure 12.2 A weak loss-of-function allele of *jafrac1* can be used to screen for enhancers. (A) The levels of *jafrac1* RNA are significantly reduced in embryos from *jafrac1^hypo^* mothers when compared to a heterozygous control as measured by RT-PCR. (B) Germ cells in embryos from wild type (w^{1118}), *jafrac1* transheterozygotes or *shg* heterozygotes rarely fail to internalize and as such are not often observed on the outside of the embryo. However, when one copy of *shg* is removed from embryos from *jafrac1* transheterozygous mutant mothers, the number of germ cells that fail to internalize increases. Data are the means $\pm$ standard error (S.E.). *jf1*, *jafrac1*.

homozygous or in combination with a null allele of *jafrac1* (Fig. 12.2B). To demonstrate that this background was sensitized enough to be dominantly modified, we crossed a null allele of *shg* (the gene that encodes DE-cadherin), which has previously been shown to interact with *jafrac1* (DeGennaro et al., 2011). The *shg* null allele substantially dominantly enhanced the germ cell migration defect, demonstrating that this background is suitable for screening (Fig. 12.2B).

If a stronger loss-of-function mutation is used, one can screen for dominant suppressors, or if an intermediate is chosen potentially, both enhancers and suppressors can be screened for. Whether one chooses to screen for enhancers, suppressors, or both depends on what one is hoping to find and on the phenotype being investigated. Typically, many genes can enhance a mutant phenotype, while only mutations in a few key regulators suppress (Jorgensen & Mango, 2002; Karim et al., 1996).

In addition to using loss-of-function mutations, sensitized backgrounds can be generated with neomorphic alleles, RNA interference (RNAi), or gene overexpression. If dominant mutations or transgenes are used to create a sensitized background, the number of crosses needed to perform the screen might be reduced, as they need not be made homozygous.

4.4. Inducing second-site mutations

Second-site mutations and alterations can be introduced in multiple ways. The way in which one chooses to induce second-site mutations will depend largely on whether the phenotype is amenable to high-throughput analysis. In *D. melanogaster*, there are three principle ways in which this is commonly done, each with its own benefits and limitations.

The first uses X-rays or chemical mutagens, most commonly ethyl methanesulfonate (EMS), which primarily induces single base changes (Lewis & Bacher, 1968; St Johnston, 2002). A major benefit of these types of mutagens is that mutations can be induced directly in the sensitized strain, and thus the progeny (F1 generation) of the mutagenized parent can be screened. This obviates the need to individually cross each second-site mutation one-by-one into the sensitized background (see, e.g., Simon et al., 1991). However, F1 screens can only be conducted for phenotypes that can be measured without killing the animal because if the animal is killed there will be no material left to map the modifier mutation. A major limitation to using X-rays or chemical mutagens is that mapping mutations induced by these agents is labor-intensive and difficult (Bokel, 2008). New genome sequence-based approaches might alleviate this somewhat (Bokel, 2008).

In the second method, second-site mutations are instead introduced into the sensitized background using engineered transposons, which when inserted into the genome frequently interfere directly with gene expression and function (Bellen et al., 2011). A major advantage of this method is that once a modifier is found, the gene mutated within it can be easily and rapidly identified by PCR using unique sequences of the transposon. New insertions can be generated by mobilizing a transposable element within the sensitized background (Mathieu, Sung, Pugieux, Soetaert, & Rorth, 2007). The utility of this approach may be limited, however, because transposons tend to be very inefficient mutagens and often exhibit considerable biases when integrating into the genome (Mathieu et al., 2007). Though possibly more labor-intensive, strains containing transposons of known location can instead be crossed one-by-one into the sensitized background (see, e.g., Karim et al., 1996). To facilitate this, the *Drosophila* gene disruption project has generated a public collection of mutant strains, each containing a single unique transposon insertion of known location, which covers nearly two-thirds of all annotated protein-coding genes (Bellen et al., 2011, 2004; Spradling et al., 1999). Many of these transposons contain upstream activating sequences of yeast which can be used to transcriptionally activate an endogenous gene next to the insertion site (Beinert et al., 2004; Bellen et al., 2004; Rorth, 1996; Staudt et al., 2005; Thibault et al., 2004). This permits gain-of-function modifier screens in which the effect of misexpressing a gene can be tested (see, e.g., Gregory et al., 2007). A major advantage of this approach is that redundant genes can be identified.

Third, large deletions (or deficiencies) that remove hundreds of genes at the same time can be used to identify modifiers. As deficiencies are equivalent to null mutations in all of the genes that are deleted, this type of deficiency screen provides a rapid way to identify regions of the genome that contain potential enhancers or suppressors (St Johnston, 2002). This approach is particularly useful if it is difficult or time consuming to screen the phenotype of interest and allows for rapid survey of the entire genome. To facilitate deficiency screens, the Bloomington stock center has assembled a collection of deficiencies that cover 98.4% of annotated euchromatic genes (Cook et al., 2012; Parks et al., 2004). Recently, microRNAs (miRNAs) have been used as an alternative means to deficiencies to identify modifiers (Szuplewski et al., 2012). miRNA overexpression can cause simultaneous reduction of the expression levels of hundreds of genes concurrently. One advantage to use miRNAs to induce second-site alterations is that they may allow access to genes not covered by the deficiency kit (Szuplewski et al., 2012). However, this approach is limited in that miRNAs do not target all protein-coding genes with equal efficiency, and identification of biologically significant targets can be problematic (Szuplewski et al., 2012).

Although chemical mutagens, transposons, and deficiencies are most commonly used, second-site alterations can be induced in other ways, for example, using RNAi. As with transposons, a major advantage of using RNAi is that the gene affected is already known, providing no other off-target genes are being silenced. Publicly available collections of transgenic flies containing conditionally expressible RNAi constructs have been generated by the Transgenic RNAi Project at Harvard Medical School (Ni et al., 2008, 2011) and the Vienna Drosophila RNAi Center (which covers 88% of *Drosophila* protein-coding genes) (Dietzl et al., 2007). A benefit and a limitation to using RNAi is that the target gene expression will in most cases be reduced by more than 50% of wild-type levels. Although removing one copy of a gene rarely leads to observable phenotypes, greater reductions in gene expression may prevent scoring the phenotype being measured, for example, due to lethality. In some instances, this can be avoided using tissue-specific or heat shock-inducible drivers or generating clones. Genes that are less dose-sensitive are more likely to be identified using this method.

Lastly, instead of searching for genes that modify a particular phenotype, a similar approach can be used to identify small molecule modifiers (Gonsalves et al., 2011; Lawal et al., 2012). Most recently, sensitized *Drosophila* strains have been used to screen for neuropsychiatric drugs (Lawal et al., 2012). Such screens could potentially be used to identify small molecules that modify redox-regulated pathways.

4.5. Crossing schemes

If second-site mutations are induced directly in the sensitized background, then crossing schemes such as those used in the original enhancer screens might be appropriate (Simon et al., 1991; St Johnston, 2002). In our case, however, because we used deficiencies, it was necessary to first cross the deficiencies into the sensitized *jafrac1* background, as outlined in Fig. 12.3.

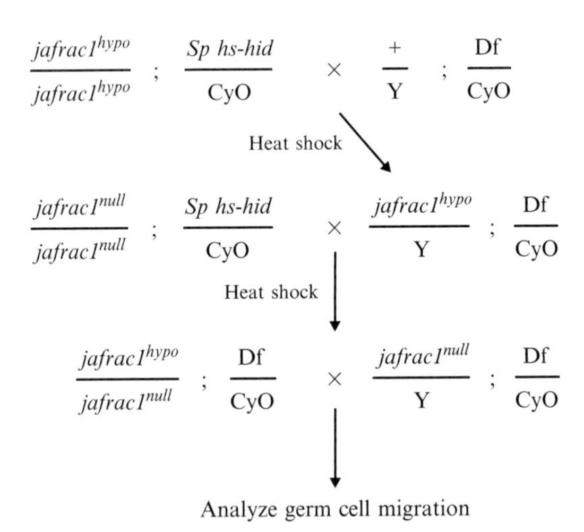

Figure 12.3 Schematic of the crosses carried out to identify dominant enhancers of the *jafrac1* germ adhesion defect. A stock homozygous for the *jafrac1^{hypo}* loss-of-function mutation that also contained one copy of *hs-hid* in trans to the balancer CyO was generated. Next, each deficiency line was crossed to females from this stock. Approximately 5 days later, when the embryos had developed into 3rd instar larvae, they were heat-shocked by immersion of the vials for 2 h in a 37 °C water bath. This induced the expression of the lethal gene *hid*, which caused all larvae that had inherited it to die. Any larvae that had inherited two copies of the balancer also died (because it is homozygous-lethal) leaving alive only flies with one copy of the deficiencies and one copy of the CyO chromosome. This process was then repeated by crossing virgin females containing the null allele of *jafrac1^{null}* to *jafrac1^{hypo}* males containing one copy of the deficiency and one copy of CyO. Df, deficiency.

We screened deficiencies on the second chromosome that are maintained in trans with a special type of chromosome called a balancer chromosome. Balancer chromosomes have three important features. First and most importantly, they contain multiple inversions, which prevent proper pairing with the homologous chromosome and suppress recovery of viable progeny after recombination. As a consequence, the deficiency-bearing chromosome remains stable through the crossing scheme. Second, balancers contain easily visualized dominant markers so that they can be followed in living flies from one generation to the next. For example, all flies with the second chromosome balancer, CyO, have curly wings. And last, most balancer chromosomes are homozygous-lethal. These features allow us to follow the balancer, and by its absence the deficiencies, so as to ensure they are faithfully crossed into the sensitized background.

To efficiently bring a set of individual deficiencies into the *jafrac1* sensitized mutant background, we first crossed deficiency strains with a stock homozygous for the $jafrac1^{hypo}$ loss-of-function mutation. This $jafrac1^{hypo}$ strain also contained one copy of the balancer CyO in trans with a chromosome that bears a transgene expressing the cell death-promoting gene *hid* under control of the heat shock promoter (*hs-hid*) (Grether, Abrams, Agapite, White, & Steller, 1995; Moore, Broihier, Van Doren, Lunsford, & Lehmann, 1998). By simply heat shocking the progeny of a cross, all flies inheriting the *hs-hid* chromosome will die. This "trick" eliminates the need for sorting correct genotypes in the next generation. This process was then repeated by crossing virgin females containing the null allele of $jafrac1^{null}$ to $jafrac1^{hypo}$ males containing one copy of the deficiency and one copy of CyO. Germ cell migration was subsequently assayed in embryos from *jafrac1* mutant flies containing one copy of a 2nd chromosome deficiency.

4.6. Assaying germ cell migration

To assay germ cell migration, germ cells were visualized in fixed embryos using an antibody that detects the germ cell-specific protein, Vasa (McDonald & Montell, 2005; Richardson & Lehmann, 2010). Briefly, 50–100 female and male adult flies were placed in a cage (a perforated 125-ml plastic beaker). Yeast paste was dabbed on a 60-mm diameter egg collection plate (2.1% (w/v) BactoAgar (Fisher Scientific), 2.4% (w/v) granulated sugar (Domino®), 23.7% (v/v) apple juice (Hansens's Natural), 0.07% (w/v) Tegosept® (Sigma), 0.7% (v/v) ethanol (Fisher Scientific)), which was placed on the mouth of the beaker and affixed with a piece of tape to prevent flies from escaping. The caged flies were then incubated overnight at 25 °C. The following morning the embryo-laden plate was removed from the cage, and the outer impermeable layer of the eggshell, the chorion, was removed by incubating the eggs in 50% Chlorox® bleach solution for 2 min. The bleach was then washed away with water, and the embryos were permeablized and fixed in 5% formaldehyde, 50% heptane, and 45% water in a glass scintillation vial with constant agitation for 40 min. Lastly, the inner impermeable layer of the eggshell, the vitelline membrane, was removed by hand using a needle (Rothwell & Sullivan, 2007). The vitelline membrane can also be removed osmotically by methanol shock (Rothwell & Sullivan, 2007).

To visualize the germ cells, the fixed embryos were incubated in blocking buffer (phosphate-buffered saline solution containing 0.1% (v/v) Tween-20 and 0.1% (w/v) bovine serum albumin) for 1 h at 23 °C. The embryos were then probed with rabbit Vasa antiserum (Moore et al., 1998) diluted in blocking buffer (1:5000) and incubated overnight at 4 °C. The next day the embryos were washed with blocking buffer three times (30 min incubation per wash) and incubated with Alexa Flour® 488 goat antirabbit IgG (A11034, Invitrogen) for 2 h at 23 °C. Embryos were washed as before in PBS to remove the secondary antibody, mounted in VectaShield mounting medium (H-1000, Vector Labs) on glass slides and covered with a 20 × 40-mm glass slip. Germ cells were imaged using an Axiovert 200M fluorescent microscope (Zeiss) equipped with a X-cite 120 mercury arc lamp (EFXO). The average number of germ cells left outside the embryo was determined in embryos which had already gastrulated and were approximately 4–11 h after egg laying old, corresponding to stages 9–14, respectively (Campos-Ortega & Hartenstein, 1985).

4.7. Identifying enhancers

To identify enhancers of *jafrac1*, we screened deficiencies covering the second chromosome. Of the deficiencies screened, two, Df(2L)ED1315 and Df(2R)BSC595, significantly enhanced the *jafrac1* phenotype (Fig. 12.4). In both cases, we observed a substantial and significant increase in the number of germ cells that failed to be internalized and were subsequently left on the outside of the embryo. If we had induced second-site mutations with a mutagen, such as EMS, then the next step would be to identify the mutated enhancer gene by standard procedures (Blumenstiel et al., 2009; Bokel,

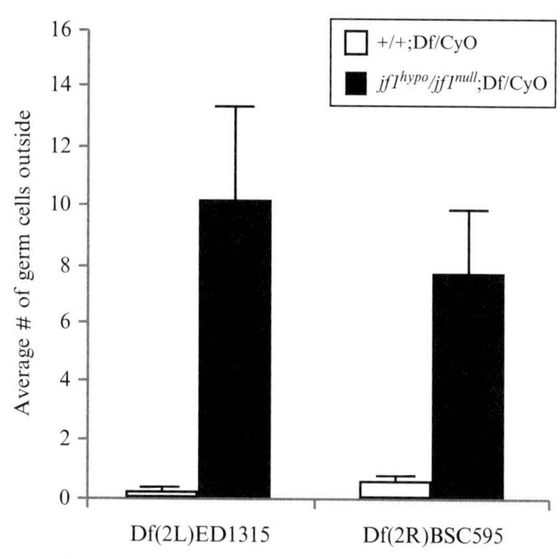

Figure 12.4 Deficiencies Df(2L)ED1315 and Df(2R)BSC595 significantly increase germ cell adhesion defects in *jafrac1* mutant embryos. Germ cell migration is normal in the deficiencies lines (white boxes). However, deficiencies Df(2L)ED1315 and Df(2R) BSC595 cause significant increases in the germ cell adhesion defects when *jafrac1* levels are reduced (black boxes). Data are the means ± S.E.

2008). However, because we used deficiencies, it was necessary to conduct a secondary screen to identify the causative genes within the enhancing deficiencies.

To do this with Df(2L)ED1315, we repeated the process described earlier with smaller deficiencies spanning the locus deleted by Df(2L)ED1315. From this analysis, we were able to further map the enhancer gene to an eight-gene region because the smaller Df(2L)Exel7079 enhanced while Df(2L)ED1378 did not (Fig. 12.5A). Since according to publicly available temporal expression data (modENCODE; Graveley et al., 2011) only five of these genes are expressed in the embryo, we tested whether mutations in any of these five genes enhanced the *jafrac1* phenotype. Mutations in one of the genes in this region, *caudal*, enhanced to a similar degree as the Df(2L)ED1315, identifying *caudal* as a enhancer of *jafrac1* (Fig. 12.5B).

Further experiments are currently underway to identify the enhancer within Df(2R)BSC595, the other deficiency we identified that enhances the *jafrac1* phenotype. Using a number of smaller deficiencies, we have been able to map the enhancer to a small region containing 13 genes, three of which are homologous to mammalian peroxiredoxin VI. Further work is underway to determine the precise enhancer gene(s).

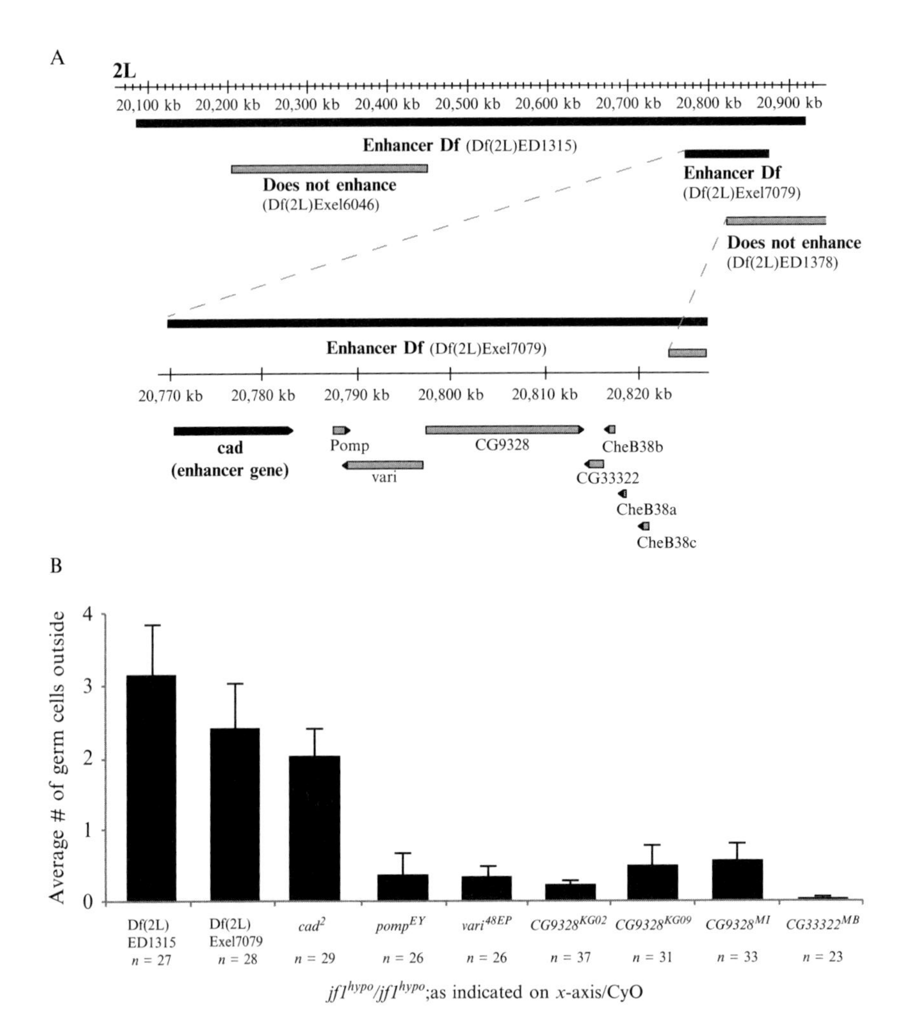

Figure 12.5 Null mutations in *caudal* dominantly enhance germ cell adhesion defects in *jafrac1* mutants. (A) A map of the deficiencies available on 2L between 20,100 and 20,900 kb. Df(2L)Exel6046 and Df(2L)ED1378 did not modify the *jafrac1* phenotype. Df(2L)ED1315 and Df(2L)Exel7079 significantly enhanced germ cell migration defects in *jafrac1* mutants uncovering an 8 gene region within which the enhancer gene must lie. (B) The effect of various alleles (EMS and P-transposon-induced) on germ cell migration in a *jafrac1* sensitized background. One copy of a null mutation in the gene *caudal*, *cad*^{2} (Macdonald & Struhl, 1986), significantly increases the number of germ cells that fail to internalize in a *jafrac1* hypomorphic mutant. Mutations in other genes in this region have no significant effect of germ cell migration. Data are the means ± S.E.

A difficult aspect to using deficiencies as modifiers is identifying the causative modifying gene within the deficiency. Although it is possible to narrow down candidates using overlapping deficiencies, loss-of-function alleles and validated RNAi constructs, there will still be instances where tools are not available to test all putative modifier genes within a deficiency. As the collections of deficiencies and number of mutations in genes improve, however, this should become less and less of a problem. A further possible issue with using deficiencies to identify modifiers is the potential for synthetic effects. For example, if both a suppressor and an enhancer are contained within one deficiency, when assayed in a sensitized background, these modifiers might not be identified because they might cancel each other out. On the flip side, if a gene is duplicated and multiple copies have redundant functions, deficiencies might be the only way to identify the gene as a modifier when all copies are included in the deficiency. For example, if the three peroxiredoxin VI homologs contained within Df(2R)BSC595 are indeed enhancing the *jafrac1* phenotype and have completely overlapping functions, it seems unlikely these genes would have been found using other methods (see also misexpression screens above). Lastly, although there are limitations to using deficiencies in modifier screens, deficiencies can always be used as a quick means of assessing the number of potential target genes for a more extensive modifier screen using other methods (St Johnston, 2002).

5. LIMITATIONS TO DOMINANT MODIFIERS SCREENS

Modifier screens are an unbiased way to uncover components in a pathway. However, there are several reasons why one might not expect to isolate mutants in all critical genes within a pathway using dominant modifier screens.

First, although interacting genes that have a direct role in the pathway can be identified, many mutants may be identified that modify the phenotype for less obvious or direct reasons (Simon et al., 1991; St Johnston, 2002). It is therefore necessary to conduct secondary screens to understand the nature of the genetic interaction. For example, some mutants might influence the phenotype by varying the expression of the allele used to create the sensitized background, whereas others might do so by perturbing unrelated or indirect processes (St Johnston, 2002). *caudal*, the enhancer of *jafrac1* we identified, might be an example of the latter. *caudal* encodes a homeodomain transcription factor that is involved in the formation of the hindgut in embryos (Wu & Lengyel, 1998). Since germ cells are directly apposed to the gut primordium during their internalization, it would not be surprising if perturbations in hindgut formation caused defects in germ cell internalization (Wu & Lengyel, 1998). Indeed, in *caudal* mutants, the hindgut (and presumably germ cells) fails to internalize (Wu & Lengyel, 1998). Therefore, while mutations in *caudal* dominantly enhance *jafrac1* mutants, the two genes are probably not directly acting within the same pathway.

Second, not all components of a pathway are dose-sensitive. As the premise of the screen is that a 50% reduction in the wild-type levels of a critical protein will be sufficient to alter the phenotype, only those proteins that meet this criterion will be found. For example, if a pathway component is in vast excess or only a small percentage of its activity is required for normal function regardless of the background, then a far greater than 50% reduction will be necessary to visibly alter the phenotype. In this case, only dominant negative or constitutive active alleles, which are relatively rare and impossible to uncover using deficiencies, may dominantly modify.

Last, mutations identified in modifier screens may not have phenotypes in the absence of the sensitized background (St Johnston, 2002). Unfortunately, enhancement or suppression of a phenotype of interest does not guarantee that a gene plays an indispensable role in the process being studied.

6. CONCLUDING REMARKS

The use of modifier screens to elucidate components of redox-regulated pathways has not, to our knowledge, been widely employed. Whether this strategy will prove fruitful depends largely on the nature of these pathways and types of modifier screens conducted. If redox signaling pathways are organized around central nodes or regulators in a manner akin to phosphorylation cascades, then modifier screens, particularly suppressor screens, should be a viable unbiased way of identifying novel components of these pathways. However, if these pathways are organized considerably differently, then the utility of such screens might be limited. Here, we have described an enhancer screen to identify components of a redox-regulated cell adhesion and migration pathway in *Drosophila*. As proof-of-principle, we have demonstrated that genetic enhancers of the phenotype could be identified. Only once further screening is concluded, will we know the true utility of this approach, and one hopes in the process novel components of a redox-regulated germ cell migration pathway will be discovered.

ACKNOWLEDGMENTS

We thank Ryan Cinalli, Allison Blum, Michael Murphy, and Brandon Cunningham for their helpful comments. Support is provided by the Canadian Institute of Health Research (T. R. H.), the National Science Foundation (M. G. L.), Howard Hughes Medical Institute (R. L.), and by National Institutes of Health Grant R01HD041900.

REFERENCES

Beinert, N., Werner, M., Dowe, G., Chung, H. R., Jackle, H., & Schafer, U. (2004). Systematic gene targeting on the X chromosome of Drosophila melanogaster. *Chromosoma, 113*, 271–275.

Bellen, H. J., Levis, R. W., He, Y., Carlson, J. W., Evans-Holm, M., Bae, E., et al. (2011). The Drosophila gene disruption project: Progress using transposons with distinctive site specificities. *Genetics, 188*, 731–743.

Bellen, H. J., Levis, R. W., Liao, G., He, Y., Carlson, J. W., Tsang, G., et al. (2004). The BDGP gene disruption project: Single transposon insertions associated with 40% of Drosophila genes. *Genetics, 167*, 761–781.

Blumenstiel, J. P., Noll, A. C., Griffiths, J. A., Perera, A. G., Walton, K. N., Gilliland, W. D., et al. (2009). Identification of EMS-induced mutations in Drosophila melanogaster by whole-genome sequencing. *Genetics, 182*, 25–32.

Bokel, C. (2008). EMS screens: From mutagenesis to screening and mapping. *Methods in Molecular Biology, 420*, 119–138.

Campos-Ortega, J. A., & Hartenstein, V. (1985). *The embryonic development of Drosophila melanogaster.* Berlin Heidelberg, Germany: Springer.

Chouchani, E. T., James, A. M., Fearnley, I. M., Lilley, K. S., & Murphy, M. P. (2011). Proteomic approaches to the characterization of protein thiol modification. *Current Opinion in Chemical Biology, 15*, 120–128.

Cook, R. K., Christensen, S. J., Deal, J. A., Coburn, R. A., Deal, M. E., Gresens, J. M., et al. (2012). The generation of chromosomal deletions to provide extensive coverage and subdivision of the Drosophila melanogaster genome. *Genome Biology, 13*, R21.

DeGennaro, M., Hurd, T. R., Siekhaus, D. E., Biteau, B., Jasper, H., & Lehmann, R. (2011). Peroxiredoxin stabilization of DE-cadherin promotes primordial germ cell adhesion. *Developmental Cell, 20*, 233–243.

Deutschbauer, A. M., Jaramillo, D. F., Proctor, M., Kumm, J., Hillenmeyer, M. E., Davis, R. W., et al. (2005). Mechanisms of haploinsufficiency revealed by genome-wide profiling in yeast. *Genetics, 169*, 1915–1925.

Dietzl, G., Chen, D., Schnorrer, F., Su, K. C., Barinova, Y., Fellner, M., et al. (2007). A genome-wide transgenic RNAi library for conditional gene inactivation in Drosophila. *Nature, 448*, 151–156.

Finkel, T. (2011). Signal transduction by reactive oxygen species. *The Journal of Cell Biology, 194*, 7–15.

Flohe, L. (2010). Changing paradigms in thiology from antioxidant defense toward redox regulation. *Methods in Enzymology, 473*, 1–39.

Gonsalves, F. C., Klein, K., Carson, B. B., Katz, S., Ekas, L. A., Evans, S., et al. (2011). An RNAi-based chemical genetic screen identifies three small-molecule inhibitors of the Wnt/wingless signaling pathway. *Proceedings of the National Academy of Sciences of the United States of America, 108*, 5954–5963.

Graveley, B. R., Brooks, A. N., Carlson, J. W., Duff, M. O., Landolin, J. M., Yang, L., et al. (2011). The developmental transcriptome of Drosophila melanogaster. *Nature, 471*, 473–479.

Gregory, S. L., Shandala, T., O'Keefe, L., Jones, L., Murray, M. J., & Saint, R. (2007). A Drosophila overexpression screen for modifiers of Rho signalling in cytokinesis. *Fly (Austin)*, *1*, 13–22.

Grether, M. E., Abrams, J. M., Agapite, J., White, K., & Steller, H. (1995). The head involution defective gene of Drosophila melanogaster functions in programmed cell death. *Genes & Development*, *9*, 1694–1708.

Harris, T. J., & Tepass, U. (2010). Adherens junctions: From molecules to morphogenesis. *Nature Reviews. Molecular Cell Biology*, *11*, 502–514.

Janssen-Heininger, Y. M., Mossman, B. T., Heintz, N. H., Forman, H. J., Kalyanaraman, B., Finkel, T., et al. (2008). Redox-based regulation of signal transduction: Principles, pitfalls, and promises. *Free Radical Biology & Medicine*, *45*, 1–17.

Jorgensen, E. M., & Mango, S. E. (2002). The art and design of genetic screens: Caenorhabditis elegans. *Nature Reviews. Genetics*, *3*, 356–369.

Kang, S. W., Chae, H. Z., Seo, M. S., Kim, K., Baines, I. C., & Rhee, S. G. (1998). Mammalian peroxiredoxin isoforms can reduce hydrogen peroxide generated in response to growth factors and tumor necrosis factor-alpha. *The Journal of Biological Chemistry*, *273*, 6297–6302.

Karim, F. D., Chang, H. C., Therrien, M., Wassarman, D. A., Laverty, T., & Rubin, G. M. (1996). A screen for genes that function downstream of Ras1 during Drosophila eye development. *Genetics*, *143*, 315–329.

Lawal, H. O., Terrell, A., Lam, H. A., Djapri, C., Jang, J., Hadi, R., et al. (2012). Drosophila modifier screens to identify novel neuropsychiatric drugs including aminergic agents for the possible treatment of Parkinson's disease and depression. *Molecular Psychiatry*.

Lee, K. S., Iijima-Ando, K., Iijima, K., Lee, W. J., Lee, J. H., Yu, K., et al. (2009). JNK/FOXO-mediated neuronal expression of fly homologue of peroxiredoxin II reduces oxidative stress and extends life span. *The Journal of Biological Chemistry*, *284*, 29454–29461.

Lewis, E. B., & Bacher, F. (1968). Methods of feeding ethyl methane sulphonate (EMS) to *Drosophila* males. *Drosophila Information Service*, *43*, 193.

Lindsley, D. L., Sandler, L., Baker, B. S., Carpenter, A. T., Denell, R. E., Hall, J. C., et al. (1972). Segmental aneuploidy and the genetic gross structure of the Drosophila genome. *Genetics*, *71*, 157–184.

Macdonald, P. M., & Struhl, G. (1986). A molecular gradient in early Drosophila embryos and its role in specifying the body pattern. *Nature*, *324*, 537–545.

Malone, J. H., Cho, D. Y., Mattiuzzo, N. R., Artieri, C. G., Jiang, L., Dale, R. K., et al. (2012). Mediation of Drosophila autosomal dosage effects and compensation by network interactions. *Genome Biology*, *13*, r28.

Mathieu, J., Sung, H. H., Pugieux, C., Soetaert, J., & Rorth, P. (2007). A sensitized PiggyBac-based screen for regulators of border cell migration in Drosophila. *Genetics*, *176*, 1579–1590.

McDonald, J. A., & Montell, D. J. (2005). Analysis of cell migration using Drosophila as a model system. *Methods in Molecular Biology*, *294*, 175–202.

Moore, L. A., Broihier, H. T., Van Doren, M., Lunsford, L. B., & Lehmann, R. (1998). Identification of genes controlling germ cell migration and embryonic gonad formation in Drosophila. *Development*, *125*, 667–678.

Ni, J. Q., Markstein, M., Binari, R., Pfeiffer, B., Liu, L. P., Villalta, C., et al. (2008). Vector and parameters for targeted transgenic RNA interference in Drosophila melanogaster. *Nature Methods*, *5*, 49–51.

Ni, J. Q., Zhou, R., Czech, B., Liu, L. P., Holderbaum, L., Yang-Zhou, D., et al. (2011). A genome-scale shRNA resource for transgenic RNAi in Drosophila. *Nature Methods*, *8*, 405–407.

Parks, A. L., Cook, K. R., Belvin, M., Dompe, N. A., Fawcett, R., Huppert, K., et al. (2004). Systematic generation of high-resolution deletion coverage of the Drosophila melanogaster genome. *Nature Genetics*, *36*, 288–292.

Rhee, S. G., Bae, Y. S., Lee, S. R., & Kwon, J. (2000). Hydrogen peroxide: A key messenger that modulates protein phosphorylation through cysteine oxidation. *Science Signaling the Signal Transduction Knowledge Environment*, *2000*, pe1.

Richardson, B. E., & Lehmann, R. (2010). Mechanisms guiding primordial germ cell migration: Strategies from different organisms. *Nature Reviews. Molecular Cell Biology*, *11*, 37–49.

Rodriguez, J., Agudo, M., Van Damme, J., Vandekerckhove, J., & Santaren, J. F. (2000). Polypeptides differentially expressed in imaginal discs define the peroxiredoxin family of genes in Drosophila. *European Journal of Biochemistry*, *267*, 487–497.

Rorth, P. (1996). A modular misexpression screen in Drosophila detecting tissue-specific phenotypes. *Proceedings of the National Academy of Sciences of the United States of America*, *93*, 12418–12422.

Rothwell, W. F., & Sullivan, W. (2007). Fixation of Drosophila embryos. *Cold Spring Harbor Protocols*, *2007*, pdb prot4827.

Rubin, G. M., Chang, H. C., Karim, F., Laverty, T., Michaud, N. R., Morrison, D. K., et al. (1997). Signal transduction downstream from Ras in Drosophila. *Cold Spring Harbor Symposia on Quantitative Biology*, *62*, 347–352.

Simon, M. A., Bowtell, D. D., Dodson, G. S., Laverty, T. R., & Rubin, G. M. (1991). Ras1 and a putative guanine nucleotide exchange factor perform crucial steps in signaling by the sevenless protein tyrosine kinase. *Cell*, *67*, 701–716.

Spradling, A. C., Stern, D., Beaton, A., Rhem, E. J., Laverty, T., Mozden, N., et al. (1999). The Berkeley Drosophila Genome Project gene disruption project: Single P-element insertions mutating 25% of vital Drosophila genes. *Genetics*, *153*, 135–177.

Springer, M., Weissman, J. S., & Kirschner, M. W. (2010). A general lack of compensation for gene dosage in yeast. *Molecular Systems Biology*, *6*, 368.

St Johnston, D. (2002). The art and design of genetic screens: Drosophila melanogaster. *Nature Reviews. Genetics*, *3*, 176–188.

Staudt, N., Molitor, A., Somogyi, K., Mata, J., Curado, S., Eulenberg, K., et al. (2005). Gain-of-function screen for genes that affect Drosophila muscle pattern formation. *PLoS Genetics*, *1*, e55.

Szuplewski, S., Kugler, J. M., Lim, S. F., Verma, P., Chen, Y. W., & Cohen, S. M. (2012). MicroRNA transgene overexpression complements deficiency-based modifier screens in Drosophila. *Genetics*, *190*, 617–626.

Thamsen, M., & Jakob, U. (2011). The redoxome: Proteomic analysis of cellular redox networks. *Current Opinion in Chemical Biology*, *15*, 113–119.

Thibault, S. T., Singer, M. A., Miyazaki, W. Y., Milash, B., Dompe, N. A., Singh, C. M., et al. (2004). A complementary transposon tool kit for Drosophila melanogaster using P and piggyBac. *Nature Genetics*, *36*, 283–287.

Wu, L. H., & Lengyel, J. A. (1998). Role of caudal in hindgut specification and gastrulation suggests homology between Drosophila amnioproctodeal invagination and vertebrate blastopore. *Development*, *125*, 2433–2442.

CHAPTER THIRTEEN

Investigating the Role of Reactive Oxygen Species in Regulating Autophagy

Spencer B. Gibson[*,†,1]
[*]Manitoba Institute of Cell Biology, University of Manitoba, Winnipeg, Manitoba, Canada
[†]Department of Biochemistry and Medical Genetics, Faculty of Medicine, University of Manitoba, Winnipeg, Manitoba, Canada
[1]Corresponding author: e-mail address: gibsonsb@cc.umanitoba.ca

Contents

1. Introduction 218
2. Regulation of Autophagy 218
3. ROS and Autophagy 220
4. Mechanisms for ROS Regulation of Autophagy 221
   4.1 mTOR regulation of ROS-induced autophagy 221
   4.2 Beclin-1 and ROS-induced autophagy 223
   4.3 p53 and autophagy 223
   4.4 ROS and p62 autophagy adaptor 224
   4.5 Oxidation of Atg4 224
   4.6 Repression of ROS-induced autophagy 225
5. Methods for the Detection of Autophagy 225
   5.1 Quantification of GFP-LC3 puncta 226
   5.2 Western blotting of Atg8/LC3 (conversion of LC3-I to LC3-II) 227
   5.3 Quantification of acidic vesicular organelles 227
   5.4 Validation of autophagosome formation by electron microscopy 228
   5.5 Autophagy detection *in vivo* 228
6. Consideration When Using Oxidative Stress and Detecting ROS Under Autophagy Conditions 229
7. Conclusions 231
Acknowledgment 231
References 232

Abstract

Autophagy is an intracellular lysosomal degradation process induced under stress conditions. Reactive oxygen species (ROS) regulate autophagy implicated in cell survival, death, development, and many human diseases. This could be through generation of ROS from intracellular compartments such as the mitochondria or an external source

Methods in Enzymology, Volume 528
ISSN 0076-6879
http://dx.doi.org/10.1016/B978-0-12-405881-1.00013-6

such as oxidative stress. Various methods have been developed for the detection of autophagy; however, the implementation of these methods and the interpretation of results often differ. In this chapter, we summarize the current understanding of autophagy and ROS regulation of autophagy. Methods available for detecting autophagy under ROS conditions are described and considerations that need to be addressed when designing experimental protocols discussed.

1. INTRODUCTION

Autophagy is referred to as a "self-eating" process involving double membrane vacuoles that degrade cytoplasmic structures (Mizushima & Levine, 2010). There are numerous subtypes of autophagy including macroautophagy, microautophagy, chaperone-mediated autophagy, aggrephagy, the yeast cytoplasm-to-vacuole targeting pathway, endoplasmic reticulum (ER)-phagy, mitophagy, pexophagy, and xenophagy (Klionsky et al., 2012). The most common is macroautophagy involving genetically regulated steps leading to the formation of double-membraned autophagosomes that engulf organelles and protein aggregates. Autophagosomes then fuse with lysosomes to form autolysosomes that lead to the degradation of these structures (Levine & Kroemer, 2008; Mizushima & Levine, 2010) (Fig. 13.1). This allows the cell to eliminate damaged organelles and provide essential nutrients under stressful conditions.

Autophagy is often activated by stress conditions. Under starvation conditions, autophagy maintains cell survival: cytoplasmic materials are degraded into amino acids and fatty acids, which are used for protein synthesis or energy (Levine & Kroemer, 2008; Mizushima & Levine, 2010). By contrast, when autophagy is prolonged by other stress conditions, the essential components for cell survival are eliminated, causing cellular collapse and death (Azad, Chen, & Gibson, 2009).

2. REGULATION OF AUTOPHAGY

Autophagy is the cellular process defined by "self-digestion" (Levine, 2005; Levine & Klionsky, 2004). Genetic screening has identified over 35 ATG (autophagy-related) genes required for autophagy (Muller & Reichert, 2011). Several upstream regulators of autophagy have been characterized (Dou et al., 2010; Ravikumar et al., 2010; Szyniarowski et al.,

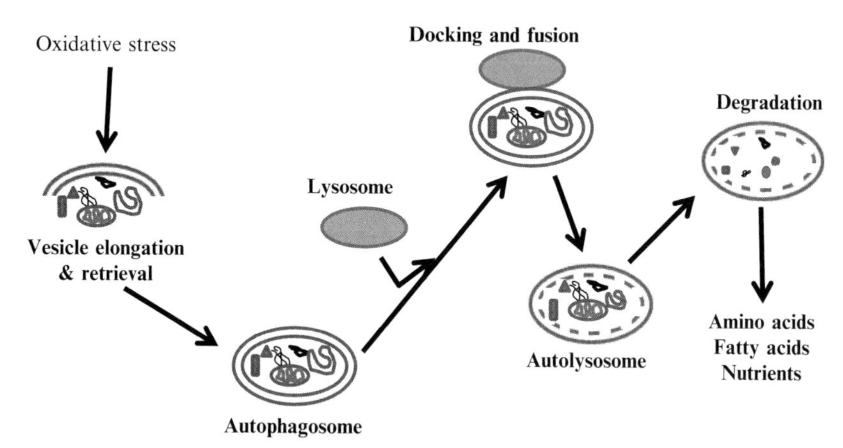

Figure 13.1 Model for autophagy and autophagic cell death. Under oxidative stress, autophagy is induced. The autophagy induction process starts with isolation membrane followed by vesicle nucleation, vesicle elongation and retrivial, and the formation of the characterized double-membraned structures, autophagosomes. Then autophagosomes fuse with lysosomes to form autolysosomes. The enclosed cytoplasmic materials and the inner membrane are degraded by the acidic proteases in the autolysosomes. So autolysosomes belong to acidic vascular organelles (AVOs). The degradation process produces amino acids and fatty acids, which can be used for protein synthesis or can be oxidized by the mitochondrial electron transport chain (mETC) to produce ATP. (See Color Insert.)

2011). The class III PI3 kinase complex including Beclin-1 is localized to *trans*-Golgi network, which controls the generation of preautophagosome structures (Kihara, Kabeya, Ohsumi, & Yoshimori, 2001; Mizushima, 2005; Periyasamy-Thandavan, Jiang, Schoenlein, & Dong, 2009). The mammalian target of rapamycin (mTOR), a nutrient-sensing kinase complex, inhibits autophagy during nutrient-rich conditions by inhibiting the formation of the class III PI3K/Beclin-1 complex (Pattingre, Espert, Biard-Piechaczyk, & Codogno, 2008). Activators of mTOR pathway (class I PI3 kinase and Akt) suppress autophagy whereas inhibitors of this pathway such as PTEN induce autophagy (Azad et al., 2009) (Fig. 13.2). Downstream of the class III PI3K/Beclin-1 complex, the Atg5–Atg12 covalent protein complex, and Atg8–phosphoethanolamine conjugates are required components of the autophagosome membrane. The Atg4 cysteine protease cleaves Atg8 at the C-terminus to facilitate its lipidation leading to autophagosome formation (Azad et al., 2009; Scherz-Shouval, 2007). Autophagosomes could also form independent of Beclin-1 mediated through Atg7 and Atg5. Reactive oxygen

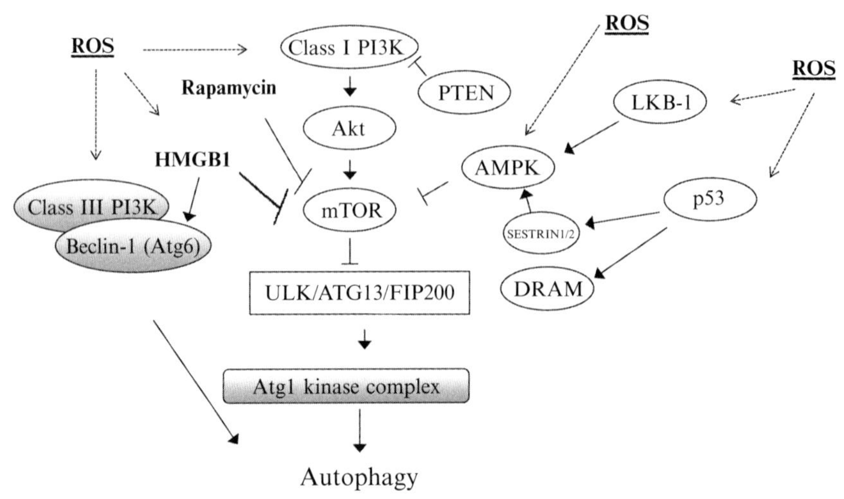

Figure 13.2 ROS regulation of autophagy. Following external stimuli such as reactive oxygen species (ROS), mTOR is inhibited either through blockage of AKT activation or activation of LKB1/AMPK leading to induction of autophagy. The mTOR-regulated Atg1 complex is involved in autophagy initiation through regulation of ULK/ATG13/FIP200 complex and the class III PI3 kinase complex including Beclin-1/Atg6 is required for generation of preautophagosome structures. Activation of p53 also increases DRAM and SESTRIN1/2 expression that increases autophagy. HMGB1 inhibits mTOR or induces Beclin-1 to increase autophagy. ROS also activates AMPK directly or indirectly through LKB-1 leading to inhibition of mTOR and increased autophagy. Rapamycin induces autophagy through inhibition of mTOR.

species (ROS) induction has been implicated in autophagy. This chapter reviews our knowledge of ROS in autophagy and how it is used to investigate ROS role in autophagy.

3. ROS AND AUTOPHAGY

Under normal physiological conditions, moderate levels of ROS can serve as a signal in various signaling pathways including autophagy (Finkel, 2003; Hamanaka & Chandel, 2010; Scherz-Shouval & Elazar, 2011). It is known that starvation can induce the production of ROS and this increase could lead to autophagy. Through studying the autophagy induced by starvation, mitochondria electron transport chain (mETC) inhibitors, and exogenous H_2O_2, we further delineated that O_2^- was selectively induced by starvation by glucose withdrawal whereas starvation of amino acids and serum induced O_2^- and H_2O_2. However, the autophagy induced by mETC inhibitors combining with the SOD inhibitor 2-methoxyestradiol (2-ME) is through increasing O_2^- levels and lowered H_2O_2 levels. The exogenous H_2O_2-induced autophagy also went through increased intracellular O_2^- but not the intracellular H_2O_2 (Chen, Azad, & Gibson, 2009). Other groups showed that the starvation-induced autophagy was through higher H_2O_2 levels (Scherz-Shouval, 2007). In addition, H_2O_2 induces autophagy in cells through both a Beclin-1-dependent and -independent mechanisms (Seo et al., 2011). Interestingly, using mitochondrial toxins that inhibit the mETC, and increased ROS, autophagy is induced in transformed HEK293, U87, and HeLa cancer cells but not in the normal mouse astrocytes (Chen & Gibson, 2008; Chen, McMillan-Ward, Kong, Israels, & Gibson, 2007). These results suggest that ROS regulates autophagy depending on the cell type and stress conditions.

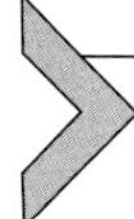

4. MECHANISMS FOR ROS REGULATION OF AUTOPHAGY

There are several signaling pathways that regulate autophagy in mammalian cells (Ravikumar et al., 2010). The classical pathway involves the inhibition of the serine/threonine protein kinase, mTOR pathway (Levin, 2008; Levine & Klionsky, 2004). In addition, there are various other pathways and small molecules regulating ROS-induced autophagy through mTOR-dependent and -independent mechanisms (Fig. 13.2).

4.1. mTOR regulation of ROS-induced autophagy

mTOR consists of two multiple protein complexes called mTOR complex 1 (mTORC1) and mTOR complex 2 (mTORC2). mTORC1 detects changes of nutrition and energy levels within a cell, whereas mTORC2 is regulated by ribosome maturation and growth factor signals (Suzuki & Inoki, 2011). The mTOR complex characterized in regulating autophagy regulation is mTORC1, which consists of the mTOR catalytic subunit, raptor (regulatory-associated protein of mTOR), GβL (G protein β-subunit-like protein; also known asmLST8), and PRAS40 (proline-rich Akt substrate of 40 kDa). Rapamycin (sirolimus), the specific inhibitor of mTORC1 activity, induces autophagy in many cell lines (Boland et al., 2008; Noda & Ohsumi, 1998; Ravikumar et al., 2004; Rubinsztein, Gestwicki, Murphy, & Klionsky, 2007) and inhibits the phosphorylation of two downstream effectors of mTORC1, ribosomal protein S6 kinase-1

(S6K1, also known as p70S6K), and translation initiation factor 4E-binding protein-1 (4E-BP1). Starvation induces autophagy through stabilizing the raptor–mTOR complex and further inhibiting mTORC1 (Kim et al., 2002). Downstream regulators of mTORC1 have been identified in mammalian autophagy in response to starvation. ULK1 (Unc51-like kinase, hATG1) is negatively regulated by mTORC1 and induces autophagy in response to starvation. Atg13 is associated with ULK1 or its homolog ULK2 to form the ULK1/2–Atg13–FIP200 stable complex. This complex plays key roles in signaling to the autophagic machinery downstream of mTOR. Conversely, ULK1 also inhibits the kinase activity of mTORC1 by binding and phosphorylating the raptor (Ganley et al., 2009; Hosokawa et al., 2009; Jung, Seo, Otto, & Kim, 2011; Jung et al., 2009). Thus mTORC1 inhibits or activates autophagy through interaction with or dissociation from the ULK1/2–Atg13–FIP200 complex, depending on the nutrient-rich or starvation conditions. Hence, the ULK1–Atg13–FIP200 complex integrates the autophagy signals downstream of mTORC1 in response to nutrition starvation. Increasing evidence strongly suggests that starvation-induced autophagy is mediated by ROS generation and downregulation of the mTOR pathway.

PI3K pathway is the major upstream signaling cascade controlling mTORC1. The binding of growth factors or insulin to cell surface receptors activates the class Ia PI3K. Activated PI3K then converts the plasma membrane lipid phosphatidylinositol-4,5-bisphosphate to phosphatidylinositol-3,4,5-trisphosphate, which then recruits pleckstrin homology domain proteins, such as Akt to the plasma membrane (Cantley, 2002). Akt activation inhibits the rictor–mTOR complex, thereby decreasing autophagy (Datta et al., 1997). Inhibiting PI3K activity with specific inhibitors (e.g., wortmannin) can downregulate AKT activation and increases autophagy (Petiot, Ogier-Denis, Blommaart, Meijer, & Codogno, 2000). mTORC1 can also be a sensor of changes in the cellular energy state via AMP-activated kinase (AMPK) (Meijer & Codogno, 2006). AMPK senses the fluctuation of the intracellular ATP/AMP ratio. We demonstrated that starvation induced AMPK activation and autophagy while Akt activation decreased. This activation of AMPK is regulated by ROS as blocking ROS formation reduces AMPK activation and starvation-induced autophagy (Li, Chen, & Gibson, 2013). Further, several other kinases have been reported to regulate autophagy through mTOR pathway. IκB kinase induces autophagy via AMPK activation and mTOR inhibition in an NF-κB-independent manner (Criollo et al., 2010).

4.2. Beclin-1 and ROS-induced autophagy

Beclin-1 is a BH3-only protein involved in the initiation of autophagy. The Beclin-1 expression level is upregulated under oxidative stress, suggesting ROS is regulating its gene expression (Chen, McMillan-Ward, Kong, Israels, & Gibson, 2008). One study demonstrated that Beclin-1 is inhibited by tumor suppressor BRCA1. BRCA1 protected cells from autophagic cell death by blocking the H_2O_2-induced upregulation of Beclin-1 (Fan, Meng, Saha, Sarkar, & Rosen, 2009). It is important to note that the level of Beclin-1 could be the biomarker to decide whether the autophagy is activated. One study demonstrated that, when Beclin-1 levels were high through sustaining MEK/ERK activation, complete disassembly of both mTORC1 and mTORC2 complexes occurs. However, autophagy cytoprotection occurred when moderate Beclin-1 levels are observed (Wang et al., 2009). Other autophagy genes such as p62/SQSTM1 and UVRAG are upregulated by MEK-/ERK-mediated activation of transcription factor TFEB (Settembre et al., 2011). Further studies are necessary to elucidate the role of ROS in the regulation of Beclin-1 and other autophagy genes.

High-mobility group box 1 (HMGB1) is a chromatin-associated nuclear protein that functions as an extracellular signaling molecule during inflammation, cell differentiation, cell migration, and tumor metastasis (Lotze & Tracey, 2005; Tang et al., 2010). HMGB1 is a Beclin 1-binding protein important in sustaining autophagy under oxidative stress. Inhibition of HMGB1 release or loss of HMGB1 decreases the number of autophagolysosomes and autophagic flux following oxidative stress (Kang, Tang, Lotze, & Zeh, 2011). The underlying mechanism is that HMGB1 competes with Bcl-2 for interaction with Beclin-1 and allows Beclin-1 to initiate autophagy. Interestingly, HMGB1 also regulates the PI3K/Akt/mTORC1 pathway. Increased expression of HMGB1 decreased the phosphorylation levels of Akt and p70S6K whereas knocking down HMGB1 caused the disassociation of raptor from mTOR and inhibited autophagy. Thus, endogenous HMGB1 is a proautophagic protein that is regulated by ROS (Kang, Livesey, Zeh, Loze, & Tang, 2010; Kang et al., 2011; Tang et al., 2010).

4.3. p53 and autophagy

p53 induces autophagy through the transcriptional control of mTOR pathway. When oxidative stress occurs, basal p53 expression activates multiple detoxifying pathways to eliminate oxidative stress. p53 induces the expression of a range of antioxidant genes such as GPX1, MnSOD, ALDH4, and

TPP53INP1 that lead to increased antioxidant activity (Bensaad et al., 2006; Budanov, Lee, & Karin, 2010; Hu et al., 2010; Pani & Galeotti, 2011; Suzuki et al., 2010). In addition, *sestrin1* and *sestrin2* have been identified as a link between p53 activation and mTORC1 activity (Budanov & Karin, 2008; Budanov et al., 2010). p53 exerts the antioxidant effect through increased expression of Sestrin in response to oxidative stress, which leads to inhibition of mTORC1 activity and increased autophagy. The Sestrins belongs to a family of stress-responsive proteins involved in the regulation of ROS (Budanov, 2011). These proteins inhibit mTOR1 activity by binding with mTOR pathway suppressors AMPK, TSC1, and TSC2 (Budanov & Karin, 2008). In contrast, cytoplasmic p53 inhibits autophagy and is degraded under autophagy conditions (Sui et al., 2011). This degradation is regulated by mTOR downstream signaling of S6K and MDM2, the major ubiquitin ligase that controls p53 stability (Lai et al., 2010). Taken together, p53 signaling regulates autophagy in response to multiple cellular stresses including oxidative stress.

4.4. ROS and p62 autophagy adaptor

Autophagic adaptors, p62 and NBR1, play important roles in initiating autophagy (Rusten & Stenmark, 2010). The p62/SQSTM1 (sequestosome 1) protein acts as a cargo receptor for ubiquitinated targets to transport them to autophagosomes for degradation. ROS increases p62 gene expression mediated by NF-E2-related factor 2 transcription factor (Jain et al., 2010). Conversely, p62 and NBR1 are degraded by autophagy despite their roles as cargo receptors for degradation of ubiquitinated substrates. This provides control of the extent of protein degradation that occurs under autophagy (Johansen & Lamark, 2011).

4.5. Oxidation of Atg4

ROS can regulate autophagy through controlling the activity of the cysteine protease Atg4 (Chen et al., 2009; Scherz-Shouval, 2007). Under nutrient starvation, there is an increase in H_2O_2. Atg4 is a direct target for oxidation by H_2O_2, specifying a cysteine residue located near the active site. Atg4 regulates the reversible conjugation of Atg8 (LC3 in mammals) to the autophagosomal membrane (Liu & Lenardo, 2007). Oxidative inactivation of Atg4 promotes lipidation of Atg8 and an increase in autophagy (Chung, Suttangkakul, & Vierstra, 2009). It remains unclear whether ROS-mediated inhibition of Atg4 is essential for autophagy.

4.6. Repression of ROS-induced autophagy

The NF-κB transcription factor could repress ROS-mediated autophagy. One study showed that tumor necrosis factor α-induced autophagy through ROS production in the absence of NF-κB activation but with NF-κB activation autophagy is repressed (Djavaheri-Mergny et al., 2007). The c subunit (ATP6L) of vacuolar H(+)-ATPase exerts its protective role against H_2O_2-induced cytotoxicity through inhibiting the MEK/ERK signaling pathway and inhibition of autophagy (Byun et al., 2007). In addition, antiapoptotic protein, Bcl-2 reduces stress caused by ROS through decreasing ROS generation (Kane et al., 1993). In addition, Bcl-2 directly associates with Beclin-1-suppressing autophagy (Luo & Rubinsztein, 2010; Pattingre et al., 2005). Bcl-2 also contributes to the regulation of the type III PI3K and autophagy through attenuating ROS production. Hence, Beclin-1 regulates autophagy through integration of signals from many pathways in response to ROS (Lipinski et al., 2010). Therefore, ROS can regulate autophagy, depending on the severity of oxidative stress, suggesting regulation of intracellular redox status that controls autophagy (Yang, Wu, Tashino, Onodera, & Ikejima, 2008). Considering these signaling factors will be critical in understanding how autophagy is regulated by ROS.

5. METHODS FOR THE DETECTION OF AUTOPHAGY

Understanding the role of ROS in autophagy requires knowledge of methods to detect autophagy and the limitations of these techniques. There are many excellent reviews that comprehensively summarized the various methods for autophagy detection (Klionsky et al., 2012; Mizushima & Levine, 2010); however, few reviews have addressed how to apply these methods to the role of ROS and oxidative stress. The important methods that need to be used to study ROS in autophagy are presented and the context of using these methods to study ROS involvement discussed. This will hopefully guide researchers studying ROS and autophagy to achieve reliable data, and to provide potential solutions for troubleshooting failed experiments.

As autophagy is characterized by the formation of double-membraned autophagosomes (Fig. 13.1), the detection of autophagosomes is the main method to evaluate the induction of autophagy. Another way to detect autophagy is through the formation of autolysosomes, which are formed by the fusion of autophagosomes and lysosomes. Listed below are some of the most common methods to detect autophagy.

5.1. Quantification of GFP-LC3 puncta

The main method for autophagy detection is the detection and quantification of GFP-LC3 puncta by fluorescence microscopy. Under normal conditions, LC3 (microtubule-associated protein-1 light chain 3, a mammalian homolog of yeast Atg8) is localized in the cytoplasm, whereas during autophagy, LC3 is recruited to autophagosome membranes. In this method, cDNA vectors containing GFP alone or GFP-LC3 are transfected into cells using standard transfection techniques. Depending on the cell type, infection of cell with viral constructs could also be used as an alternative method. The transfected cells are then treated with a stress for a period of time (from several hours to several days). In the case of oxidative stress, H_2O_2 or a superoxide dismutase (SOD) inhibitor can be used. It is important to determine the level of intracellular H_2O_2 following treatment to ensure cells have increased ROS levels. The cells are visualized under a fluorescent microscope and the amount of cells with GFP-LC3 puncta (green dots) is quantified. GFP-LC3 protein aggregates could form under overexpression conditions. This could lead to large clump structures that are not autophagosomes. To avoid this, stable transfection of GFP-LC3 could be used as an alternative approach. This illustrates the need to use multiple techniques to detect autophagy and not rely on just one method.

To quantify the GFP-LC3 puncta in control and treated cells, there are two approaches: the first approach is to quantify the amount of cells containing GFP-LC3 puncta (number of cells with green dots/total number of transfected cells) (Klionsky et al., 2012; Mizushima & Levine, 2010). The second approach is to quantify the number of GFP-LC3 puncta in an individual cell (Chen et al., 2008; Klionsky et al., 2012). This means counting the number of green dots in each of approximately 200 cells per condition and calculating the average number of green dots per transfected cell. This procedure is a time-consuming process and errors could be due to variations in transfection efficiency and counting styles. We recommend transfecting cells at 50–70% confluency without antibiotics in the medium. At least 200 green fluorescent cells should be counted for each sample to obtain statistically significant results and be repeated by another researcher to eliminate counting bias.

It is often detected that green puncta are present in GFP-LC3 "control" cells, especially when cells are incubated for more than 24 h after treatment. This is not surprising as there is a constant level of basal autophagy occurring in cells and it is the differences between control and treatment conditions that need to be monitored. Over time in tissue culture, cells use nutrients to a point where starvation condition could exist thereby inducing an autophagy response. Determining cell destiny and time under tissue culture condition is important to control for basal autophagy. Measuring ROS levels over time in control conditions is also important to eliminate any effects of changes in autophagy due to tissue culture conditions.

5.2. Western blotting of Atg8/LC3 (conversion of LC3-I to LC3-II)

The second method for detecting autophagy is to determine the level of LC3-II protein. During autophagy, the cytosolic form of LC3 (LC3-I, 18 kDa) is lipidated and translocates to autophagosome membranes. The differences between LC3-I and LC3-II can be separated by western blotting. Unfortunately, LC3-I protein is not always detectable using the currently available antibodies (Mizushima & Yoshimori, 2007). Using different antibodies could give better results depending on conditions. Further, the relatively low-molecular weight of LC3 means that a higher percentage of SDS-PAGE gels needs to be considered. Another issue to be considered for the detection of LC3 is autophagosome trafficking. This process involved autophagosome fusing with lysosome to form autolysosome where components from autophagosome including LC3-II will be cleaved or degraded. When the clearance of autophagosome components in the autolysosomes is blocked, LC3-II protein can be detected. Using standard Western blotting for LC3-II, steady-state autophagy can be measured (Klionsky et al., 2008). However, when autophagic flux needs to be determined, an additional control is required: a lysosomal inhibitor or an inhibitor of autophagosome/ lysosome fusion. This is important because LC3-II is degraded during autophagy, so an accumulation of LC3-II following blockage of autolysosome formation could indicate an increase in autophagy flux. In case of measuring autophagy following oxidative stress, the measurement of autophagy flux is required. Some lysosomal inhibitors that could be used in autophagy studies include NH_4Cl, chloroquine, bafilomycin A1, E64d, and pepstatin A (Chen et al., 2009; Mizushima & Yoshimori, 2007).

5.3. Quantification of acidic vesicular organelles

The formation of autolysosomes can be determined by a number of different methods (Chen & Gibson, 2008; Chen et al., 2007, 2009; Mizushima, 2005).

The pH-sensitive dye, acridine orange (AO), can be used to detect acidic vesicular organelle (AVO) formation in live cells. AO is a cell-permeable fluorescent dye that, when it enters acidic compartments, such as lysosomes and autolysosomes, emits red fluorescence. This can be measured by flow cytometry/fluorescence-activated cell sorting or under fluorescent microscope (Chen & Gibson, 2008; Chen et al., 2007, 2009; Mizushima, 2005). Another stain called monodansylcadaverine (MDC) can also be used to detect AVO formation (Klionsky et al., 2012) that emits a blue fluorescence under a microscope. One consideration that needs to be appreciated is that detection of AVO is not a direct measure of autolysosomes but correlates increased AVO with increased autophagosomes indicating the formation of autolysosomes. Finally, the AVOs can also be confirmed by immunostaining for LAMP1 and LAMP2, which are membrane-bound proteins found in lysosomes. An increase in LAMP1 or LAMP2 punctate staining in cells indicates an increase in AVO and is another indicator for formation of autolysosomes (Eskelinen et al., 2002). These methods need to be correlated with autophagosome detection as the detection of other AVOs may lead to inaccurate conclusions.

5.4. Validation of autophagosome formation by electron microscopy

Measurement of GFP-LC3 puncta and LC3-II protein is only an indirect measure of autophagosome formation. The absolute validation of autophagosome formation requires direct detection by using electron microscopy (EM). Although this technique is labor intensive for frequent or repeated experiments, it is required when a stress is first investigated to determine whether autophagy occurred. To identify true double-membraned autophagosomes as other types of vacuole formation can also be detected by EM, multiple EM pictures showing intracellular structure such as the nucleus must be taken to ensure double-membraned autophagosomes are detected correctly (Klionsky et al., 2008). EM pictures should also be taken over different times to ensure detection of autophagosomes. One problem is that artifacts may be easily generated during fixation and staining processes. Thus, repeating the experiments is also needed.

5.5. Autophagy detection *in vivo*

The above-mentioned methods are based on studies using cell cultures. As the *in vivo* studies of autophagy link autophagy to human diseases, it is important to discuss the methods for *in vivo* studies of autophagy. Generally, the methods used for cell cultures can also be used for mammalian tissues *in vivo* studies but some differences exist. Analysis of autophagy using transgenic mice that expressed GFP-LC3 autophagic marker revealed GFP-LC3 puncta under a fluorescent microscope and by Western blotting LC3II protein levels following starvation (Mizushima, 2009). Interestingly, in some tissues such as thymic epithelial cells and lens epithelial cells, basal autophagy was detected without starvation. When the lysosomal inhibitor chloroquine or bafilomycin A1 was concurrently administrated with the autophagy inducer rapamycin or exogenous H_2O_2 in GFP-LC3 transgenic mice, the increased number of GFP-LC3 puncta per cardiac myocyte cell was detected indicating autophagy flux. Using fixed tissue, staining for LC3 puncta structures was detected in both normal and cancer tissues. Similar results were obtained using frozen tissue through western blotting for LC3-II.

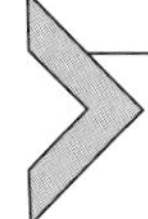

6. CONSIDERATION WHEN USING OXIDATIVE STRESS AND DETECTING ROS UNDER AUTOPHAGY CONDITIONS

The use of oxidative stress to induce autophagy requires the use of multiple autophagy detection techniques but also requires the use of a chemical inhibitor of autophagy such as 3-methyladenine (3-MA) or wortmannin and/or to knockdown autophagy genes (e.g., *beclin-1/atg-6*, *atg-5*, and *atg-7*) using siRNAs or shRNAs. This will distinguish the observed autophagic characteristics (GFP-LC3 puncta, LC3-II protein, AVO formation, etc.) from nonautophagic characteristics (Klionsky, Cuervo, & Seglen, 2007).

There are many chemical inhibitors of autophagy such as 3-MA, wortmannin, NH_4Cl, chloroquine, bafilomycin A1, E64d, and pepstatin A (Chen & Gibson, 2008; Chen et al., 2007, 2008; Mizushima & Yoshimori, 2007). 3-MA and chloroquine are the most frequently used, where 3-MA blocks the formation of autophagosomes by inhibiting the class III PI3K. Chloroquine blocks the formation of autolysosomes and the activity of lysosomes in cells. However, this inhibitor is known to have other effects, such as blocking ATP production and indirectly increasing ROS levels, so they should be used in combination with siRNA against autophagy genes (described below) to confirm results.

As the available chemical inhibitors for autophagy are "unspecific" in that they have additional known or unknown effects, it is better to study the relationships between autophagy and cell death by manipulating the expression of specific autophagy genes. In a particular context, if knocking down or knocking out autophagy genes decreases cell survival whereas overexpressing autophagy genes increases cell survival, then autophagy contributes to cell survival. The role of autophagy in cell survival and death has been demonstrated in many studies where specific autophagy genes were knocked down by siRNA or knocked out altogether (Chen & Gibson, 2008; Chen et al., 2007, 2008; Klionsky et al., 2007; Levine & Yuan, 2005; Mizushima & Yoshimori, 2007). Autophagy genes *atg-5*, *atg-6/beclin-1*, and *atg-7* are commonly knocked down by siRNAs, in cells to block autophagy. Indeed, we found that these genes effectively block autophagy induced by exogenous H_2O_2 or 2-ME (Chen et al., 2008). Pyo et al. (2008) showed that exogenous H_2O_2-induced autophagy was reduced in *atg-5*-deficient mouse embryo fibroblasts (*atg-5*$^{-/-}$ MEFs) as well as in the murine hippocampal neuronal cell line HT22 (Pyo et al., 2008). We have also demonstrated that siRNAs against *beclin-1* and *atg-5* inhibited electron transport chain inhibitors rotenone- or TTFA-induced autophagy mediated by ROS (Chen et al., 2007).

An important control in determining the role ROS plays in autophagy is the use of ROS scavengers. Two commonly used scavengers, Tiron and *N*-acetyl-L-cysteine, are added to cells prior to autophagy stimulation and effects observed (Chen et al., 2008, 2009). In addition, overexpression of SOD protein is also effective at reduced ROS levels and effect on autophagy determined. Conversely, inhibitors of SOD could be used to induce ROS and study autophagy (Chen et al., 2008, 2009). These experiments are critical to determine whether ROS is involved in autophagy and should be included in any experimental design where ROS is investigated.

Timing for detection of autophagy following oxidative stress can be critical: When AVOs were detected with AO or MDC, weak signals were usually observed. This could be attributed to delayed formation of autolysosomes compared to autophagosomes. Indeed, autophagosome formation is increased at times when there is little or no autolysosome formation (Chen, Azad, & Gibson, 2010). In addition, ROS levels could be low at times when autophagosome and autolysosomes are formed. This is dependent on the type of autophagy stimulation. For example, direct increase in ROS by H_2O_2 would lead to high levels of ROS over prolonged times whereas using starvation will lead to transient increase in ROS levels (Chen et al., 2009). It is very important that data are obtained from different methods over time courses, following autophagy induction. This will prevent misinterpretation of data due to events happening at different times.

The source of ROS in regulating autophagy is also important. As mitochondria are the main ROS production sites in mammalian cells, it has been proposed that ROS produced within mitochondria may account for most of the ROS induction of autophagy. When mitochondrial DNA is eliminated from cells thereby blocking oxidative phosphorylation and mitochondrial ROS production, the ability of cells to induce autophagy following starvation was impaired (Li, Chen, & Gibson, 2013). Further, ROS within mitochondria can be induced by using a mitochondrial-targeted photosensitizer, mitochondrial KillerRed (mtKR). Using mtKR resulted in the loss of membrane potential and the subsequent activation of autophagy (Wang, Nartiss, Steipe, McQuibban, & Kim, 2012). ER stress produces ROS that leads to autophagy, and blocking the ER stress response signaling blocks autophagy (Park et al., 2008). There is also evidence of cross talk between the ER and mitochondria during autophagy (Hamed et al., 2010). Thus, the source of ROS dependent upon the autophagy stimuli needs to be considered to understand the regulatory pathways involved.

7. CONCLUSIONS

As ROS regulates many aspects of autophagy, the study of ROS in autophagy will become increasingly important in many fields of biological sciences. Various methods for autophagy detection have been developed. However, there is no single method that is specific to autophagy. There needs to be more than one method applied to detect autophagy under oxidative stress or for the role of ROS in autophagy regulation. As ROS and autophagy are dynamic cellular signals that change over time, it is also essential that time courses and source of ROS are evaluated to ensure that cellular events are not missed due to the time points or the localization of ROS production. Thus, the use and verification of these autophagy methods in the context of the redox environment of the cell will be extremely important to understand autophagy and maintenance of the balance of ROS levels in biological systems.

ACKNOWLEDGMENT

SBG is a Manitoba Research Chair supported by MHRC.

REFERENCES

Azad, M. B., Chen, Y., & Gibson, S. B. (2009). Regulation of autophagy by reactive oxygen species (ROS): Implications for cancer progression and treatment. *Antioxidants & Redox Signaling*, *11*(4), 777–790.

Bensaad, K., Tsuruta, A., Selak, M. A., Vidal, M. N., Nakano, K., Bartrons, R., et al. (2006). TIGAR, a p53-inducible regulator of glycolysis and apoptosis. *Cell*, *126*(1), 107–120.

Boland, B., Kumar, A., Lee, S., Platt, F. M., Wegiel, J., Yu, W. H., et al. (2008). Autophagy induction and autophagosome clearance in neurons: Relationship to autophagic pathology in Alzheimer's disease. *The Journal of Neuroscience*, *28*(27), 6926–6937.

Budanov, A. V. (2011). Stress-responsive sestrins link p53 with redox regulation and mammalian target of rapamycin signaling. *Antioxidants & Redox Signaling*, *15*(6), 1679–1690.

Budanov, A. V., & Karin, M. (2008). p53 target genes sestrin1 and sestrin2 connect genotoxic stress and mTOR signaling. *Cell*, *134*(3), 451–460.

Budanov, A. V., Lee, J. H., & Karin, M. (2010). Stressin' Sestrins take an aging fight. *EMBO Molecular Medicine*, *2*(10), 388–400.

Byun, Y. J., Lee, S. B., Kim, D. J., Lee, H. O., Son, M. J., Yang, C. W., et al. (2007). Protective effects of vacuolar H+-ATPase c on hydrogen peroxide-induced cell death in C6 glioma cells. *Neuroscience Letters*, *425*(3), 183–187.

Cantley, L. C. (2002). The phosphoinositide 3-kinase pathway [Review]. *Science*, *296*(5573), 1655–1657.

Chen, Y., Azad, M. B., & Gibson, S. B. (2009). Superoxide is the major reactive oxygen species regulating autophagy. *Cell Death and Differentiation*, *16*(7), 1040–1052.

Chen, Y., Azad, M. B., & Gibson, S. B. (2010). Methods for detecting autophagy and determining autophagy-induced cell death. *Canadian Journal of Physiology and Pharmacology*, *88*(3), 285–295.

Chen, Y., & Gibson, S. B. (2008). Is mitochondrial generation of reactive oxygen species a trigger for autophagy? *Autophagy*, *4*(2), 246–248.

Chen, Y., McMillan-Ward, E., Kong, J., Israels, S. J., & Gibson, S. B. (2007). Mitochondrial electron-transport-chain inhibitors of complexes I and II induce autophagic cell death mediated by reactive oxygen species. *Journal of Cell Science*, *120*, 4155–4166.

Chen, Y., McMillan-Ward, E., Kong, J., Israels, S. J., & Gibson, S. B. (2008). Oxidative stress induces autophagic cell death independent of apoptosis in transformed and cancer cells. *Cell Death and Differentiation*, *15*(1), 171–182.

Chung, T., Suttangkakul, A., & Vierstra, R. D. (2009). The ATG autophagic conjugation system in maize: ATG transcripts and abundance of the ATG8-lipid adduct are regulated by development and nutrient availability. *Plant Physiology*, *149*(1), 220–234.

Criollo, A., Senovilla, L., Authier, H., Maiuri, M. C., Morselli, E., Vitale, I., et al. (2010). The IKK complex contributes to the induction of autophagy. *The EMBO Journal*, *29*(3), 619–631.

Datta, S. R., Dudek, H., Tao, X., Masters, S., Fu, H., Gotoh, Y., et al. (1997). Akt phosphorylation of BAD couples survival signals to the cell-intrinsic death machinery. *Cell*, *91*(2), 231–241.

Djavaheri-Mergny, M., Amelotti, M., Mathieu, J., Besancon, F., Bauvy, C., & Codogno, P. (2007). Regulation of autophagy by NFkappaB transcription factor and reactives oxygen species. *Autophagy*, *3*(4), 390–392.

Dou, Z., Chattopadhyay, M., Pan, J. A., Guerriero, J. L., Jiang, Y. P., Ballou, L. M., et al. (2010). The class IA phosphatidylinositol 3-kinase p110-beta subunit is a positive regulator of autophagy. *The Journal of Cell Biology*, *191*(4), 827–843.

Eskelinen, E. L., Illert, A. L., Tanaka, Y., Schwarzmann, G., Blanz, J., Von Figura, K., et al. (2002). Role of LAMP-2 in lysosome biogenesis and autophagy. *Molecular Biology of the Cell*, *13*(9), 3355–3368.

Fan, S., Meng, Q., Saha, T., Sarkar, F. H., & Rosen, E. M. (2009). Low concentrations of diindolylmethane, a metabolite of indole-3-carbinol, protect against oxidative stress in a BRCA1-dependent manner. *Cancer Research*, *69*(15), 6083–6091.

Finkel, T. (2003). Oxidant signals and oxidative stress. *Current Opinion in Cell Biology*, *15*(2), 247–254.

Ganley, I. G., Lam du, H., Wang, J., Ding, X., Chen, S., & Jiang, X. (2009). ULK1.ATG13. FIP200 complex mediates mTOR signaling and is essential for autophagy. *The Journal of Biological Chemistry*, *284*(18), 12297–12305.

Hamanaka, R. B., & Chandel, N. S. (2010). Mitochondrial reactive oxygen species regulate cellular signaling and dictate biological outcomes. *Trends in Biochemical Sciences*, *35*(9), 505–513.

Hamed, H. A., Yacoub, A., Park, M. A., Eulitt, P., Sarkar, D., Dimitrie, I. P., et al. (2010). OSU-03012 enhances Ad.7-induced GBM cell killing via ER stress and autophagy and by decreasing expression of mitochondrial protective proteins. *Cancer Biology & Therapy*, *9*(7), 526–536.

Hosokawa, N., Hara, T., Kaizuka, T., Kishi, C., Takamura, A., Miura, Y., et al. (2009). Nutrient-dependent mTORC1 association with the ULK1-Atg13-FIP200 complex required for autophagy. *Molecular Biology of the Cell*, *20*(7), 1981–1991.

Hu, W., Zhang, C., Wu, R., Sun, Y., Levine, A., & Feng, Z. (2010). Glutaminase 2, a novel p53 target gene regulating energy metabolism and antioxidant function. *Proceedings of the National Academy of Sciences of the United States of America*, *107*(16), 7455–7460.

Jain, A., Lamark, T., Sjottem, E., Bowitz Larsen, K., Atesoh Awuh, J., Overvatn, A., et al. (2010). p62/SQSTM1 is a target gene for transcription factor NRF2 and creates a positive feedback loop by inducing antioxidant response element-driven gene transcription. *The Journal of Biological Chemistry*, *285*(29), 22576–22591.

Johansen, T., & Lamark, T. (2011). Selective autophagy mediated by autophagic adapter proteins. *Autophagy*, 7(3), 279–296.

Jung, C. H., Jun, C. B., Ro, S. H., Kim, Y. M., Otto, N. M., Cao, J., et al. (2009). ULK-Atg13-FIP200 complexes mediate mTOR signaling to the autophagy machinery. *Molecular Biology of the Cell*, *20*(7), 1992–2003.

Jung, C. H., Seo, M., Otto, N. M., & Kim, D. H. (2011). ULK1 inhibits the kinase activity of mTORC1 and cell proliferation. *Autophagy*, 7(10), 1212–1221.

Kane, D. J., Sarafian, T. A., Anton, R., Hahn, H., Gralla, E. B., Valentine, J. S., et al. (1993). Bcl-2 inhibition of neural death: Decreased generation of reactive oxygen species. *Science*, *262*(5137), 1274–1277.

Kang, R., Livesey, K. M., Zeh, H. J., Loze, M. T., & Tang, D. (2010). HMGB1: A novel Beclin 1-binding protein active in autophagy. *Autophagy*, *6*(8), 1209–1211.

Kang, R., Tang, D., Lotze, M. T., & Zeh, H. J., 3rd. (2011). RAGE regulates autophagy and apoptosis following oxidative injury. *Autophagy*, 7(4), 442–444.

Kihara, A., Kabeya, Y., Ohsumi, Y., & Yoshimori, T. (2001). Beclin-phosphatidylinositol 3-kinase complex functions at the trans-Golgi network. *EMBO Reports*, *2*(4), 330–335.

Kim, D. H., Sarbassov, D. D., Ali, S. M., King, J. E., Latek, R. R., Erdjument-Bromage, H., et al. (2002). mTOR interacts with raptor to form a nutrient-sensitive complex that signals to the cell growth machinery. *Cell*, *110*(2), 163–175.

Klionsky, D. J., Abdalla, F. C., Abeliovich, H., Abraham, R. T., Acevedo-Arozena, A., Adeli, K., et al. (2012). Guidelines for the use and interpretation of assays for monitoring autophagy. *Autophagy*, *8*(4), 445–544.

Klionsky, D. J., Abeliovich, H., Agostinis, P., Agrawal, D. K., Aliev, G., Askew, D. S., et al. (2008). Guidelines for the use and interpretation of assays for monitoring autophagy in higher eukaryotes. *Autophagy*, *4*(2), 151–175.

Klionsky, D. J., Cuervo, A. M., & Seglen, P. O. (2007). Methods for monitoring autophagy from yeast to human. *Autophagy*, *3*(3), 181–206.

Lai, K. P., Leong, W. F., Chau, J. F., Jia, D., Zeng, L., Liu, H., et al. (2010). S6K1 is a multifaceted regulator of Mdm2 that connects nutrient status and DNA damage response. *The EMBO Journal*, *29*(17), 2994–3006.

Levine, B. (2005). Eating oneself and uninvited guests: Autophagy-related pathways in cellular defense. *Cell*, *120*(2), 159–162.

Levine, B., & Klionsky, D. J. (2004). Development by self-digestion: Molecular mechanisms and biological functions of autophagy. *Developmental Cell*, *6*(4), 463–477.

Levine, B., & Kroemer, G. (2008). Autophagy in the pathogenesis of disease. *Cell*, *132*(1), 27–42.

Levine, B., & Yuan, J. (2005). Autophagy in cell death: An innocent convict? *The Journal of Clinical Investigation*, *115*(10), 2679–2688.

Li, L., Chen, Y., & Gibson, S. B. (2013). Starvation-induced autophagy is regulated by mitochondrial reactive oxygen species leading to AMPK activation. *Cellular Signalling*, *25*(1), 50–65.

Lipinski, M. M., Zheng, B., Lu, T., Yan, Z., Py, B. F., Ng, A., et al. (2010). Genome-wide analysis reveals mechanisms modulating autophagy in normal brain aging and in Alzheimer's disease. *Proceedings of the National Academy of Sciences*, *107*(32), 14164–14169.

Liu, Z., & Lenardo, M. J. (2007). Reactive oxygen species regulate autophagy through redox-sensitive proteases. *Developmental Cell*, *12*(4), 484–485.

Lotze, M. T., & Tracey, K. J. (2005). High-mobility group box 1 protein (HMGB1): Nuclear weapon in the immune arsenal. *Nature Reviews. Immunology*, *5*(4), 331–342.

Luo, S., & Rubinsztein, D. C. (2010). Apoptosis blocks Beclin 1-dependent autophagosome synthesis: An effect rescued by Bcl-xL. *Cell Death and Differentiation*, *17*(2), 268–277.

Meijer, A. J., & Codogno, P. (2006). Signalling and autophagy regulation in health, aging and disease. *Molecular Aspects of Medicine*, *27*(5–6), 411–425.

Mizushima, N. (2005). The pleiotropic role of autophagy: From protein metabolism to bactericide. *Cell Death and Differentiation*, *12*(Suppl. 2), 1535–1541.

Mizushima, N. (2009). Methods for monitoring autophagy using GFP-LC3 transgenic mice. *Methods in Enzymology*, *452*, 13–23.

Mizushima, N., & Levine, B. (2010). Autophagy in mammalian development and differentiation. *Nature Cell Biology*, *12*(9), 823–830.

Mizushima, N., & Yoshimori, T. (2007). How to interpret LC3 immunoblotting. *Autophagy*, *3*(6), 542–545.

Muller, M., & Reichert, A. S. (2011). Mitophagy, mitochondrial dynamics and the general stress response in yeast. *Biochemical Society Transactions*, *39*(5), 1514–1519.

Noda, T., & Ohsumi, Y. (1998). Tor, a phosphatidylinositol kinase homologue, controls autophagy in yeast. *The Journal of Biological Chemistry*, *273*(7), 3963–3966.

Pani, G., & Galeotti, T. (2011). Role of MnSOD and p66shc in mitochondrial response to p53. *Antioxidants & Redox Signaling*, *15*(6), 1715–1727.

Park, M. A., Zhang, G., Martin, A. P., Hamed, H., Mitchell, C., Hylemon, P. B., et al. (2008). Vorinostat and sorafenib increase ER stress, autophagy and apoptosis via ceramide-dependent CD95 and PERK activation. *Cancer Biology & Therapy*, 7(10), 1648–1662.

Pattingre, S., Espert, L., Biard-Piechaczyk, M., & Codogno, P. (2008). Regulation of macroautophagy by mTOR and Beclin 1 complexes. *Biochimie*, *90*(2), 313–323.

Pattingre, S., Tassa, A., Qu, X., Garuti, R., Liang, X. H., Mizushima, N., et al. (2005). Bcl-2 antiapoptotic proteins inhibit Beclin 1-dependent autophagy. *Cell*, *122*(6), 927–939.

Periyasamy-Thandavan, S., Jiang, M., Schoenlein, P., & Dong, Z. (2009). Autophagy: Molecular machinery, regulation, and implications for renal pathophysiology. *American Journal of Physiology. Renal Physiology*, *297*(2), F244–F256.

Petiot, A., Ogier-Denis, E., Blommaart, E. F., Meijer, A. J., & Codogno, P. (2000). Distinct classes of phosphatidylinositol 3′-kinases are involved in signaling pathways that control macroautophagy in HT-29 cells. *The Journal of Biological Chemistry*, *275*(2), 992–998.

Pyo, J. O., Nah, J., Kim, H. J., Lee, H. J., Heo, J., Lee, H., et al. (2008). Compensatory activation of ERK1/2 in Atg5-deficient mouse embryo fibroblasts suppresses oxidative stress-induced cell death. *Autophagy*, *4*(3), 315–321.

Ravikumar, B., Sarkar, S., Davies, J. E., Futter, M., Garcia-Arencibia, M., Green-Thompson, Z. W., et al. (2010). Regulation of mammalian autophagy in physiology and pathophysiology. *Physiological Reviews*, *90*(4), 1383–1435.

Ravikumar, B., Vacher, C., Berger, Z., Davies, J. E., Luo, S., Oroz, L. G., et al. (2004). Inhibition of mTOR induces autophagy and reduces toxicity of polyglutamine expansions in fly and mouse models of Huntington disease. *Nature Genetics*, *36*(6), 585–595.

Rubinsztein, D. C., Gestwicki, J. E., Murphy, L. O., & Klionsky, D. J. (2007). Potential therapeutic applications of autophagy. *Nature Reviews. Drug Discovery*, *6*(4), 304–312.

Rusten, T. E., & Stenmark, H. (2010). p62, an autophagy hero or culprit? *Nature Cell Biology*, *12*(3), 207–209.

Scherz-Shouval, R., & Elazar, Z. (2011). Regulation of autophagy by ROS: Physiology and pathology. *Trends in Biochemical Sciences*, *36*(1), 30–38.

Scherz-Shouval, R., Shvets, E., & Elazar, Z. (2007). Oxidation as a post-translational modification that regulates autophagy. *Autophagy*, *3*(4), 371–373.

Seo, G., Kim, S. K., Byun, Y. J., Oh, E., Jeong, S. W., Chae, G. T., et al. (2011). Hydrogen peroxide induces Beclin 1-independent autophagic cell death by suppressing the mTOR pathway via promoting the ubiquitination and degradation of Rheb in GSH-depleted RAW 264.7 cells. *Free Radical Research*, *45*(4), 389–399.

Settembre, C., Di Malta, C., Polito, V. A., Garcia Arencibia, M., Vetrini, F., Erdin, S., et al. (2011). TFEB links autophagy to lysosomal biogenesis. *Science*, *332*(6036), 1429–1433.

Sui, X., Jin, L., Huang, X., Geng, S., He, C., & Hu, X. (2011). p53 signaling and autophagy in cancer: A revolutionary strategy could be developed for cancer treatment. *Autophagy*, 7(6), 565–571.

Suzuki, T., & Inoki, K. (2011). Spatial regulation of the mTORC1 system in amino acids sensing pathway. *Acta Biochimica et Biophysica Sinica*, *43*(9), 671–679.

Suzuki, S., Tanaka, T., Poyurovsky, M. V., Nagano, H., Mayama, T., Ohkubo, S., et al. (2010). Phosphate-activated glutaminase (GLS2), a p53-inducible regulator of glutamine metabolism and reactive oxygen species. *Proceedings of the National Academy of Sciences of the United States of America*, *107*(16), 7461–7466.

Szyniarowski, P., Corcelle-Termeau, E., Farkas, T., Hoyer-Hansen, M., Nylandsted, J., Kallunki, T., et al. (2011). A comprehensive siRNA screen for kinases that suppress macroautophagy in optimal growth conditions. *Autophagy*, 7(8), 892–903.

Tang, D., Kang, R., Livesey, K. M., Cheh, C. W., Farkas, A., Loughran, P., et al. (2010). Endogenous HMGB1 regulates autophagy. *The Journal of Cell Biology*, *190*(5), 881–892.

Wang, Y., Nartiss, Y., Steipe, B., McQuibban, G. A., & Kim, P. K. (2012). ROS-induced mitochondrial depolarization initiates PARK2/PARKIN-dependent mitochondrial degradation by autophagy. *Autophagy*, *8*(10), 1462–1476.

Wang, J., Whiteman, M. W., Lian, H., Wang, G., Singh, A., Huang, D., et al. (2009). A non-canonical MEK/ERK signaling pathway regulates autophagy via regulating Beclin 1. *The Journal of Biological Chemistry*, *284*(32), 21412–21424.

Yang, J., Wu, L.-J. , Tashino, S.-I. , Onodera, S., & Ikejima, T. (2008). Reactive oxygen species and nitric oxide regulate mitochondria-dependent apoptosis and autophagy in evodiamine-treated human cervix carcinoma HeLa cells. *Free Radical Research*, *42*(5), 492–504.

CHAPTER FOURTEEN

H_2O_2: A Chemoattractant?

Balázs Enyedi, Philipp Niethammer[1]
Cell Biology Program, Memorial Sloan-Kettering Cancer Center, New York, USA
[1]Corresponding author: e-mail address: niethamp@mskcc.org

Contents

1. Introduction 238
2. The Zebrafish Tail Fin Wounding Assay 242
   2.1 Protocol for leukocyte recruitment assay 242
   2.2 Required materials 243
   2.3 Setting up the experiment 243
   2.4 Image analysis and quantification 245
3. Measuring H_2O_2 Signals in Zebrafish 246
   3.1 Considerations on choosing measurement techniques 246
   3.2 Preparing tools to use HyPer in zebrafish 246
   3.3 Preparing and injecting HyPer RNA 247
   3.4 Generating transgenic lines expressing HyPer 247
4. Imaging H_2O_2 Production by Wide-Field Microscopy 248
   4.1 Wounding the fish 248
   4.2 Microscope settings, fluorescence imaging 248
   4.3 Image processing and data analysis 249
5. Imaging H_2O_2 Production by Confocal Microscopy 249
   5.1 Wounding and image acquisition 250
   5.2 Image processing and data analysis 251
Acknowledgments 253
References 253

Abstract

H_2O_2 is a relatively stable, rapidly diffusing reactive oxygen species that has been recently implicated as a mediator of leukocyte recruitment to epithelial wounds and transformed cells in zebrafish. Whether H_2O_2 activates the innate immune response by acting as a *bona fide* chemoattractant, enhancing chemoattractant sensing, or triggering production of other chemoattractive ligands remains largely unclear. Here, we describe the basic experimental procedures required to study these questions. We present a detailed protocol of the zebrafish tail fin wounding assay and explain how to use it for analyzing leukocyte chemotaxis *in vivo*. We further outline a method for H_2O_2 measurement in live zebrafish larvae using the genetically encoded sensor HyPer on a wide-field and a spinning disk confocal microscope. These methods provide a basis for dissecting the role of H_2O_2 in leukocyte chemotaxis in a vertebrate animal.

Methods in Enzymology, Volume 528
ISSN 0076-6879
http://dx.doi.org/10.1016/B978-0-12-405881-1.00014-8

1. INTRODUCTION

Chemotaxis is fundamental for a plethora of biological processes including host defense, embryogenesis, or pathological conditions such as cancer metastasis. The protective function of innate immune responses strongly depends on the capability of leukocytes to detect and migrate along chemical gradients emitted by pathogens, wounds, or damaged cells. Due to their central role during host defense, leukocytes, especially neutrophil granulocytes, are a prime focus of chemotaxis research. It has been known for a long time that these cells produce reactive oxygen species (ROS) to kill pathogens. Recent evidence suggests that ROS also control other crucial aspects of leukocyte biology such as directional migration.

Cellular ROS are derived mainly from two sources: (i) mitochondrial metabolism (Murphy, 2009) or (ii) regulated NADPH oxidase (NOX) activity (Bedard & Krause, 2007). There are 7 NOX enzymes (NOX1-5 and DUOX1-2) in mammals that catalyze the transfer of electrons from NADPH to molecular oxygen. The initial product of this reaction is superoxide ($O_2^{\cdot-}$), or in case of DUOX enzymes, hydrogen peroxide (H_2O_2) (Geiszt, 2006). Pathogen killing by leukocytes requires the phagocytic NADPH oxidase enzyme, NOX2, to produce large amounts of $O_2^{\cdot-}$ (Rada et al., 2008). Lower, nontoxic concentrations of ROS can modulate inflammation by activating proinflammatory signaling cascades such as the NF-κB pathway, leading, for example, to chemokine production (Hayden & Ghosh, 2011). Hydrogen peroxide is a particularly stable ROS with an extracellular half-life of ~20 s (Barnard & Matalon, 1992). It diffuses approximately as fast as water (Bienert, Schjoerring, & Jahn, 2006). Due to its biophysical properties, hydrogen peroxide seems well suited to convey biological signals over larger distances (i.e., hundreds of μm) through the extracellular space of tissues. Indeed, recent evidence suggests that H_2O_2 may function in a paracrine fashion to mediate recruitment of leukocytes to injury sites (Niethammer, Grabher, Look, & Mitchison, 2009) and transformed cells (Feng, Santoriello, Mione, Hurlstone, & Martin, 2010).

Klyubin and colleagues were the first to propose that H_2O_2 acts as a leukocyte chemoattractant. Using an *in vitro* under-agarose chemotaxis assay, these authors showed that mouse peritoneal neutrophils migrate toward H_2O_2 at concentrations as low as 10 μ*M* (Klyubin, Kirpichnikova, & Gamaley, 1996). In addition, H_2O_2 produced by the extracellular enzyme lysyl oxidase (LOX) has been shown to be required for chemotaxis of vascular smooth muscle cells (VSMCs; Li, Liu, Chou, & Kagan, 2000). LOX catalyzes oxidation of specific lysine residues in collagen and elastin fibers of the extracellular matrix. H_2O_2 and ammonia are by-products of this reaction. Catalase administration abolished the chemotactic effect of LOX, demonstrating that H_2O_2 mediates VSMC chemotaxis.

To act as a leukocyte chemoattractant *in vivo*, H_2O_2 must form a concentration gradient in the extracellular space that persists long enough and reaches far enough to convey spatial information to distant leukocytes. In principle, spatially graded production of H_2O_2 (i.e., through a NOX activity gradient in the tissue), reaction–diffusion of H_2O_2, or a combination of both mechanisms could create such a pattern. Indeed, tissue culture experiments underline the paracrine signaling capabilities of H_2O_2. Specifically, it has been shown that H_2O_2 produced by DUOX-expressing cells diffuses into the cytoplasm of neighboring, nonexpressing cells, where it oxidizes cysteine residues (Enyedi, Zana, Donko, & Geiszt, 2012).

The first evidence that H_2O_2 promotes leukocyte chemotaxis in intact tissues came from Niethammer and colleagues. Using the zebrafish tail fin wounding assay, they demonstrated that injury activates DUOX at the wound site, which generates a H_2O_2 gradient that extends $\sim$100–200 μm into the tissue (Niethammer et al., 2009). This signal was measured by cytosolic HyPer, a genetically encoded, ratiometric H_2O_2 sensor (Belousov et al., 2006). The maximal HyPer-signal correlated in time with the arrival of the first leukocytes. Genetic or pharmacological inhibition of DUOX abolished the H_2O_2 gradient and impaired normal leukocyte recruitment. This study demonstrated that H_2O_2 is required for wound recruitment of leukocytes in zebrafish, but left open the question of how H_2O_2 precisely mediates leukocyte recruitment.

Particularly, it remains unclear whether H_2O_2 *in vivo* mainly acts (i) as a primary chemoattractant, (ii) as a permissive signal that enhances chemoattractant sensing of leukocytes, or (iii) as an inducer of chemoattractant production/release in nonmyeloid cells. While all these ideas are consistent with a paracrine signaling role of H_2O_2 and supported by published data, only the first model implements H_2O_2 as a *bona fide* chemoattractant. Chemotactic receptors are typically activated by noncovalent binding of peptide- or lipid ligands to G-protein coupled receptors (GPCRs) or growth factor receptors on the plasma membrane. Cells exposed to a localized source of chemoattractant polarize and directionally migrate in response to subtle spatial differences in receptor occupancy. This allows cells to utilize shallow concentration gradients of chemotactic ligands as guidance cues

(Iglesias & Devreotes, 2012). Stable, noncovalent interactions of H_2O_2 with chemotactic plasma membrane receptors have not been reported to date. Thus, it is difficult to draw direct mechanistic analogies between the mode of action of classic chemoattractants (e.g., chemokines, oxidized lipids) and H_2O_2. It is, however, known that H_2O_2 can modulate intracellular signaling through reversible oxidation of cysteine residues that posttranslationally regulate protein function (Nathan, 2003). Cysteine residues on the extracellular side of transmembrane proteins are predominantly oxidized. Thus, H_2O_2 has to enter the cells to elicit reversible oxidation and modulate signaling. H_2O_2 can pass through the plasma membrane via aquaporin channels or—more slowly—by passive diffusion (Bienert et al., 2006). Due to its short cytoplasmic half-life, H_2O_2's activity radius is limited to the vicinity of the plasma membrane, where many of the proteins involved in chemotactic signaling reside. Known targets of reversible cysteine oxidation and potential chemotactic regulators are protein tyrosine kinases and phosphatases. Cysteine oxidation of the myeloid-specific tyrosine kinase Lyn has recently been demonstrated by Yoo and colleagues to be required for H_2O_2-dependent leukocyte migration to wound sites in zebrafish (Yoo, Starnes, Deng, & Huttenlocher, 2011). The exact mechanism, however, by which Lyn instructs directional migration and regulates cell polarity remains to be clarified.

The lipid phosphatase PTEN has also been described as a direct target of H_2O_2-mediated oxidation, leading to its inhibition (Lee et al., 2002). Along with SHIP1, PTEN regulates the levels of the polarization factor PtdIns(3,4,5)P3 (Weiner, 2002), a lipid product of phosphatidylinositol 3-kinase (PI3K) that asymmetrically accumulates on the plasma membrane of leukocytes upon exposure to a chemoattractant gradient. Oxidative inhibition of PTEN may be required to generate an anterior–posterior gradient of PI3K lipid products during stimulation with chemoattractants such as fMLP (Kuiper, Sun, Magalhaes, & Glogauer, 2011). While PTEN is crucial for effective directional sensing of the social amoeba *Dictyostelium discoideum*, genetic disruption of PTEN only has minor effects on neutrophil chemotaxis (Sarraj et al., 2009). By contrast, SHIP1 seems to have a more pronounced role in regulating neutrophil chemotaxis (Mondal, Subramanian, Sakai, Bajrami, & Luo, 2012); however, it is unclear if this enzyme is also regulated by H_2O_2-mediated oxidation.

In tissues, H_2O_2 may also stimulate leukocyte recruitment by affecting the production of chemoattractants. For example, H_2O_2 may enhance production of GPCR ligands through activating the complement cascade

(Shingu & Nobunaga, 1984), increase enzymatic or nonenzymatic lipid oxidation (Grant et al., 2011; Perez, Weksler, & Goldstein, 1980), or stimulate transcription of proinflammatory chemokines in wounded or infected tissue (Roebuck et al., 1999). Chemotactic signaling by GPCR ligands is well established and may provide a simple explanation for at least some of the proinflammatory effects of H_2O_2.

To understand how H_2O_2 promotes leukocyte recruitment *in vivo*, it is crucial to know when and where it is generated, how far and fast it propagates through the tissue, whether it reaches potential target cells such as leukocytes in a relevant time interval, and whether these target cells directly sense H_2O_2. The most straightforward way to study this is to directly visualize the spatiotemporal patterns of H_2O_2 in intact tissues, probing its propagation through intra- and extracellular spaces, and also the oxidative responses it elicits in different target cells.

Zebrafish (*Danio rerio*) is excellently suited for this type of *in vivo* imaging approach. Zebrafish larvae have an immune system that closely resembles that of mammals (Lieschke & Trede, 2009). Their larvae are thin and transparent, which make their tissues uniquely accessible for fluorescence microscopy. Ectopic expression of fluorescent proteins is achieved through mRNA injection into embryos at the one-cell stage, or through transgenesis using tissue-specific promoters to drive protein expression. The contribution of individual signaling pathways can be conveniently interrogated by reverse genetics using morpholinos (Chen & Ekker, 2004), or addition of pharmacological antagonists/agonists to the fish-bathing medium. While mammalian intravital imaging experiments are typically laborious and expensive to repeat, sample throughputs of over 150–200 zebrafish/day can be achieved in wide-field imaging assays by using a motorized stage that drastically increases statistical confidence in the *in vivo* imaging results.

Wounding of zebrafish larvae at the tail fin generates a local, endogenous source of H_2O_2 in the tissue that is required for recruitment of distant, perivascular leukocytes (Niethammer et al., 2009). Notably, the currently available protein-based H_2O_2 sensors detect H_2O_2 levels through cysteine oxidation, which is only reversible inside cells. This restricts steady-state measurements of H_2O_2 to the cytoplasm, while direct imaging of extracellular, paracrine H_2O_2 patterns in tissues remains a technical challenge to date. However, as extracellular H_2O_2 rapidly diffuses into cells, cytoplasmic HyPer is likely to indirectly report on extracellular H_2O_2 levels. *In vivo* measurements of cytoplasmic H_2O_2 in response to tail fin wounding are feasible with HyPer, a genetically encoded H_2O_2 sensor of high dynamic range.

Combined imaging of HyPer and leukocyte recruitment allows addressing two related set of questions: First, is H_2O_2 production sufficient to recruit leukocytes? The sufficiency hypothesis could be challenged through identification of experimental conditions that perturb leukocyte chemotaxis (but not general migration), while leaving the wound-induced H_2O_2 gradient intact. Second, is H_2O_2 directly sensed by leukocytes? Direct sensing of an extracellular H_2O_2 gradient should gradually increase H_2O_2 levels inside leukocytes as they move toward the wound. Confirmation of direct sensing would strongly argue for a role of H_2O_2 as spatially instructive chemoattractant *in vivo* as long as the sufficiency hypothesis is not rejected. Rejection of the sufficiency hypothesis, but confirmation of the direct-sensing hypothesis, would suggest a permissive role during leukocyte chemotaxis. Rejection of both hypotheses would point to a role of H_2O_2 as an inducer of chemoattractant release/production in tissues.

In the following section, we outline the basic experimental procedures to set up and evaluate these types of experiments.

2. THE ZEBRAFISH TAIL FIN WOUNDING ASSAY

Wound healing and regeneration typically proceeds on the timescale of hours to days. The first protective tissue responses (e.g., H_2O_2 production and leukocyte recruitment) are initiated only seconds to minutes after injury. We term this initial time window the "wound-detection phase". Within the first ~40 min post-wounding (pw), cyclohexamide does not inhibit leukocyte migration (our unpublished observations). This suggests that leukocyte chemotaxis during the wound-detection phase does not require new protein synthesis. It may rather depend on release of preformed chemoattractants, or their rapid, enzymatic production. By contrast, the regulation of later stages of wound inflammation and healing requires expression of new proteins (e.g., chemokines, growth factors, etc.). The H_2O_2 signal at the wound peaks ~20 min after injury and largely diminishes over the course of ~40–60 min. To investigate the effects of wound margin H_2O_2 production on leukocytes, a measurement time window of ~40 min after injury is generally sufficient.

2.1. Protocol for leukocyte recruitment assay

Time-lapse imaging by transmission microscopy allows easy identification of all moving leukocytes in the larval tail fin. To retain the transparency of the embryos, pigment formation should be inhibited by maintaining the larvae in medium containing 0.2 m*M* *N*-phenylthiourea (PTU; Sigma). Alternatively, pigmentation mutant lines such as the *casper* can be used (White et al., 2008). Retaining transparency is particularly important for assays involving fluorescence imaging, while it may be dispensable for assays that solely rely on transmission microscopy. An alternative approach to visualize leukocytes is to use transgenic zebrafish lines expressing fluorescent proteins under the control of different leukocyte-specific promoters (mpx, lysC, fli1, etc.) (Hall, Flores, Storm, Crosier, & Crosier, 2007; Lawson & Weinstein, 2002; Renshaw et al., 2006). This method can be of advantage if there is a need to differentiate between various leukocyte types. Silencing of the transgene, however, may result in mosaic expression of the fluorescent marker, leaving only a proportion of the leukocytes detectable. To avoid this type of error, we prefer to use transmission microscopy to visualize all leukocytes that migrate to the wound. The leukocyte population that migrates to tail fin wounds within the first 40 min after injury mainly consists of neutrophils and macrophages. Even without using transgenic markers, these two cell types can often be differentiated by inspecting their morphology (e.g., migrating neutrophils appear compact, triangular shaped, while macrophages appear more branched) or quantifying their dynamic parameters (neutrophils move approximately double as fast as macrophages). Slightly offsetting the focus enhances the contrast of moving leukocytes and makes them appear as black blobs allowing easy tracking (Fig. 14.1A).

2.2. Required materials

- E3 embryo medium (5 m*M* NaCl, 0.17 m*M* KCl, 0.33 m*M* $CaCl_2$, 0.33 m*M* $MgSO_4$), PTU, Tricaine (Sigma), low melting agarose (Lonza)
- Plastic and glass petri dishes, sterile plastic transfer pipettes
- Dissecting microscope and inverted microscope equipped with a camera and motorized stage
- Tungsten needle (Fine Science Tools, 0.25 mm) and hair loop tool (a short loop of hair squeezed into a pipette tip)

2.3. Setting up the experiment

- Maintain zebrafish at 28 °C in E3 embryo medium supplemented with 0.2 m*M* PTU.
- Dechorionate 2.5–3 dpf zebrafish embryos and anesthetize them ~5 min prior to wounding in a glass petri dish in E3 medium containing 0.2 mg/ml Tricaine (it is advised to keep the anesthetized larvae in glass

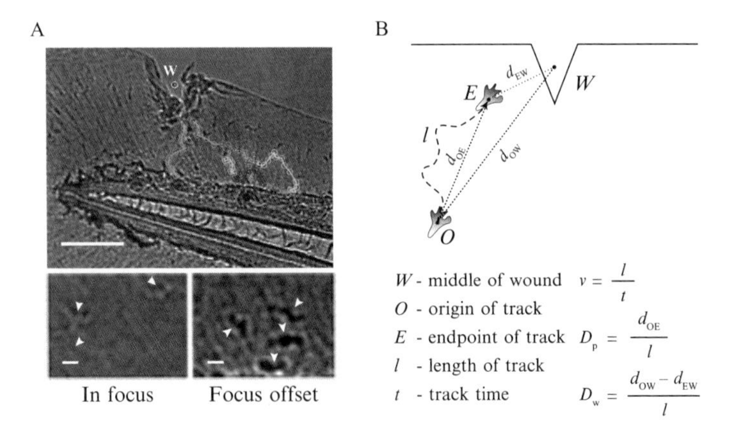

Figure 14.1 Quantitative analysis of leukocyte recruitment to incisional tail fin wounds. (A) Transmitted-light images of wounded caudal tail fins of zebrafish. Upper panel: Low magnification image showing tracks of leukocyte migration over the course of 40 min post-wounding (scale bar 100 μm). Lower panel: Zoom into tail fin region. In focus (left) or slightly out of focus (right) images of migrating leukocytes (marked by arrows). Scale bars, 10 μm. (B) Scheme of parameters measured for in-depth analysis of leukocyte trajectories. (See Color Insert.)

dishes as they easily stick to plastic surfaces leading to a series of unintentional micro-wounds along the fin). Pharmacological compounds can be added to the E3 medium for a typical preincubation time of ~45–60 min.

- Transfer larvae to a glass-bottom dish (Matek Corporation) using a plastic transfer pipette. If a motorized stage is available and multiple embryos can be imaged during the course of one experiment, it is advised to align the embryos and flatten all tail fins before wounding. A hair loop tool is most suitable for this task. After aligning the embryos, immobilize them by carefully adding 1% low melting agarose (Lonza, dissolved in E3, prepared fresh and kept at 42 °C) to the medium in a ratio of approx. 1:1 using a plastic transfer pipette and realign them if they moved as fast as possible.
- Make an incision on the ventral tail fin of the embryos using a sterile tungsten needle. In our experience, the size of the wound does not significantly affect the extent of leukocyte recruitment if the length of the wound margin is kept between 50 and 150 μm. For the sake of reproducibility, it is advised to wound the tail fin between the tip of the notochord and the loop of the caudal artery and vein as shown in

Fig. 14.1A. The embryos should be wounded before the agarose solidifies, within ~2 min.

- Place the glass-bottom dish on the inverted microscope and gently add E3 medium containing 0.2 mg/ml Tricaine to prevent the sample from drying out. If pharmacological compounds are used, they should be added to the agarose and the covering medium as well.
- Set the position of each embryo for multisample imaging and choose an appropriate focal plane in which leukocytes appear as black blobs in the light transmission images (Fig. 14.1A). Acquire an image every minute for a period of 40 min keeping the temperature at ~26–28 °C. The time between wounding and the first acquisition should be noted and kept at ~3 min for each experiment.

2.4. Image analysis and quantification

Leukocyte recruitment can be imaged on any inverted microscope equipped with a camera. We use a Nikon Eclipse Ti microscope equipped with a 20 × plan-apochromat NA 0.75 air objective lens, an Andor Clara CCD camera and a LED light source (Lumencor). Leukocyte recruitment is determined by counting all migrating cells that arrive at the wound margin within 43 min pw. Cells that already reside at the wound margin at the beginning of the time-lapse sequence (~3 min pw) are not counted.

In-depth analysis of leukocyte trajectories can be performed on the same time-lapse data. To generate tracks, open the acquired time-lapse data set in Fiji (Schindelin et al., 2012). Mark the center of the wound and the approximate center of mass of the cells that move in the imaged tail fin using the MTrackJ plugin. Include only the cells that describe a path of at least 50 μm in the statistical path analysis. Furthermore, do not analyze tracks or part of tracks within a radius of 50 μm around the center of mass of the triangular wound region in order to avoid tracking of cells that had already reached the wound and merely move along the wound margins. Average velocity (v) is calculated as $v = l/t\ track$, with l being the length of the track and $t\ track$ being the total track time. Path linearity (which is frequently also termed "directionality") is calculated as $D_p = d_{OE}/l$, with d_{OE} being the Euclidian distance between origin (O) and endpoint (E) of the track. Wound directionality is calculated as $D_w = (d_{OW} - d_{EW})/l$, with d_{OW} being the distance between track origin and center of mass of the wound (W) and d_{EW} being the distance between track endpoint and W (Fig. 14.1B).

It is worth noting that the average values obtained from the tracking data also represent the proportion of the different cell types that are dominantly migrating. Macrophages typically move slower (~4 μm/min) while granulocytes migrate faster (~8 μm/min). The values of D_p and D_w typically range from 0.5 to 0.8, being lower when cell are not migrating and having higher values during directional migration.

3. MEASURING H_2O_2 SIGNALS IN ZEBRAFISH

3.1. Considerations on choosing measurement techniques

The most straightforward way of assessing paracrine H_2O_2 would be to image extracellular H_2O_2 concentration in living tissues. A number of H_2O_2-specific, membrane-impermeable dyes exist that principally allow such extracellular measurements. In practice, however, their usefulness for imaging H_2O_2 in tissues is limited. Varying tissue thickness/structure demands for signal normalization in order to avoid false-positive signals. Thus, ratiometric assays are preferred over single-wavelength approaches. Unfortunately, most of the currently available redox dyes are nonratiometric and react with H_2O_2 in an irreversible manner. Therefore, they only allow cumulative endpoint measurements without signal normalization. In addition, diffusion of irreversibly converted dye may create artifactual diffusion patterns that likely obscure actual spatial differences in extracellular H_2O_2 concentration.

Because of these complications, the best current option for H_2O_2 measurements in tissues is the use of fluorescent protein-based intracellular probes. Several groups have developed genetically encoded H_2O_2 sensors in the past years, such as the single-fluorescent protein-based HyPer and roGFP2-Orp1 (Morgan, Sobotta, & Dick, 2011) or the FRET-based OxyFRET and PerFRET probes (Enyedi et al., 2012). Other chapters in this volume extensively describe how these probes work, and what their advantages and drawbacks are. In our experience, fluorescent sensors with high dynamic range such as HyPer are best suited to measure signals in live animals. The following section will thus provide detailed description on using HyPer to assess H_2O_2 levels.

3.2. Preparing tools to use HyPer in zebrafish

Since the first report by Niethammer et al. on the measurement of H_2O_2 gradients in live zebrafish larvae (Niethammer et al., 2009), several groups have successfully used HyPer in this system (Pase et al., 2012; Yoo, Freisinger, LeBert, & Huttenlocher, 2012). The following section will describe how to prepare and inject HyPer RNA into one-cell stage embryos for ubiquitous expression and how to create transgenic zebrafish lines expressing the probe in different tissues.

3.3. Preparing and injecting HyPer RNA

HyPer should be subcloned from the commercially available plasmid (Evrogen) into a vector containing an SP6 RNA polymerase promoter site such as pCS2+. For *in vitro* RNA transcription, the mMESSAGE mMACHINE® SP6 kit (Life Technologies) can be used after linearizing ~1–2 μg of the pCS2+ vector with NotI. We suggest using the MEGAclear™ Kit (Life Technologies) to purify the capped RNA from the reaction mix. A typical reaction yields ~20 μg of RNA in a concentration of 0.3–0.5 μg/μl. Run an aliquot of the RNA on an agarose gel to ensure the integrity of the product and store at −80 °C until further use. For expression, inject ~2.3 nl of RNA (0.5 μg/μl) into the 1–2-cell stage embryos of wild-type AB or *casper* fish.

3.4. Generating transgenic lines expressing HyPer

The Tol2kit system is a convenient tool to generate stable transgenic zebrafish lines (Kwan et al., 2007). The multisite Gateway technology (Life Technologies) allows quick, site-specific recombination-based cloning and modular assembly of promoter-coding sequence-3′ tag constructs in a Tol2 transposon backbone. To assemble an expression clone of HyPer, subclone it into a middle entry clone available in the Tol2kit system, such as pME-MCS. Various promoters can be used to create expression vectors allowing HyPer expression in different tissues, such as the ubiquitous beta-actin promoter (from the p5E-*bactin2* entry clone of the Tol2kit system) or leukocyte-specific promoter such as lysC (Hall et al., 2007). The expression vector is created in an LR reaction catalyzed by the LR Clonase II Plus enzyme mix (Life Technologies) using the 5′, the middle and the SV40 late polyA 3′ entry clones along with the pDestTol2CG2 destination vector. The resultant expression clone also drives the expression of a cardiac green fluorescent protein (cmlc2:EGFP transgenesis marker) allowing for the easy identification of transgenic embryos.

To create transgenic fish, a solution containing 25–25 pg of the expression plasmid and transposase mRNA (made by *in vitro* transcription from the pCS2FA-transposase plasmid) is injected into the cytosol of one-cell stage *casper* or AB embryos. Injected larvae with mosaic cardiac EGFP expression are raised to sexual maturity and screened by crossing with wild-type fish to identify founders. F1 embryos are identified by EGFP expression and raised to sexual maturity. Experiments should be performed on the progeny of F1 outcross with wild-type fish.

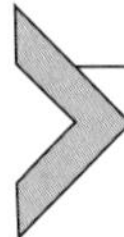

4. IMAGING H_2O_2 PRODUCTION BY WIDE-FIELD MICROSCOPY

Wound-induced H_2O_2 signals are generated on the minute-to-hour timescale. Capturing these signals does not require fast imaging and high sampling rates. Low-resolution wide-field microscopy is suitable to assess epithelial H_2O_2 production. Wide-field imaging at low magnification (e.g., 20 ×) provides the large field of view necessary to capture tissue-scale patterns and allows simultaneous leukocyte tracking in the transmitted light channel. Multiple specimens can be followed at the same time using a motorized stage that increases the sample size and statistical confidence.

4.1. Wounding the fish

Incisional wounds result in H_2O_2 production along the wound margin. To achieve consistent and comparable results, it is advised to injure the larvae by tail fin tip amputation using a needle knife (Fine Science Tools). For imaging purposes, larvae should be embedded in 1% low melting agarose in a glass-bottom dish as previously described for the leukocyte recruitment assay (see earlier).

4.2. Microscope settings, fluorescence imaging

The optimal excitation wavelengths for ratiometric imaging of HyPer are 500 and 420 nm. Emission is acquired in the YFP channel around 520 nm. Strong illumination of the sample may lead to irreversible photo-oxidation. It is thus advised to optimize the imaging setup for minimal light exposure by using neutral density filters, or tuning down the illuminating light power. Using high-sensitivity CCD cameras reduces the need for strong illumination, and if needed, exposure time can be increased to ~300 ms/wavelength to gain a signal significantly higher than the background. Since subcellular resolution is not required, it also helps to use high camera gain and binning (4–8 × bin). Similar to the leukocyte recruitment assay, imaging may be started at ~3 min pw. We typically acquire an image every 1–2 min to follow the course of H_2O_2 production.

4.3. Image processing and data analysis

The following steps are required to calculate HyPer ratio images from the raw YFP_{500} and YFP_{420} images using Fiji:

- Apply a one-pass median filter to remove noise (Process → Filters → Median filter → 1 pixel).
- Subtract background by selecting a ROI outside the tail fin and subtracting the average intensity of this ROI from the images frame by frame (e.g., by using the "BG Subtraction from ROI" plugin (www.uhnresearch.ca/facilities/wcif/imagej) by Cammer and Collins).
- Create a background mask by thresholding either image using the "dark background" setting and calculating the threshold for every image using the "triangle" method (Image → Adjust → Threshold). Divide the 8-bit binary image with a fix value of 255 to set the mask values to "1" and the background values to "0" and finally convert the image to 32-bits.
- Calculate the ratio (rat) image by dividing the background corrected YFP_{500} and YFP_{420} images (Process → Image calculator → divide (32-bit)) and apply the 16_colors LUT.
- Remove the background noise by multiplying the ratio image with the background mask.

To standardize the measurement of H_2O_2 upregulation across different samples, the mean ratio acquired in a region of interest ~100 μm wide directly at the wound margin (rat_{wound}) is divided by the mean basal ratio acquired in a region of interest inside the body (rat_{body}, ~300–400 μm distant from the wound margin; Fig. 14.2).

5. IMAGING H_2O_2 PRODUCTION BY CONFOCAL MICROSCOPY

Confocal imaging allows acquisition of high-resolution 3D images. Traditional laser scanning techniques provide superior background rejection but are rather slow and insensitive due to the use of low quantum efficiency photomultiplier tube detection. As a result, samples are often heavily illuminated. In live samples, this may affect the biological process of interest. We therefore use laser spinning disk confocal microscopy, which is more suited to observe live samples due to its speed and sensitive EMCCD detection. Imaging a complete fin tissue at 1.5 μm Z-resolution is possible within

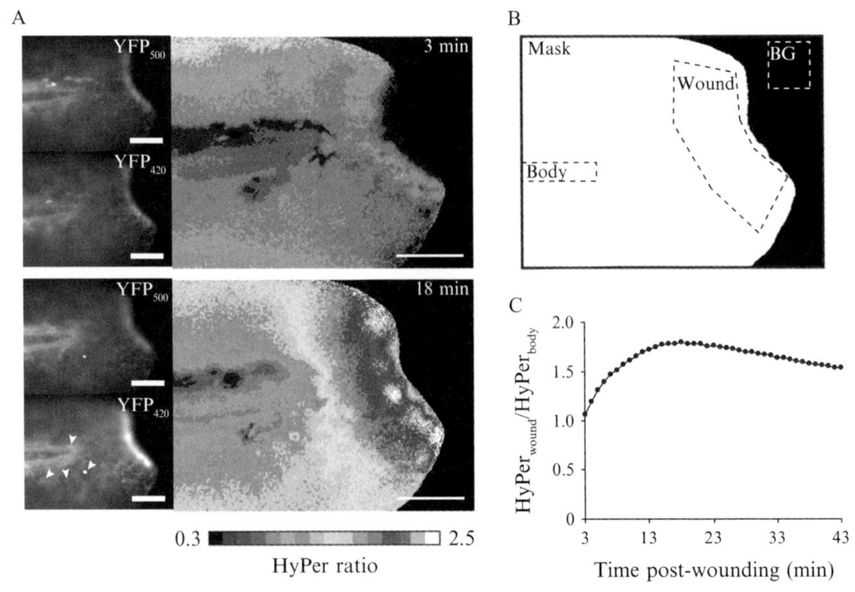

Figure 14.2 Wide-field imaging of wound margin H_2O_2 production in TG (bactin2: HyPer) zebrafish. (A and B) HyPer ratio images at indicated time points calculated from YFP_{500} and YFP_{420} images after applying a one-pass median filter, performing background (BG) subtraction and multiplying the ratio images with a mask derived from the thresholded YFP images. (C) Normalized H_2O_2 production is calculated by dividing the ratio values of a ROI along the wound margin by the ratio values of a ROI in the body of the embryo, distant from the wound margin. Scale bars, 100 μm. (See Color Insert.)

~1 s (single color) or ~2 s (two colors) at low light exposure resulting in minimal photo toxicity. We use a Nikon Eclipse FN1 upright microscope equipped with a 25 × LWD Apo/NA1.1 water-dipping objective, a Yokogawa spinning disk scan head (CSU-X1) and an Andor iXon3 897 electron multiplying CCD (EMCCD) camera with ~98% quantum efficiency. This imaging system achieves cellular resolution, which allows measuring signals in individual epithelial cells or leukocytes.

5.1. Wounding and image acquisition

Confocal microscopy can be combined with laser ablation that can be performed prior to or during image acquisition. Laser injury allows us to visualize rapid responses that happen within seconds after injury. We use a 435-nm UV Micropoint laser (Andor) to wound the tail fin of anesthetized larvae embedded in 1% low melting agarose. With the Micropoint laser, best wounding results are achieved by focusing the laserblast on tissue boundaries.

For ratiometric HyPer measurements, we use the 405- and 488-nm diode laser lines (Andor Revolution XD). Emission is collected with a 535/20 bandpass filter (Chroma). Up to 30 Z-stack slices with a resolution of 2 μm are acquired for every time point with the NIS-Elements software (Nikon) to obtain images of the whole tail fin.

5.2. Image processing and data analysis

To process the 3D confocal series as wide-field images, project the YFP_{488} and YFP_{405} Z-stack images using the "average intensity" projection function and continue with the image processing steps described above to create a projected ratio image. This is useful to evaluate H_2O_2 upregulation over larger sections of the tail fin (Fig. 14.3A and B), or for samples in which individual cells such as leukocytes express the HyPer probe. In the latter case, moving cells often change their position along the *Z*-axis during the time-lapse experiment, which makes it hard to follow them using wide-field microscopy. This problem can be avoided by projecting all Z-slices of a 3D-stack.

The following protocol describes how to retain the original 3D data:

- Apply a one-pass 3D median filter on the YFP_{488} and YFP_{405} z-stacks to eliminate noise (Process → Filters → Median 3D filter → 1 pixel).
- Subtract background by selecting a ROI outside the tail fin and subtract the average intensity of this ROI from the images frame by frame (e.g., by using the "BG Subtraction from ROI" plugin (www.uhnresearch.ca/facilities/wcif/imagej) by Cammer and Collins).
- Create a background mask by thresholding either image using the "dark background" setting and calculating the threshold using a fix selected value (e.g., Image → Adjust → Huang Threshold). Divide the 8-bit binary image with a fix value of 255 to set the mask values to "1" and the background values to "0" and finally convert the image to 32-bits.
- Calculate the ratio (rat) images by dividing the background corrected YFP_{488} and YFP_{405} images (Process → Image calculator → divide (32-bit)) and apply the 16_colors LUT.
- Remove background noise by multiplying the ratio image with the background mask.

Orthogonal views of the 3D image reveal the ratio values of cells underneath the superficial epithelial layer, such as leukocytes (Fig. 14.3C). 3D ratio analysis is particularly useful for measuring H_2O_2 production of single cells when HyPer is expressed in the whole tissue.

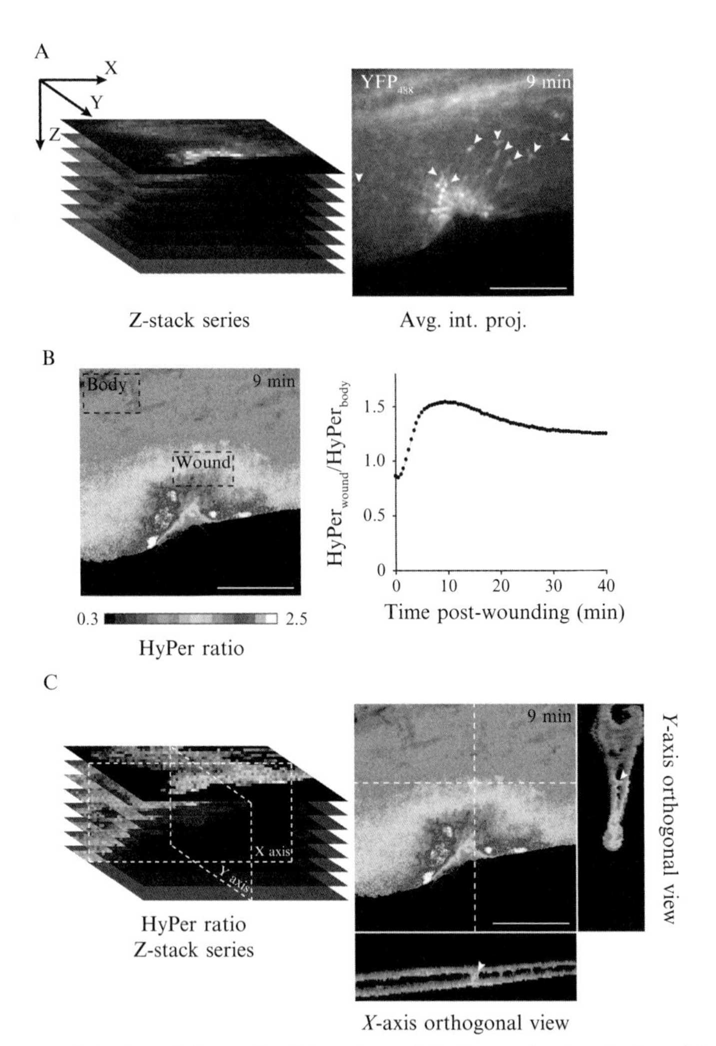

Figure 14.3 Spinning disk confocal imaging of H_2O_2 production induced by laser wounding in TG (bactin2:HyPer) zebrafish. (A) Z-stack series (~30 slices/fin, with a resolution of 2 μm/slice) of the YFP_{488} channel of HyPer taken at the indicated time point on a spinning disk confocal microscope. Average intensity projection of the Z-stack series collapses the 3D data into 2D images and allows leukocytes, marked by arrows to be clearly visualized. (B) HyPer ratio image calculated from the average intensity projection of median filtered, background subtracted and thresholded YFP_{488} and YFP_{405} Z-stack series. Normalized H_2O_2 production is calculated by dividing the ratio values of a ROI along the wound margin by the ratio values of a ROI in the body of the embryo. (C) Z-stack series of HyPer ratio images with *X*- and *Y*-axis orthogonal views derived from sections of the 3D image marked by white dashed lines. A leukocyte, marked by an arrow, is visible between the epithelial layers. Scale bars, 100 μm. (See Color Insert.)

ACKNOWLEDGMENTS

This work was supported by NIH Grant GM099970 and a Louis V. *Gerstner*, Jr. *Young Investigator award.*

REFERENCES

Barnard, M. L., & Matalon, S. (1992). Mechanisms of extracellular reactive oxygen species injury to the pulmonary microvasculature. *Journal of Applied Physiology*, *72*(5), 1724–1729.

Bedard, K., & Krause, K. H. (2007). The NOX family of ROS-generating NADPH oxidases: Physiology and pathophysiology. *Physiological Reviews*, *87*(1), 245–313. http://dx.doi.org/10.1152/physrev.00044.2005.

Belousov, V. V., Fradkov, A. F., Lukyanov, K. A., Staroverov, D. B., Shakhbazov, K. S., Terskikh, A. V., et al. (2006). Genetically encoded fluorescent indicator for intracellular hydrogen peroxide. *Nature Methods*, *3*(4), 281–286. http://dx.doi.org/10.1038/nmeth866.

Bienert, G. P., Schjoerring, J. K., & Jahn, T. P. (2006). Membrane transport of hydrogen peroxide. *Biochimica et Biophysica Acta*, *1758*(8), 994–1003. http://dx.doi.org/10.1016/j.bbamem.2006.02.015.

Chen, E., & Ekker, S. C. (2004). Zebrafish as a genomics research model. *Current Pharmaceutical Biotechnology*, *5*(5), 409–413.

Enyedi, B., Zana, M., Donko, A., & Geiszt, M. (2012). Spatial and temporal analysis of Nadph oxidase-generated hydrogen peroxide signals by novel fluorescent reporter proteins. *Antioxidants & Redox Signaling*. [Epub ahead of print]. http://dx.doi.org/10.1089/ars.2012.4594.

Feng, Y., Santoriello, C., Mione, M., Hurlstone, A., & Martin, P. (2010). Live imaging of innate immune cell sensing of transformed cells in zebrafish larvae: Parallels between tumor initiation and wound inflammation. *PLoS Biology*, *8*(12), e1000562. http://dx.doi.org/10.1371/journal.pbio.1000562.

Geiszt, M. (2006). NADPH oxidases: New kids on the block. *Cardiovascular Research*, *71*(2), 289–299. http://dx.doi.org/10.1016/j.cardiores.2006.05.004.

Grant, G. E., Gravel, S., Guay, J., Patel, P., Mazer, B. D., Rokach, J., et al. (2011). 5-oxo-ETE is a major oxidative stress-induced arachidonate metabolite in B lymphocytes. *Free Radical Biology & Medicine*, *50*(10), 1297–1304. http://dx.doi.org/10.1016/j.freeradbiomed.2011.02.010.

Hall, C., Flores, M. V., Storm, T., Crosier, K., & Crosier, P. (2007). The zebrafish lysozyme C promoter drives myeloid-specific expression in transgenic fish. *BMC Developmental Biology*, 7, 42. http://dx.doi.org/10.1186/1471-213X-7-42.

Hayden, M. S., & Ghosh, S. (2011). NF-kappaB in immunobiology. *Cell Research*, *21*(2), 223–244. http://dx.doi.org/10.1038/cr.2011.13.

Iglesias, P. A., & Devreotes, P. N. (2012). Biased excitable networks: How cells direct motion in response to gradients. *Current Opinion in Cell Biology*, *24*(2), 245–253. http://dx.doi.org/10.1016/j.ceb.2011.11.009.

Klyubin, I. V., Kirpichnikova, K. M., & Gamaley, I. A. (1996). Hydrogen peroxide-induced chemotaxis of mouse peritoneal neutrophils. *European Journal of Cell Biology*, *70*(4), 347–351.

Kuiper, J. W., Sun, C., Magalhaes, M. A., & Glogauer, M. (2011). Rac regulates PtdInsP(3) signaling and the chemotactic compass through a redox-mediated feedback loop. *Blood*, *118*(23), 6164–6171. http://dx.doi.org/10.1182/blood-2010-09-310383.

Kwan, K. M., Fujimoto, E., Grabher, C., Mangum, B. D., Hardy, M. E., Campbell, D. S., et al. (2007). The Tol2kit: A multisite gateway-based construction kit for Tol2 transposon transgenesis constructs. *Developmental Dynamics*, *236*(11), 3088–3099. http://dx.doi.org/10.1002/dvdy.21343.

Lawson, N. D., & Weinstein, B. M. (2002). In vivo imaging of embryonic vascular development using transgenic zebrafish. *Developmental Biology*, *248*(2), 307–318.

Lee, S. R., Yang, K. S., Kwon, J., Lee, C., Jeong, W., & Rhee, S. G. (2002). Reversible inactivation of the tumor suppressor PTEN by H2O2. *The Journal of Biological Chemistry*, *277*(23), 20336–20342. http://dx.doi.org/10.1074/jbc.M111899200.

Li, W., Liu, G., Chou, I. N., & Kagan, H. M. (2000). Hydrogen peroxide-mediated, lysyl oxidase-dependent chemotaxis of vascular smooth muscle cells. *Journal of Cellular Biochemistry*, *78*(4), 550–557.

Lieschke, G. J., & Trede, N. S. (2009). Fish immunology. *Current Biology*, *19*(16), R678–R682. http://dx.doi.org/10.1016/j.cub.2009.06.068.

Mondal, S., Subramanian, K. K., Sakai, J., Bajrami, B., & Luo, H. R. (2012). Phosphoinositide lipid phosphatase SHIP1 and PTEN coordinate to regulate cell migration and adhesion. *Molecular Biology of the Cell*, *23*(7), 1219–1230. http://dx.doi.org/10.1091/mbc.E11-10-0889.

Morgan, B., Sobotta, M. C., & Dick, T. P. (2011). Measuring E(GSH) and H2O2 with roGFP2-based redox probes. *Free Radical Biology & Medicine*, *51*(11), 1943–1951. http://dx.doi.org/10.1016/j.freeradbiomed.2011.08.035.

Murphy, M. P. (2009). How mitochondria produce reactive oxygen species. *The Biochemical Journal*, *417*(1), 1–13. http://dx.doi.org/10.1042/BJ20081386.

Nathan, C. (2003). Specificity of a third kind: Reactive oxygen and nitrogen intermediates in cell signaling. *The Journal of Clinical Investigation*, *111*(6), 769–778. http://dx.doi.org/10.1172/JCI18174.

Niethammer, P., Grabher, C., Look, A. T., & Mitchison, T. J. (2009). A tissue-scale gradient of hydrogen peroxide mediates rapid wound detection in zebrafish. *Nature*, *459*(7249), 996–999. http://dx.doi.org/10.1038/nature08119.

Pase, L., Layton, J. E., Wittmann, C., Ellett, F., Nowell, C. J., Reyes-Aldasoro, C. C., et al. (2012). Neutrophil-delivered myeloperoxidase dampens the hydrogen peroxide burst after tissue wounding in zebrafish. *Current Biology*, *22*(19), 1818–1824. http://dx.doi.org/10.1016/j.cub.2012.07.060.

Perez, H. D., Weksler, B. B., & Goldstein, I. M. (1980). Generation of a chemotactic lipid from a arachidonic acid by exposure to a superoxide-generating system. *Inflammation*, *4*(3), 313–328.

Rada, B., Hably, C., Meczner, A., Timar, C., Lakatos, G., Enyedi, P., et al. (2008). Role of Nox2 in elimination of microorganisms. *Seminars in Immunopathology*, *30*(3), 237–253. http://dx.doi.org/10.1007/s00281-008-0126-3.

Renshaw, S. A., Loynes, C. A., Trushell, D. M., Elworthy, S., Ingham, P. W., & Whyte, M. K. (2006). A transgenic zebrafish model of neutrophilic inflammation. *Blood*, *108*(13), 3976–3978. http://dx.doi.org/10.1182/blood-2006-05-024075.

Roebuck, K. A., Carpenter, L. R., Lakshminarayanan, V., Page, S. M., Moy, J. N., & Thomas, L. L. (1999). Stimulus-specific regulation of chemokine expression involves differential activation of the redox-responsive transcription factors AP-1 and NF-kappaB. *Journal of Leukocyte Biology*, *65*(3), 291–298.

Sarraj, B., Massberg, S., Li, Y., Kasorn, A., Subramanian, K., Loison, F., et al. (2009). Myeloid-specific deletion of tumor suppressor PTEN augments neutrophil transendothelial migration during inflammation. *The Journal of Immunology*, *182*(11), 7190–7200. http://dx.doi.org/10.4049/jimmunol.0802562.

Schindelin, J., Arganda-Carreras, I., Frise, E., Kaynig, V., Longair, M., Pietzsch, T., et al. (2012). Fiji: An open-source platform for biological-image analysis. *Nature Methods*, *9*(7), 676–682. http://dx.doi.org/10.1038/nmeth.2019.

Shingu, M., & Nobunaga, M. (1984). Chemotactic activity generated in human serum from the fifth component of complement by hydrogen peroxide. *The American Journal of Pathology, 117*(2), 201–206.

Weiner, O. D. (2002). Regulation of cell polarity during eukaryotic chemotaxis: The chemotactic compass. *Current Opinion in Cell Biology, 14*(2), 196–202.

White, R. M., Sessa, A., Burke, C., Bowman, T., LeBlanc, J., Ceol, C., et al. (2008). Transparent adult zebrafish as a tool for in vivo transplantation analysis. *Cell Stem Cell, 2*(2), 183–189. http://dx.doi.org/10.1016/j.stem.2007.11.002.

Yoo, S. K., Freisinger, C. M., LeBert, D. C., & Huttenlocher, A. (2012). Early redox, Src family kinase, and calcium signaling integrate wound responses and tissue regeneration in zebrafish. *The Journal of Cell Biology, 199*(2), 225–234. http://dx.doi.org/10.1083/jcb.201203154.

Yoo, S. K., Starnes, T. W., Deng, Q., & Huttenlocher, A. (2011). Lyn is a redox sensor that mediates leukocyte wound attraction in vivo. *Nature, 480*(7375), 109–112. http://dx.doi.org/10.1038/nature10632.

CHAPTER FIFTEEN

Measuring Mitochondrial Uncoupling Protein-2 Level and Activity in Insulinoma Cells

Jonathan Barlow*, Verena Hirschberg*, Martin D. Brand†, Charles Affourtit*,1

*School of Biomedical and Biological Sciences, Plymouth University, Drake Circus, Plymouth, United Kingdom
†Buck Institute for Research on Aging, Novato, California, USA
[1]Corresponding author: e-mail address: charles.affourtit@plymouth.ac.uk

Contents

1. Introduction 258
2. Tissue Culture 258
3. UCP2 Protein Detection 259
   3.1 Sample preparation 259
   3.2 Western analysis 259
4. UCP2 Protein Knockdown 260
5. UCP2 Activity 261
   5.1 Mitochondrial bioenergetics: Glucose-stimulated respiration and coupling efficiency 261
   5.2 Mitochondrial bioenergetics: Basal respiratory activity 263
   5.3 Mitochondrial reactive oxygen species 264
   5.4 Insulin secretion 266
Acknowledgments 267
References 267

Abstract

Mitochondrial uncoupling protein-2 (UCP2) regulates glucose-stimulated insulin secretion (GSIS) by pancreatic beta cells—the physiological role of the beta cell UCP2 remains a subject of debate. Experimental studies informing this debate benefit from reliable measurements of UCP2 protein level and activity. In this chapter, we describe how UCP2 protein can be detected in INS-1 insulinoma cells and how it can be knocked down by RNA interference. We demonstrate briefly that UCP2 knockdown lowers glucose-induced rises in mitochondrial respiratory activity, coupling efficiency of oxidative phosphorylation, levels of mitochondrial reactive oxygen species, and insulin secretion. We provide protocols for the detection of the respective UCP2 phenotypes, which are indirect, but invaluable measures of UCP2 activity. We also introduce a convenient

Methods in Enzymology, Volume 528
ISSN 0076-6879
http://dx.doi.org/10.1016/B978-0-12-405881-1.00015-X

method to normalize cellular respiration to cell density allowing measurement of UCP2 effects on specific mitochondrial oxygen consumption.

1. INTRODUCTION

Pancreatic beta cells contribute to glucose homeostasis by secreting insulin when the blood glucose level is high. Mitochondria are essential for this glucose-stimulated insulin secretion (GSIS) as they generate signals that couple oxidative glucose catabolism to insulin release (Barlow, Hirschberg, & Affourtit, 2013). According to the canonical GSIS model, the ATP/ADP ratio is the key signal that triggers insulin secretion (Rutter, 2001), but other mitochondria-derived GSIS mediators are of likely importance (MacDonald, 2004). For example, hydrogen peroxide has recently been suggested as a novel GSIS signal (Pi et al., 2007). In mouse pancreatic islets, mitochondrial uncoupling protein-2 (UCP2) can dampen glucose-induced increases in ATP (Zhang et al., 2001) and lower reactive oxygen species (ROS) (Robson-Doucette et al., 2011), and UCP2 thus likely plays an important part in beta cell biology. Studies aimed at clarifying the still debated physiological role of the beta cell UCP2 benefit from reliable information on protein levels and activity. In this chapter, we describe how to measure and knockdown UCP2 protein in INS-1 insulinoma cells, and we provide methods to determine UCP2 activity in this widely used pancreatic beta cell model.

2. TISSUE CULTURE

INS-1 cells are grown in a humidified carbogen atmosphere (5% CO_2, 95% air) in RPMI-1640 medium (Sigma R0883) that contains 11 m*M* glucose and 2 g/L sodium bicarbonate—buffering the growth medium at pH 7.2 under carbogen—and is supplemented with 5% (v/v) heat-inactivated fetal calf serum (Sigma F9665), 10 m*M* HEPES, 1 m*M* sodium pyruvate, 2 m*M* glutamine, 50 μ*M* β-mercaptoethanol, 100 U/mL penicillin, and 100 μg/mL streptomycin. Note that glutamine is an absolute requirement in all growth media and assay buffers to ensure full translation of UCP2 mRNA. Cells are kept in 75-cm^2 BD Falcon™ flasks and passaged by trypsinization: at 85–90% confluence, cells are washed twice with 10 mL Dulbecco's phosphate-buffered saline (DPBS, Invitrogen 14190-185) after which 2 mL trypsin-EDTA (Invitrogen 15630-056) is spread equally across the monolayer and then removed immediately. Cells are detached by gentle tapping and resuspended in 10 mL supplemented RPMI. For UCP2 activity assays, cells are counted with an "Improved Neubauer" hemocytometer (Weber Scientific International Ltd.) and seeded at appropriate density (see details below) on Seahorse Bioscience XF24 V7 (Part #100777-004) or 96-well Corning (Costar® 3595) cell culture microplates.

3. UCP2 Protein Detection

3.1. Sample preparation

Cells grown to roughly 80% confluence on Seahorse or Corning microplates are washed by adding 500 μL DPBS to the 200 μL growth medium in each well and removing 650 μL, leaving a notional volume of 50 μL. Another 500 μL DPBS is added to a single well and cells are resuspended by pipetting up and down vigorously—the cell suspension is then transferred to obtain cells from another well and this process is repeated until cells from an entire plate have been collected. Pooled cells are harvested and counted in DPBS following a 5-min spin at maximum speed in a microfuge. After another 2-min centrifugation, cells are solubilized in gel-loading buffer (10% (w/v) SDS, 250 m*M* Tris–HCl (pH 6.8), 5 m*M* EDTA, 50% (v/v) glycerol, 5% (v/v) β-mercaptoethanol, 0.05% (w/v) bromophenol blue) at 5.6×10^3 cells/μL. Complete solubilization is ensured by thorough pipetting, 30-s vortexing and a 5 min incubation at 100 °C. After being allowed to cool down for 2 min, samples are divided into 25 μL aliquots (1.4×10^5 cells each) that can be used for analysis immediately or may be stored at −80 °C for at least several weeks.

3.2. Western analysis

Cell lysate equivalent to 1.12×10^5 cells (20 μL sample) is separated on a 12% SDS polyacrylamide gel by running it at 150 V for 90 min. Proteins are transferred to a nitrocellulose membrane (Whatman Protan BA85) using a semi-dry Trans-Blot SD (Biorad) transfer cell set at 20 V for 30 min. After one rinse with Tris-buffered saline (TBS), the membrane is blocked for 2 h at room temperature in TBS containing 3% (w/v) bovine serum albumin (BSA) and then probed overnight at 4 °C with 1° antibodies diluted 1:2500 in TBS containing 0.1% (v/v) Tween-20 (TBST). Previously, we used goat anti-UCP2 antibodies from Santa Cruz for this purpose

(Affourtit & Brand, 2009), but we obtain more consistent results with rabbit anti-UCP2 antibodies from Calbiochem (catalogue #662047). Following incubation with 1° antibodies, the membrane is washed (1 × quickly, 1 × 15 min, 3 × 5 min) with TBST and then incubated for 1 h—at room temperature and protected from light—with ECL Plex goat antirabbit 2° antibodies (GE Healthcare Life Sciences, catalogue #GZPA45012) that are diluted 1:2500 in TBST and are coupled to Cy5, a fluorescent dye that exhibits maximum excitation/emission at 650/670 nm. After washing the membrane with TBST (1 × quickly, 1 × 15 min, 2 × 5 min) and TBS (2 × 5 min), cross-reacted proteins are visualized with a Typhoon™ FLA 9500 biomolecular imager (GE Healthcare Life Sciences) applying a long pass red filter to detect Cy5 fluorescence. Following 5-min rehydration in distilled water, the membrane is stained for 5 min with GelCode® Blue reagent (Pierce Biotechnology) diluted 2 × in water, destained via two brief washes with a mix of 50% (v/v) methanol and 1% (v/v) acetic acid, and then rescanned with the Typhoon™ imager to visualize total protein. Images are analyzed with ImageQuant TL software (GE Healthcare Life Sciences), and Cy5 fluorescence is normalized to total protein allowing quantitative comparison between different experiments.

4. UCP2 PROTEIN KNOCKDOWN

UCP2 protein can be knocked down in INS-1 cells with siRNA oligo-nucleotides predesigned by Ambion (Huntingdon, UK). In our hands, the most effective siRNA is targeted at exon 8 of the rat *Ucp2* (5′–3′ sense sequence: CGUAGUAAUGUUUGUCACCtt) and typically achieves more than 80% protein knockdown (Affourtit & Brand, 2009). Cells are seeded on Seahorse XF24 or Corning 96-well cell culture plates (4×10^4 cells in 200 μL per well) and cultured overnight in RPMI lacking antibiotics to roughly 50% confluence. Cells are transfected with 200 n*M* *Ucp2* siRNA that has been allowed to complex with 1.7 μg/mL Lipofectamine™ 2000 (Invitrogen, UK) for 20 min at room temperature in RPMI lacking fetal calf serum and antibiotics. In parallel, cells are transfected with scrambled siRNA (Ambion, Silencer® Negative Control 1) to control for possible nonspecific transfection effects. Following 2–3 days' further growth, cells are collected for Western analysis to confirm UCP2 knockdown or subjected to functional UCP2 assays.

5. UCP2 ACTIVITY

Uncoupling proteins catalyze a proton leak across the mitochondrial inner membrane and thus dissipate the protonmotive force as heat (Esteves & Brand, 2005). Ideally, comparison of proton leak activity between experimental systems requires simultaneous measurement of oligomycin-resistant mitochondrial respiratory activity *and* the membrane potential so that proton fluxes can be compared at identical driving forces (Affourtit & Brand, 2009). Although a novel method to quantify membrane potentials in intact cells has been reported recently (Gerencser et al., 2012), its relatively complicated nature impedes routine application at present. Our current inability to readily compare proton leak kinetics (i.e., the dependency of proton leak on membrane potential) in INS-1 cells with and without UCP2 thus stands in the way of a *direct* beta cell UCP2 activity assay. However, we have shown that UCP2 increases the basal mitochondrial INS-1E respiratory activity (Affourtit, Jastroch, & Brand, 2011). Moreover, UCP2 attenuates glucose-induced increases in mitochondrial respiration, coupling efficiency of oxidative phosphorylation, mitochondrial ROS, and insulin secretion (Affourtit et al., 2011). These UCP2-dependent phenotypes allow *indirect* measurement of UCP2 activity.

5.1. Mitochondrial bioenergetics: Glucose-stimulated respiration and coupling efficiency

The glucose sensitivity of mitochondrial respiration and the coupling efficiency of oxidative phosphorylation—defined as the proportion of mitochondrial respiration used to make ATP—are calculated from mitochondrial oxygen uptake rates, which are conveniently measured in intact INS-1 cells using Seahorse extracellular flux (XF) technology (Affourtit & Brand, 2009). Cells are seeded and cultured on XF24 cell culture plates and then washed into a glucose-free Krebs-Ringer buffer (KRH) containing 135 m*M* NaCl, 3.6 m*M* KCl, 10 m*M* HEPES (pH 7.4), 0.5 m*M* $MgCl_2$, 1.5 m*M* $CaCl_2$, 0.5 m*M* NaH_2PO_4, 2 m*M* glutamine, and 0.1% (w/v) BSA. Note that this KRH formulation contains glutamine to ensure that UCP2 mRNA is translated fully but lacks the usually included sodium bicarbonate to prevent alkalinization of the medium during respiratory and other functional assays. Washed cells are incubated at 37 °C under air for 30 min and then transferred to a Seahorse XF24 XF analyzer (controlled at 37 °C) for a 10-min calibration and three measurement cycles comprising a 1-min mix, 2-min wait, and 3-min measure period each. Glucose is then added and stimulated oxygen uptake is monitored for eight further measurement cycles. At this point, 1 μ*M* oligomycin and a mixture of 1 μ*M* rotenone and 2 μ*M* antimycin A are added sequentially to, respectively, inhibit the mitochondrial ATP synthase and to determine non-mitochondrial respiration.

As illustrated in Fig. 15.1, glucose sensitivity of mitochondrial respiratory activity is calculated by dividing total oxygen uptake activity in the presence of glucose (J_{G15}) by the basal respiratory rate (J_B). The coupling efficiency of oxidative phosphorylation is approximated (cf. Affourtit & Brand, 2009) by expressing oligomycin-sensitive respiration (J_{G15}–J_{OLI}) as a fraction of total respiration (J_{G15}). Rotenone and antimycin A-resistant respiratory activity (J_{RA}) is subtracted from all other activities to correct for non-mitochondrial oxygen consumption. Thus, glucose sensitivity$=(J_{G15}-J_{RA})/(J_B-J_{RA})$ and coupling efficiency$=1-((J_{OLI}-J_{RA})/(J_{G15}-J_{RA}))$. The glucose sensitivity and coupling efficiency calculated from Fig. 15.1 are 1.6 ± 0.046 and 0.54 ± 0.0057, respectively. In other words, 15 m*M* glucose causes a 60%

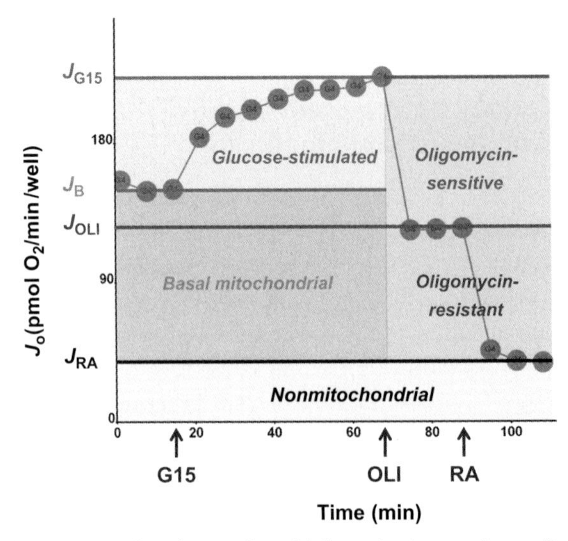

Figure 15.1 Glucose-stimulated mitochondrial respiration and coupling efficiency of oxidative phosphorylation. Cells were seeded in XF24 plates at 6×10^4 cells per well and grown for 48 h, washed into glucose-free KRH, and assayed in a Seahorse XF analyser (see text). J_B, J_{G15}, J_{OLI}, and J_{RA} reflect basal respiration or the respiratory rates observed in the cumulative presence of 15 m*M* glucose, 1 μ*M* oligomycin, and a mixture of 1 μ*M* rotenone and 2 μ*M* antimycin A, respectively. Compounds were added at times indicated by arrows. Data represent oxygen uptake rate (J_o) means of five wells. (See Color Insert.)

increase in basal INS-1 respiratory activity and the cells used just over half of this stimulated activity to make ATP. Importantly, in substrate-starved INS-1E cells, UCP2 knockdown stimulates the sensitivity of mitochondrial respiration to glucose (Fig. 15.2A) as well as the coupling efficiency at 30 m*M* glucose (Fig. 15.2B). These UCP2 effects are relatively modest—28% and 25% stimulation, respectively—but they are statistically significant and demonstrate that respiratory sensitivity to glucose and coupling efficiency are valuable indicators of UCP2 activity.

5.2. Mitochondrial bioenergetics: Basal respiratory activity

In addition to the bioenergetic phenotype shown in Fig. 15.2, UCP2 has a stimulatory effect on basal mitochondrial oxygen uptake in INS-1 cells

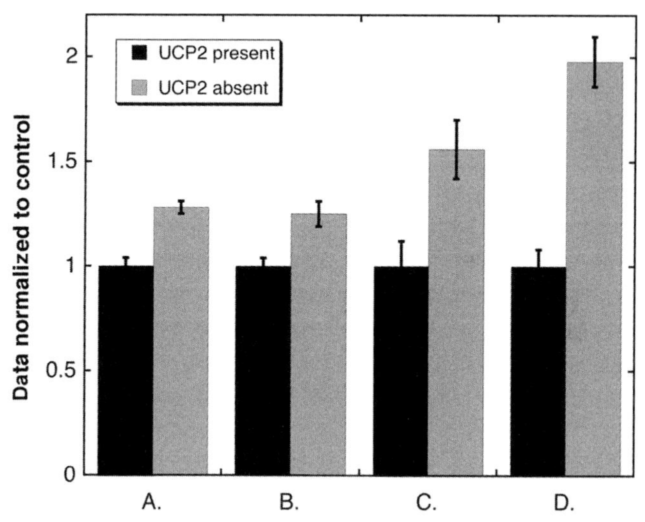

Figure 15.2 UCP2 activity. Cells were seeded in XF24 (A and B) or 96-well (C and D) plates at 4×10^4 cells per well, grown overnight, and then transfected with *Ucp2* or scrambled siRNA. Prior to all functional assays, cells were starved for 2 h in RPMI lacking glucose and pyruvate and containing only 1% (v/v) fetal calf serum. Cells were washed twice with glucose-free KRH, incubated in this buffer for 30 min at 37 °C (in a Seahorse XF analyzer or a shaking plate incubator procured from Labnet International), and then subjected to 30 m*M* glucose. UCP2 knockdown increases glucose sensitivity of mitochondrial respiration (A), coupling efficiency of oxidative phosphorylation (B), MitoSOX oxidation (C), and insulin secretion (D). Data were derived from published results (Affourtit et al., 2011) and were normalized to the means calculated from experiments with UCP2-containing cells. Controls ± SEM: glucose sensitivity = 1.3 ± 0.06; coupling efficiency = 0.32 ± 0.01; MitoSOX oxidation = 0.00118 ± 0.000145 relative fluorescence units per second; insulin secretion = 8.97 ± 0.76 ng insulin per well per 30 min. (For color version of this figure, the reader is referred to the online version of this chapter.)

(Affourtit et al., 2011). Basal oxygen uptake is an *a priori* non-normalized bioenergetic parameter and may thus only be compared between experimental systems (e.g., cells ± UCP2) if normalized, for example, to cell number. Densities of attached INS-1 cells can be determined conveniently in XF24 plates by probing the metabolic activity in each well with C_{12}-resazurin, a cell viability probe that is reduced in metabolically active cells to the fluorescent C_{12}-resorufin. Lyophilized C_{12}-resazurin powder is bought as part of a commercial kit (Invitrogen V-23110), dissolved at 10 μ*M* in dimethylsulfoxide (Fisher BPE231-100), and stored at −20 °C in 500 μL aliquots. On the day of use, C_{12}-resazurin is diluted in PBS (Sigma P4417) to 300 n*M* and protected from light. Immediately following Seahorse measurement of basal oxygen uptake, KRH buffer is replaced with 100 μL per well supplemented RPMI to which 20 μL diluted C_{12}-resazurin is added achieving a working concentration of 50 n*M*. The XF24 plate is then wrapped in aluminum foil and incubated at 37 °C under carbogen for 50 min. Total C_{12}-resorufin produced in metabolically active cells is measured by detecting the total fluorescence in each well using a PHERAstar FS microplate reader (BMG LABTECH) in fluorescence intensity, bottom-reading and well-scanning mode. C_{12}-resorufin is excited at 540 nm, and emitted light is detected at 590 nm.

Figure 15.3A shows the dependency of total well C_{12}-resorufin fluorescence on seeding density, which is described by a relation that is linear for densities between 1×10^4 and 1×10^5 cells per well. Typical seeding densities fall in the middle of this range, and the resazurin assay is, therefore, well suited to quantitatively detect positive and negative effects of a treatment on cell growth and any functional cell-normalized parameter. Figures 15.3B and C demonstrate, for example, clearly that normalization of cell respiration to C_{12}-resorufin fluorescence yields specific oxygen uptake rates that are consistent within a wide range of cell densities. It is generally possible, however, that parameters normalized to cell number using the resazurin assay are confounded by effects of a particular treatment on the specific metabolic activity of cells—the lack of such effects needs to be confirmed by independent control experiments.

5.3. Mitochondrial reactive oxygen species

Cells are seeded and cultured on 96-well plates and then starved and washed as described in the legend of Fig. 15.2. Subsequently, cells are incubated with 5 μ*M* MitoSOX (Invitrogen M36008), a mitochondria-targeted hydroethidine derivative that is oxidized by superoxide. Fluorescent MitoSOX

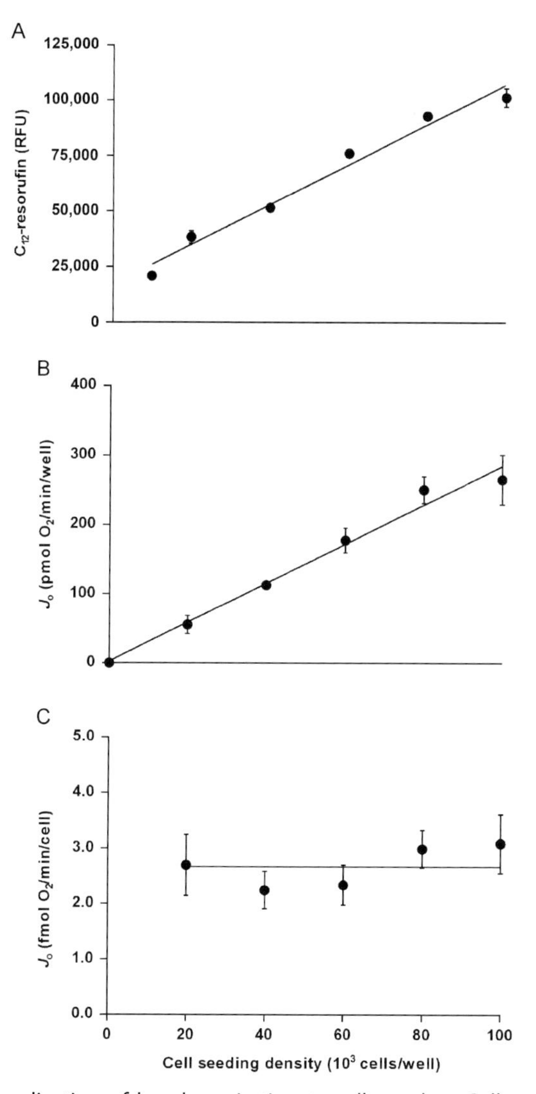

Figure 15.3 Normalization of basal respiration to cell number. Cells were seeded on XF24 plates at densities ranging from 1×10^4 to 1×10^5 cells in 200 μL RPMI per well and grown for 24 or 48 h (A and B, respectively). (A) Total C_{12}-resorufin formed during a 50-min incubation with 50 n*M* C_{12}-resazurin was measured as described in the text. Data represent fluorescence means ± SEM expressed in relative fluorescence units (RFU) corrected for cell-free background fluorescence. (B, C) Basal respiration was measured 4× after which the cells were incubated for 50 min with 50 n*M* C_{12}-resazurin. Respiratory rates are plotted against applied seeding density and expressed as time-resolved oxygen uptake per well (B) or per cell (C). Cell densities to which basal respiration was normalized were derived from the measured total C_{12}-resorufin fluorescence using data shown in panel A. The plate reader's focal height was set at 2 mm, and its gain was fixed between different experiments. Data represent means ± SEM calculated from 6 to 12 individual wells sampled from 2 to 3 separate plates. Data were fitted to linear expressions with slope kept variable (A and B) or set to zero (C).

oxidation products are excited at 510 nm, and emission is recorded at 580 nm twice every minute for 1 h. The slope of the resultant progress curve is expressed in relative fluorescent units per second and is proportional to the mitochondrial superoxide level. The mitochondrial origin of superoxide is confirmed by repeating the experiment with 1 μ*M* antimycin A or 1 μ*M* of the chemical uncoupling agent carbonyl cyanide *p*-trifluoromethoxyphenylhydrazone, which increases and decreases the MitoSOX oxidation rate, respectively. Although MitoSOX is used widely to report superoxide, it should be stressed that this probe is also oxidized, albeit to a lesser extent, by hydrogen peroxide (in the presence of peroxidases) and intracellular oxidoreductases. Additionally, since MitoSOX is charged, its uptake into the mitochondria and subsequent oxidation are affected by mitochondrial membrane potential as well as by ROS. Importantly, UCP2 knockdown increases the MitoSOX oxidation rate seen with starved INS-1 cells incubated at 30 m*M* glucose by almost 60% (Fig. 15.2C). Keeping in mind the incomplete specificity of MitoSOX and the complicating effects of possible changes in membrane potential, it may be concluded that UCP2 activity lowers mitochondrial ROS at high glucose. The UCP2 effect on mitochondrial ROS is larger than the UCP2 effects on mitochondrial respiration (Fig. 15.2), and MitoSOX oxidation differences between systems ± UCP2 are thus a useful additional qualitative indication of UCP2 activity.

5.4. Insulin secretion

We observe the largest effect of UCP2 knockdown in INS-1 cells on GSIS: insulin secretion at 30 m*M* glucose is almost twice as high in cells lacking UCP2 as in cells containing UCP2 (Fig. 15.2D). The relative magnitude of the UCP2 phenotype renders GSIS an attractive assay to gauge UCP2 activity, but it should be kept in mind that insulin secretion is a less direct reflector of UCP2 function (i.e., partial dissipation of the proton motive force) than mitochondrial respiratory activity, coupling efficiency, or mitochondrial ROS. For GSIS experiments, cells are grown, starved, and washed the same way as for the MitoSOX assay. Following a 30-min glucose exposure, 50 μL assay medium is collected from each well and centrifuged for 1 min at maximum microfuge speed to pellet any detached cells. The supernatants can be stored at −20 °C for several weeks or their insulin content can be quantified immediately by ELISA (Mercodia, Uppsala, Sweden).

ACKNOWLEDGMENTS

Our research is funded by the Medical Research Council (C. A., V. H.), the National Institutes of Health (M. D. B.), and Plymouth University (C. A., J. B.).

REFERENCES

Affourtit, C., & Brand, M. D. (2009). Measuring mitochondrial bioenergetics in INS-1E insulinoma cells. *Methods in Enzymology*, *457*, 405–424.

Affourtit, C., Jastroch, M., & Brand, M. D. (2011). Uncoupling protein-2 attenuates glucose-stimulated insulin secretion in INS-1E insulinoma cells by lowering mitochondrial reactive oxygen species. *Free Radical Biology & Medicine*, *50*, 609–616.

Barlow, J., Hirschberg, V., & Affourtit, C. (2013). On the role of mitochondria in pancreatic beta cells. In *Research on diabetes*. (edited by: i.Press) iConcept Press. ISBN: 978-14775550-1-9.

Esteves, T. C., & Brand, M. D. (2005). The reactions catalysed by the mitochondrial uncoupling proteins UCP2 and UCP3. *Biochimica et Biophysica Acta*, *1709*, 35–44.

Gerencser, A. A., Chinopoulos, C., Birket, M. J., Jastroch, M., Vitelli, C., Nicholls, D. G., et al. (2012). Quantitative measurement of mitochondrial membrane potential in cultured cells: Calcium-induced de- and hyperpolarization of neuronal mitochondria. *The Journal of Physiology*, *590*, 2845–2871.

MacDonald, M. J. (2004). Perspective: Emerging evidence for signaling roles of mitochondrial anaplerotic products in insulin secretion. *American Journal of Physiology. Endocrinology and Metabolism*, *288*, E1–E15.

Pi, J., Bai, Y., Zhang, Q., Wong, V., Floering, L. M., Daniel, K., et al. (2007). Reactive oxygen species as a signal in glucose-stimulated insulin secretion. *Diabetes*, *56*, 1783–1791.

Robson-Doucette, C. A., Sultan, S., Allister, E. M., Wikstrom, J. D., Koshkin, V., Bhattacharjee, A., et al. (2011). Uncoupling protein-2 regulates reactive oxygen species production, which influences both insulin and glucagon secretion. *Diabetes*, *60*, 2710–2719.

Rutter, G. A. (2001). Nutrient-secretion coupling in the pancreatic islet beta-cell: Recent advances. *Molecular Aspects of Medicine*, *22*, 247–284.

Zhang, C.-Y. , Baffy, G., Perret, P., Krauss, S., Peroni, O., Grujic, D., et al. (2001). Uncoupling protein-2 negatively regulates insulin secretion and is a major link between obesity, beta cell dysfunction, and type 2 diabetes. *Cell*, *105*, 745–755.

CHAPTER SIXTEEN

Effects of H_2O_2 on Insulin Signaling the Glucose Transport System in Mammalian Skeletal Muscle

Erik J. Henriksen[1]

Department of Physiology, Muscle Metabolism Laboratory, University of Arizona College of Medicine, Tucson, Arizona, USA

[1]Corresponding author: e-mail address: ejhenrik@u.arizona.edu

Contents

Abstract

Hydrogen peroxide (H_2O_2) is an important regulator of cellular events leading to glucose transport activation in mammalian skeletal muscle. In the absence of insulin, H_2O_2 in the low micromolar range engages the canonical IRS-1/PI3K/Akt-dependent insulin signaling pathway, as well as other signaling elements (AMPK and p38 MAPK), to increase basal glucose transport activity. In contrast, in the presence of insulin, H_2O_2 antagonizes insulin signaling by recruitment of various deleterious serine/threonine kinases, producing a state of insulin resistance. Here, we describe the H_2O_2 enzymatic-generating system, utilizing glucose oxidase, that has been used to investigate the impact of H_2O_2 on cellular signaling mechanisms related to glucose transport activity in isolated rat skeletal muscle preparations, such as the soleus. By varying the glucose oxidase

Methods in Enzymology, Volume 528
ISSN 0076-6879
http://dx.doi.org/10.1016/B978-0-12-405881-1.00016-1

concentration in the medium, target ranges of steady-state H_2O_2 concentrations (30–90 μM) can be attained for up to 6 h, with subsequent assessment of cellular signaling and glucose transport activity.

1. INTRODUCTION

The transmembrane transport of glucose in mammalian skeletal muscle cells is considered the rate-limiting step for the intracellular metabolism of this sugar (Holloszy & Hansen, 1996). This glucose transport process in myocytes is regulated, on the one hand, by the peptide hormone insulin through a series of intracellular signaling events (Henriksen, 2002; Henriksen, Diamond-Stanic, & Marchionne, 2011; Shepherd & Kahn, 1999; Zierath, Krook, & Wallberg-Henriksson, 2000), and, on the other hand, by a distinct set of signaling molecules that are activated during muscle contractions (Jessen & Goodyear, 2005; Shepherd & Kahn, 1999). These signaling events are briefly summarized below.

1.1. Regulation of the skeletal muscle glucose transport system

The canonical insulin-dependent signaling pathway includes activation of the insulin receptor and intracellular insulin receptor substrate (IRS) by tyrosine phosphorylation, engagement of phosphatidylinositol 3-kinase (PI3K) producing phosphatidylinositol-3,4,5-trisphosphate, which allosterically activates phosphoinositide-dependent kinases (PDKs). PDKs can stimulate Akt directly by Thr^{308} phosphorylation. Activation of the mTOR complex 2 (Ikenoue, Inoki, Yang, Zhou, & Guan, 2008; Kleiman, Carter, Ghansah, Patel, & Cooper, 2009) can also stimulate Akt by a mechanism involving Ser^{473} phosphorylation. An important substrate for Akt is the Rab-GTPase-containing AS160, also known as TBC1D4 (Sano et al., 2003). Akt phosphorylation inactivates AS160, facilitating translocation of GLUT-4-sequetering vesicles to the plasma membrane, thereby enhancing glucose transport activity (Cartee & Wojtaszewski, 2007). Defects in this IRS/PI3K/Akt/AS160 signaling pathway are responsible for many conditions of insulin resistance in skeletal muscle (Shepherd & Kahn, 1999; Zierath et al., 2000), and this insulin resistance is critical for the transition to a state of overt type 2 diabetes (American Diabetes Association, 2013).

The contraction-dependent pathway engages a distinct set of signaling elements, with a critical role of the activation of 5′-AMP-dependent protein kinase (AMPK) (Hardie, 2011; Winder & Thomson, 2007). This AMPK-dependent pathway also causes phosphorylation and inactivation of AS160, with associated increases in GLUT-4 translocation and glucose transport activity (Cartee & Wojtaszewski, 2007). This pathway for activation of glucose transport is obviously crucial for supplying glucose as a substrate to contracting skeletal muscle during an exercise bout (Jessen & Goodyear, 2005).

1.2. Evidence for H_2O_2 as a modulator of cellular signaling

Ample evidence exists in the literature for an important role of H_2O_2 as a modulator of cellular signaling in the context of glucose transport regulation in mammalian skeletal muscle (Henriksen et al., 2011). An important site of H_2O_2 production in myocytes is the mitochondrion (Stone & Yang, 2006). This H_2O_2 production is likely crucial for many cellular functions (Stone & Yang, 2006), including activation of the basal glucose transport process by engagement of AMPK-dependent and IRS/PI3K/Akt-dependent pathways (Kim, Saengsirisuwan, Sloniger, Teachey, & Henriksen, 2006). However, overproduction of H_2O_2 from myocellular mitochondria under conditions of nutrient excess or in the context of obesity is associated with impaired insulin signaling and glucose transport activity, that is, insulin resistance (Anderson et al., 2009; Hey-Mogensen, Jeppesen, Madsen, Kiens, & Franch, 2012). Although H_2O_2 is only a moderately reactive oxidant, it can be readily reduced via the Fenton reaction (using Cu^{2+} or Fe^{2+}) to the highly reactive hydroxyl radical (OH), which can markedly impair IRS/PI3K/Akt-dependent insulin signaling, leading to insulin resistance of glucose transport activity (Evans, Goldfine, Maddux, & Grodsky, 2002, 2003; Henriksen et al., 2011).

The remainder of this chapter will focus primarily on methods for modulating H_2O_2 exposure of isolated skeletal muscle strips from rodents (primarily rats). Subsequently, the impact of this oxidant exposure on the regulation of the glucose transport system in these isolated muscle preparations will be discussed, in order to underscore the physiological and pathophysiological significance of H_2O_2 as a modulator of glucose metabolism.

2. *IN VITRO* EXPOSURE TO H_2O_2

2.1. Simple addition of H_2O_2 to medium

The most basic approach for determining the impact of H_2O_2 on cellular signaling for regulation of glucose transport activity is the simple addition of H_2O_2 directly to the medium in order to attain some given H_2O_2 concentration. Early studies were restricted to exposure of cells to high concentrations (1 m*M* or more) of H_2O_2 (Hadari et al., 1992; Hayes & Lockwood, 1987; Heffetz, Bushkin, Dror, & Zick, 1990), which did not reflect the concentrations typically seen in the plasma (Bonnard et al., 2008). Later studies utilized more reasonable H_2O_2 concentrations. For example, the study of Blair, Hajduch, Litherland, and Hundal (1999) used the approach of exposing L6 myotubes for short durations (30 min) to H_2O_2 concentrations ranging from 50 μ*M* to 1 m*M*, with subsequent assessment of glucose transport activity or cellular signaling (such as Akt phosphorylation). A similar approach was used more recently by Bonnard et al. (2008), in which C_2C_{12} muscle cells were incubated for extended periods of time (up to 96 h) with 100 μ*M* H_2O_2 in order to assess the effect of this oxidant on mitochondrial protein expression and function. While this approach is certainly straight forward and easily implemented, the actual concentration of H_2O_2 in the medium, unless explicitly determined, is likely to be less than that targeted due to the removal of H_2O_2 by the endogenous catalase activity possessed by the cells. For this reason, methods for continuously producing H_2O_2 in the medium to attain a target value or range have been developed and used, as described in the following section.

2.2. Glucose oxidase method of *in vitro* H_2O_2 generation

A well-defined H_2O_2 enzymatic-generating system for investigations of H_2O_2 on the cellular regulation of glucose transport activity was used initially in studies utilizing cultured cells by the research group of Nava Bashan in Israel (Kozlovsky, Rudich, Potashnik, & Bashan, 1997; Rudich, Kozlovsky, Potashnik, & Bashan, 1997; Rudich, Tirosh, Potashnik, Khamaisi, & Bashan, 1999; Rudich et al., 1998). In these experiments, 3T3-L1 adipocytes were incubated in medium containing 5 m*M* glucose and 25–50 mU/ml glucose oxidase, which produced a steady-state concentration of 8–15 μ*M* H_2O_2 in the medium (Rudich et al., 1997, 1998), whereas incubation with 5 m*M* glucose and 100 mU/ml glucose oxidase produced ~27 μ*M* H_2O_2 (Rudich et al., 1999). Experiments in L6 myotubes using 5 m*M* glucose and 50 mU/ml glucose oxidase produced 20–40 μ*M* H_2O_2 for up to 24 h (Kozlovsky et al., 1997). Maddux et al. (2001) subsequently investigated this approach in L6 myotubes overexpressing GLUT-4 protein, and reported ~35 μ*M* H_2O_2 for up to 18 h when cells were incubated in 5 m*M* glucose and 100 mU/ml glucose oxidase.

This H_2O_2 enzymatic-generating system has been applied more recently to isolated rat skeletal muscle preparations (Archuleta et al., 2009; Diamond-Stanic et al., 2011; Dokken, Saengsirisuwan, Kim, Teachey, & Henriksen, 2008; Kim et al., 2006; Santos, Diamond-Stanic, Prasannarong, & Henriksen, 2012; Vichaiwong et al., 2009). These studies have utilized both isolated epitrochlearis muscle (predominantly fast-twitch glycolytic fibers) and soleus muscle strips (mainly slow-twitch oxidative fibers) (Henriksen et al., 1990). In these studies, 8 m*M* glucose and 100 mU/ml glucose oxidase produced 50–90 μ*M* H_2O_2 for up to 2 h in the absence (Kim et al., 2006) or presence (Dokken et al., 2008; Vichaiwong et al., 2009) of insulin. Extended exposures to these same incubation conditions for 4 h (Archuleta et al., 2009) or 6 h (Diamond-Stanic et al., 2011; Santos et al., 2012) produced similar medium concentrations of H_2O_2 (~70 μ*M*). It should be noted that the incubation medium is changed every hour to provide fresh medium containing glucose oxidase.

H_2O_2 in the medium of these *in vitro* studies was determined using the method of Thurman, Ley, and Scholz (1972) as described by Kozlovsky et al. (1997) and Maddux et al. (2001). In this method, 1 ml of the medium is acidified with 0.1 ml trichloroacetic acid (50%, w/v) on ice and centrifuged for 10 min at 5000 × *g*. The supernatant is then added to 0.2 ml of 10 m*M* ferrous ammonium sulfate and 0.1 ml of 2.5 *M* potassium thiocyanate. The absorbance of the ferrithiocyanate complex at 491 n*M* is then measured spectrophotometrically, and H_2O_2 concentration is determined using a *tert*-butyl hydroperoxide standard curve.

2.3. Other methods for assessment of H_2O_2 in plasma and skeletal muscle

While the spectrophotometric approach for assessing H_2O_2 in the medium is effective and useful, other approaches for determining cellular exposure to H_2O_2 exist and should be briefly mentioned. In the study of Anderson et al. (2009), the H_2O_2-emitting potential of mitochondria, which reflects the equilibrium between the formation of superoxide ion (and electron leak) in the electron transport chain and the matrix scavenging of H_2O_2, was determined *in vitro*. Interestingly, both overnutrition of rats (high fat feeding) and human obesity are associated with an increased H_2O_2-emitting potential of mitochondria and can be linked mechanistically with impaired insulin signaling in skeletal muscle (Anderson et al., 2009).

Other methods for assessing H_2O_2 exposure are of utility. For example, in Bonnard et al. (2008), a commercially available kit (Invitrogen, Grand

Island, NY) using Amplex Red hydrogen peroxide assessment was used to demonstrate that a high-fat/high-sucrose diet in mice leads to elevations in plasma H_2O_2 and to specific mitochondrial dysfunctions in skeletal muscle. A similar approach (Molecular Probes, Carlsbad, CA) was used recently to show that mitochondrial H_2O_2 release is elevated in insulin-resistant skeletal muscle from obese Zucker rats compared to lean Zucker rats (Hey-Mogensen et al., 2012). Finally, numerous novel methods, such as H_2O_2-selective fluorescent probes (e.g., HyPer), have been developed by Rhee, Chang, Jeong, and Kang (2010) for the assessment of H_2O_2 in biological fluids.

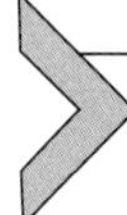

3. EFFECTS OF H_2O_2 ON THE GLUCOSE TRANSPORT SYSTEM IN ISOLATED SKELETAL MUSCLE

3.1. Measurement of glucose transport in isolated skeletal muscle preparations

Some basic information regarding the use of isolated skeletal muscle preparations would be prudent in this discussion. Except for very small rats (~70 g or less), most isolated muscles, such as the soleus and extensor digitorum longus, must be prepared into strips of smaller size (typically 25–35 mg) for use in *in vitro* incubations (Henriksen & Holloszy, 1991). Detailed information regarding the incubation conditions can be found in Henriksen et al. (1990) and Henriksen and Jacob (1995). In brief, following exposure of the muscles to the desired factors (e.g., H_2O_2) in Krebs-Henseleit buffer, the muscles are rinsed briefly (10 min) in glucose-free medium. Glucose transport activity is then assessed (typically for a 20-min period) by determining the specific intracellular accumulation of the glucose analogue 2-deoxy-D-glucose, using 1 m*M* 2-deoxy-[1,2-^{3}H] glucose (2-DG) (300 μCi/mmol) and 39 m*M* [U-^{14}C] mannitol (0.8 μCi/mmol), with mannitol being used to measure the extracellular space of the incubated muscle preparation. This approach has been validated as a reliable method for measuring the rate of transmembrane glucose transport (not glucose phosphorylation) (Hansen, Gulve, & Holloszy, 1994).

3.2. H_2O_2 engagement of insulin signaling factors in the absence of insulin

The H_2O_2 enzymatic-generating system has been used in isolated rat skeletal muscle to make some important observations regarding the impact of H_2O_2 on regulation of basal glucose transport activity. Exposure of rat soleus muscle to ~50–70 μ*M* H_2O_2 for 2 h results in enhanced glucose transport activity associated with engagement of IR tyrosine phosphorylation and Akt Ser^{473} phosphorylation (Kim et al., 2006), with additional (and likely redundant) roles of the activation of AMPKα, p38 MAPK, and c-jun N-terminal kinase (JNK) (Kim et al., 2006; Vichaiwong et al., 2009).

3.3. Impairment of insulin-stimulated insulin signaling by H_2O_2

The H_2O_2 enzymatic-generating system has also been used to delineate the deleterious effects of H_2O_2 on insulin action in the regulation of the glucose transport system in isolated rat skeletal muscle. Exposure of muscle to ~70 μ*M* H_2O_2 for 2–4 h or to ~35 μ*M* H_2O_2 for 6 h results in almost complete insulin resistance of glucose transport activity, due to engagement of various serine/threonine kinases known to negatively impact IRS-1-dependent insulin signaling, including glycogen synthase kinase-3 (Dokken et al., 2008), p38 MAPK (Archuleta et al., 2009; Blair et al., 1999; Diamond-Stanic et al., 2011; Maddux et al., 2001; Vichaiwong et al., 2009), and JNK (Santos et al., 2012). It is likely that the collective effects of these oxidant-activated kinases (and perhaps others yet to be investigated in this context) account for the complete impairment by H_2O_2 of insulin-dependent glucose transport activity in isolated rat skeletal muscle (Henriksen et al., 2011; Santos et al., 2012).

4. SUMMARY

This chapter presents an overview of the basic approach for exposing isolated mammalian skeletal muscle preparations to low micromolar concentrations of the oxidant H_2O_2 for up to 6 h, using the H_2O_2 enzymatic-generating system set up previously for cultured fat and muscle cells (Kozlovsky et al., 1997; Maddux et al., 2001; Rudich et al., 1997, 1998, 1999). Steady-state concentrations of H_2O_2 in the range of 30–90 μ*M* can be produced by varying the amount of glucose oxidase added to the incubation medium. The exposure of isolated skeletal muscle to H_2O_2 in this manner can be used to investigate the molecular underpinnings for the development of insulin resistance of glucose transport activity in this tissue, a critical defect for the development of type 2 diabetes.

REFERENCES

American Diabetes Association, (2013). Diagnosis and classification of diabetes mellitus. *Diabetes Care, 36*, S67–S74.

Anderson, E. J., Lustig, M. E., Boyle, C. E., Woodlief, T. L., Kane, D. A., Lin, C.-T. , et al. (2009). Mitochondrial H_2O_2 emission and cellular redox state link excess fat intake to insulin resistance in both rodents and humans. *The Journal of Clinical Investigation, 119*, 573–581.

Archuleta, T. L., Lemieux, A. M., Saengsirisuwan, V., Teachey, M. K., Lindborg, K. A., Kim, J. S., et al. (2009). Oxidant stress-induced loss of IRS-1 and IRS-2 proteins in rat skeletal muscle: Role of p38 MAPK. *Free Radical Biology & Medicine, 47*, 1486–1493.

Blair, A. S., Hajduch, E., Litherland, G. J., & Hundal, H. S. (1999). Regulation of glucose transport and glycogen synthesis in L6 muscle cells during oxidative stress. Evidence for cross-talk between the insulin and SAPK2/p38 mitogen-activated protein kinase signaling pathways. *The Journal of Biological Chemistry, 274*, 36293–36299.

Bonnard, C., Durand, A., Peyrol, S., Chanseaume, E., Chauvin, M.-A. , Morio, B., et al. (2008). Mitochondrial dysfunction results from oxidative stress in the skeletal muscle of diet-induced insulin-resistant mice. *The Journal of Clinical Investigation, 118*, 789–800.

Cartee, G. D., & Wojtaszewski, J. F. (2007). Role of Akt substrate of 160 kDa in insulin-stimulated and contraction-stimulated glucose transport. *Applied Physiology, Nutrition, and Metabolism, 32*, 557–566.

Diamond-Stanic, M. K., Marchionne, E. M., Teachey, M. K., Durazo, D. E., Kim, J. S., & Henriksen, E. J. (2011). Critical role of transient p38 MAPK activation in skeletal muscle insulin resistance caused by low-level in vitro oxidant stress. *Biochemical and Biophysical Research Communications, 405*, 439–444.

Dokken, B. B., Saengsirisuwan, V., Kim, J. S., Teachey, M. K., & Henriksen, E. J. (2008). Oxidative stress-induced insulin resistance in skeletal muscle: Role of glycogen synthase kinase-3. *American Journal of Physiology. Endocrinology and Metabolism, 294*, E615–E621.

Evans, J. L., Goldfine, I. D., Maddux, B. A., & Grodsky, G. M. (2002). Oxidative stress and stress-activated signaling pathways: A unifying hypothesis of type 2 diabetes. *Endocrine Reviews, 23*, 599–622.

Evans, J. L., Goldfine, I. D., Maddux, B. A., & Grodsky, G. M. (2003). Are oxidative stress-activated signaling pathways mediators of insulin resistance and β-cell dysfunction? *Diabetes, 52*, 1–8.

Hadari, Y., Tzahar, E., Nadiv, O., Roberts, C. T., Jr., LeRoith, D., Yarden, Y., et al. (1992). Insulin and insulinomimetic agents induce activation of phosphatidylinositol 3-kinase upon its association with pp 185 (IRS-1) in intact rat livers. *The Journal of Biological Chemistry, 267*, 17483–17486.

Hansen, P. A., Gulve, E. A., & Holloszy, J. O. (1994). Suitability of 2-deoxyglucose for in vitro measurement of glucose transport activity in skeletal muscle. *Journal of Applied Physiology, 76*, 1862–1867.

Hardie, D. G. (2011). Energy sensing by the AMP-activated protein kinase and its effects on muscle metabolism. *The Proceedings of the Nutrition Society, 70*, 92–99.

Hayes, G. R., & Lockwood, D. H. (1987). Role of insulin receptor phosphorylation in the insulinomimetic effects of hydrogen peroxide. *Proceedings of the National Academy of Sciences of the United States of America, 84*, 8115–8119.

Heffetz, D., Bushkin, I., Dror, R., & Zick, Y. (1990). The insulinomimetic agents H_2O_2 and vanadate stimulate protein tyrosine phosphorylation in intact cells. *The Journal of Biological Chemistry, 265*, 2896–2902.

Henriksen, E. J. (2002). Invited review: Effects of acute exercise and exercise training on insulin resistance. *Journal of Applied Physiology, 93*, 788–796.

Henriksen, E. J., Bourey, R. E., Rodnick, K. J., Koranyi, L., Permutt, M. A., & Holloszy, J. O. (1990). Glucose transporter protein content and glucose transport capacity in rat skeletal muscles. *American Journal of Physiology–Endocrinology and Metabolism, 259*, E593–E598.

Henriksen, E. J., Diamond-Stanic, M. K., & Marchionne, E. M. (2011). Oxidative stress and the etiology of insulin resistance and type 2 diabetes. *Free Radical Biology & Medicine, 51*, 993–999.

Henriksen, E. J., & Holloszy, J. O. (1991). Effect of diffusion distance on measurement of glucose transport in rat skeletal muscles in vitro. *Acta Physiologica Scandinavica, 143*, 381–386.

Henriksen, E. J., & Jacob, S. (1995). Effects of captopril on glucose transport activity in skeletal muscle of obese Zucker rats. *Metabolism, 44*, 267–272.

Hey-Mogensen, M., Jeppesen, M. J., Madsen, K., Kiens, B., & Franch, J. (2012). Obesity augments the age-induced increase in mitochondrial capacity for H_2O_2 release in Zucker fatty rats. *Acta Physiologica, 204*, 354–361.

Holloszy, J. O., & Hansen, P. A. (1996). Regulation of glucose transport into skeletal muscle. In M. P. Blaustein, H. Grunicke, E. Habermann, D. Pette, G. Schultz, & M. Schweiger (Eds.), *Reviews of physiology, biochemistry and pharmacology* (pp. 99–193). Berlin: Springer Verlag.

Ikenoue, T., Inoki, K., Yang, Q., Zhou, X., & Guan, K. L. (2008). Essential function of TORC2 in PKC and Akt turn motif phosphorylation, maturation and signaling. *The EMBO Journal, 27*, 1919–1931.

Jessen, N., & Goodyear, L. J. (2005). Contraction signaling to glucose transport in skeletal muscle. *Journal of Applied Physiology, 99*, 330–337.

Kim, J. S., Saengsirisuwan, V., Sloniger, J. A., Teachey, M. K., & Henriksen, E. J. (2006). Oxidant stress and skeletal muscle glucose transport: Roles of insulin signaling and p38 MAPK. *Free Radical Biology & Medicine, 41*, 818–824.

Kleiman, E., Carter, G., Ghansah, T., Patel, N. A., & Cooper, D. R. (2009). Developmentally spliced PKCbetaII provides a possible link between mTORC2 and Akt kinase to regulate 3 T3-L1 adipocyte insulin-stimulated glucose transport. *Biochemical and Biophysical Research Communications, 388*, 554–559.

Kozlovsky, N., Rudich, A., Potashnik, R., & Bashan, N. (1997). Reactive oxygen species activate glucose transport in L6 myotubes. *Free Radical Biology & Medicine, 23*, 859–869.

Maddux, B. A., See, W., Lawrence, J. C., Jr., Goldfine, A. L., Goldfine, I. D., & Evans, J. L. (2001). Protection against oxidative stress-induced insulin resistance in rat L6 muscle cells by micromolar concentrations of α-lipoic acid. *Diabetes, 50*, 404–410.

Rhee, S. G., Chang, T. S., Jeong, W., & Kang, D. (2010). Methods for detection and measurement of hydrogen peroxide inside and outside of cells. *Molecules and Cells, 29*, 539–549.

Rudich, A., Kozlovsky, N., Potashnik, R., & Bashan, N. (1997). Oxidant stress reduces insulin responsiveness in 3 T3-L1 adipocytes. *American Journal of Physiology. Endocrinology and Metabolism, 272*, E935–E940.

Rudich, A., Tirosh, A., Potashnik, R., Hemi, R., Kanety, H., & Bashan, N. (1998). Prolonged oxidative stress impairs insulin-induced GLUT4 translocation in 3 T3-L1 adipocytes. *Diabetes, 47*, 1562–1569.

Rudich, A., Tirosh, A., Potashnik, R., Khamaisi, M., & Bashan, N. (1999). Lipoic acid protects against oxidative stress induced impairment in insulin stimulation of protein kinase B and glucose transport in 3 T3-L1 adipocytes. *Diabetologia, 42*, 949–957.

Sano, H., Kane, S., Sano, E., Miinea, C. P., Asara, J. M., Lane, W. S., et al. (2003). Insulin-stimulated phosphorylation of a Rab GTPase-activating protein regulates GLUT4 translocation. *The Journal of Biological Chemistry, 278*, 14599–14602.

Santos, F. R., Diamond-Stanic, M. K., Prasannarong, M., & Henriksen, E. J. (2012). Contribution of serine kinase c-Jun N-terminal kinase (JNK) to oxidant-induced insulin resistance in isolated rat skeletal muscle. *Archives of Physiology and Biochemistry, 118*, 231–236.

Shepherd, P. R., & Kahn, B. B. (1999). Glucose transporters and insulin action. Implications for insulin resistance and diabetes mellitus. *The New England Journal of Medicine, 341*, 248–257.

Stone, J. R., & Yang, S. (2006). Hydrogen peroxide: A signaling messenger. *Antioxidants & Redox Signaling, 8*, 243–270.

Thurman, R. G., Ley, H. G., & Scholz, R. (1972). Hepatic microsomal ethanol oxidation. *European Journal of Biochemistry, 25*, 420–430.

Vichaiwong, K., Henriksen, E. J., Toskulkao, C., Prasannarong, M., Bupha-Intr, T., & Saengsirisuwan, V. (2009). Attenuation of oxidant-induced muscle insulin resistance and p38 MAPK by exercise training. *Free Radical Biology & Medicine, 47*, 593–599.

Winder, W. W., & Thomson, D. M. (2007). Cellular energy sensing and signaling by AMP-activated protein kinase. *Cell Biochemistry and Biophysics, 47*, 332–347.

Zierath, J. R., Krook, A., & Wallberg-Henriksson, H. (2000). Insulin action and insulin resistance in human skeletal muscle. *Diabetologia, 43*, 821–835.

CHAPTER SEVENTEEN

Monitoring of Hydrogen Peroxide and Other Reactive Oxygen and Nitrogen Species Generated by Skeletal Muscle

Malcolm J. Jackson[1]

MRC-Arthritis Research UK Centre for Integrated Research into Musculoskeletal Ageing, Institute of Ageing and Chronic Disease, University of Liverpool, Liverpool, United Kingdom

[1]Corresponding author: e-mail address: m.j.jackson@liverpool.ac.uk

Contents

1. Introduction 280
   1.1 Nature of the free radicals and ROS that are generated by skeletal muscle 280
   1.2 Physiological roles mediated by the H_2O_2 and other ROS generated by contractions in skeletal muscle 282
   1.3 Developments in approaches to study ROS in skeletal muscle 284
2. Monitoring Extracellular ROS Using Microdialysis Techniques 284
   2.1 Use of microdialyis to measure extracellular ROS in skeletal muscle 286
   2.2 Experimental procedures 286
3. Assessment of Intracellular ROS Activities 290
   3.1 Choice of model system 290
4. Concluding Remarks 296

Acknowledgments 297

References 297

Abstract

Understanding the roles and functions of reactive oxygen and nitrogen species in skeletal muscle requires the ability to monitor specific species at rest and during muscle use. These species are generated at a variety of sites in muscle fibers, and approaches to their analysis are becoming available. We utilize microdialysis approaches to sample the interstitial space of skeletal muscle *in vivo* to allow continuous monitoring of nitric oxide and some reactive oxygen species. The approach to monitor intracellular species that we currently favor utilizes isolated single muscle fibers to allow the use of fluorescent probes and epifluorescence microscopy. Methods are described that illustrate these approaches.

Methods in Enzymology, Volume 528
ISSN 0076-6879
http://dx.doi.org/10.1016/B978-0-12-405881-1.00017-3

1. INTRODUCTION

1.1. Nature of the free radicals and ROS that are generated by skeletal muscle

The nature of the molecules generated by skeletal muscle that are known as reactive oxygen species (ROS) or reactive nitrogen species (RNS) and which include various free radicals has been extensively reviewed in recent years (see Jackson, 2011; Powers & Jackson, 2008). In brief, skeletal muscle fibers generate superoxide and nitric oxide (NO) as the primary species, and these parent molecules lead to formation of several secondary ROS and RNS. Both superoxide and NO are generated from various sources within muscle fibers and in addition superoxide (Reid, Shoji, Moody, & Entman, 1992; McArdle et al., 2001), hydrogen peroxide (Vasilaki, Mansouri, et al., 2006), and NO (Vasilaki, Mansouri, et al., 2006) are released into the interstitial space from muscle fibers or generated on the extracellular side of the muscle plasma membrane.

The intracellular contents and activities of superoxide, hydrogen peroxide, and NO are increased by contractile activity (Palomero, Pye, Kabayo, Spiller, & Jackson, 2008; Pye, Kabayo, Palmero, & Jackson, 2007; Reid et al., 1992), and superoxide, hydrogen peroxide, hydroxyl radical activity, and NO are increased in the muscle interstitial space by contractile activity (McArdle, Pattwell, Vasilaki, Griffiths, & Jackson, 2001; Pattwell, McArdle, Morgan, Patridge, & Jackson, 2004; Vasilaki, Mansouri, et al., 2006). Since the initial observations in the 1980s (Davies, Quintanilha, Brooks, & Packer, 1982; Jackson, Edwards, & Symons, 1985), most authors have assumed that the ROS generated by contractions are predominantly generated by mitochondria, but recent data argue against this possibility (for discussion, see Jackson, 2011). Nonmitochondrial sources for the generation of ROS within skeletal muscle have not been extensively studied, but some potential nonmitochondrial sources including NAD(P)H oxidase(s) have been described in skeletal muscle (Javesghani, Magder, Barreiro, Quinn, & Hussain, 2002). NAD(P)H oxidases localized to skeletal muscle plasma membrane (Javesghani et al., 2002), sarcoplasmic reticulum (Xia, Webb, Gnall, Cutler, & Abramson, 2003), T-tubules (Espinosa et al., 2006), and mitochondria (Sakellariou et al., 2013) have been reported. The activity of the T-tubule localized enzyme has also been claimed to be activated by contractions (Espinosa et al., 2006), and inhibitor studies published recently support the possibility that NADP(H) oxidases contribute to the increase in cytosolic superoxide generation during contractile activity. Both Nox2 and Nox4 isoforms of NADPH oxidase have been reported to be present in skeletal muscle (Sakellariou et al., 2013).

The sources of the extracellular ROS that are released from skeletal muscle cells in culture or isolated muscle preparations are also relatively obscure. Both hydrogen peroxide and NO can theoretically diffuse through the plasma membrane, and hence, intracellular sources for these species may play a role (but see later for a further discussion of hydrogen peroxide). Although superoxide has been frequently detected in the extracellular medium surrounding muscle cells and isolated muscles (e.g., see McArdle et al., 2001; Reid et al., 1992), substantial diffusion of this species (or its protonated form) through a plasma membrane seems extremely unlikely (Halliwell & Gutteridge, 1989). In intact muscle preparations, it appears that xanthine oxidase enzymes in the endothelium associated with the muscle play an important role in contraction-induced release of superoxide (Gomez-Cabrera, Close, Kayani, McArdle, & Jackson, 2010). This source for ROS generation has been claimed to play important roles in adaptations of muscle to contractile activity (Gomez-Cabrera et al., 2005) but has been relatively sparsely studied in recent years. In studies of isolated muscle fibers and myotubes, the role of xanthine oxidase is unclear. Javesghani et al. (2002) have reported release of ROS derived from a plasma membrane-localized NAD(P)H oxidase. Other NAD(P)H-dependent systems have also been implicated (for discussion, see Jackson, Pye, & Palomero, 2006). An updated scheme depicting the various sites that have been identified for generation of ROS and NO in skeletal muscle has been recently published (Jackson, 2011).

Relatively little is known about the factors that control ROS production by skeletal muscle. Initial studies suggested that superoxide generation by skeletal muscle was essentially a by-product of oxygen consumption by mitochondria and many early authors quoted reports that 2–5% of the total oxygen consumed by mitochondria undergoes one electron reduction and generates substantial amounts of superoxide (Boveris & Chance, 1973; Loschen, Azzi, Richter, & Flohe, 1974). This assumption was related to exercise with the assumption that the increased ROS generation that occurs during contractile activity was directly related to the elevated oxygen consumption that occurs with increased mitochondrial activity, implying potentially a 50- or 100-fold increase in superoxide generation by skeletal muscle during aerobic contractions (e.g., see Kanter, 1994; Urso & Clarkson, 2003). As previously reviewed (Jackson, 2008;

Powers & Jackson, 2008), data now argue against such a substantial formation of superoxide within mitochondria. In particular, Brand and colleagues reassessed the rate of production of ROS by mitochondria and indicated that the upper estimate of the proportion of the electron flow giving rise to ROS was an order of magnitude lower than the original minimum estimate (St-Pierre, Buckingham, Roebuck, & Brand, 2002). In more recent data, investigators have targeted redox-sensitive probes to mitochondria of muscle fibers and have been unable to demonstrate any change in mitochondrial redox potential during contractile activity which they argue implies a lack of mitochondrial ROS generation during contractions (Michaelson, Shi, Ward, & Rodney, 2010). Recent inhibitor studies have also supported this conclusion (Sakellariou et al., 2013).

The control of superoxide production by putative NAD(P)H oxidase sources in skeletal muscle is unknown. In neutrophils and other phagocytic cells, the NAD(P)H oxidase complex is assembled on membranes following a stimulus for activation, but different mechanisms apply in many non-phagocytic cells (Arora, Vaishya, Dabla, & Singh, 2010). Espinosa et al. (2006) hypothesized that the T-tubule localized NAD(P)H oxidase might be activated by depolarization of the T-tubules, but this has not been confirmed.

Xanthine oxidase has been recognized to contribute to superoxide generation in ischemia and reperfusion, but recent data also indicate that the xanthine oxidase pathway is important in superoxide formation in the extracellular fluid following a nondamaging protocol of muscle contractions (Gomez-Cabrera et al., 2010). Most studies argue that in relatively hypoxic tissues, anaerobic metabolism leads to proteolytic modification of xanthine dehydrogenase to form xanthine oxidase (Nishino, Okamoto, Eger, Pai, & Nishino, 2008) and to the increased availability of the xanthine oxidase substrates, hypoxanthine, and xanthine (Pacher, Nivorozhkin, & Szabó, 2006). This has led some workers to argue that superoxide generation by contracting muscle during exercise is greatest at exhaustion (Viña et al., 2000).

1.2. Physiological roles mediated by the H_2O_2 and other ROS generated by contractions in skeletal muscle

While it is clear that oxidative damage to lipids, DNA, and protein may contribute to tissue dysfunction in various situations, it seems likely that oxidative damage is not induced to any substantial extent by the modest changes in ROS concentrations/activities that occur during normal contractions but would be likely to require higher concentrations/activities or more sustained exposure. There has been increasing recognition that ROS mediate physiological processes in tissues, and these molecules have been recognized as important signaling molecules with regulatory functions that modulate changes in cell and tissue homeostasis and gene expression (Dröge, 2002; Haddad, 2002; Jackson et al., 2002). Signaling by these reactive molecules is mainly carried out by targeted modifications of specific residues in proteins (Janssen-Heininger et al., 2008).

ROS appear to modulate a number of physiological responses in skeletal muscle. A single period of contractile activity in mouse muscle was found to increase the activity of muscle antioxidant defense enzymes such as superoxide dismutase (SOD) and catalase together with HSP60 and HSP70 content (McArdle et al., 2001), changes that were replicated in human muscle studies. Other studies have implicated redox signaling in diverse processes in muscle such as maintenance of force production during contractions, glucose uptake, and insulin signaling (Powers & Jackson, 2008).

ROS have become increasingly recognized to mediate some adaptive responses of skeletal muscle to contractile activity through the activation of redox-sensitive transcription factors, such as NFκB, Activator Protein-1, and Heat Shock Factor 1 (Cotto & Morimoto, 1999). NFκB is a redox-regulated factor and ROS have been proposed to be principal regulators of NFκB activation in many situations (Moran, Gutteridge, & Quinlan, 2001). In skeletal muscle, NFκB modulates expression of a number of genes associated with myogenesis (Bakkar et al., 2008; Dahlman, Wang, Bakkar, & Guttridge, 2009), catabolism-related genes (Bar-Shai, Carmeli, & Reznick, 2005; Peterson & Guttridge, 2008; Van Gammeren, Damrauer, Jackman, & Kandarian, 2009), and cytoprotective proteins during adaptation to contractile activity (Vasilaki, McArdle, et al., 2006). Moreover, skeletal muscle has been identified as an endocrine organ producing cytokines via NFκB activation following a number of stresses including systemic inflammation or physical strain (Febbraio & Pedersen, 2005). Activation of NFκB by ROS appears to involve oxidation of key cysteine residues in the upstream activators of NFκB, and in many situations, the process is inhibited by antioxidants or reducing agents. Thus, the increased ROS generated by skeletal muscle during contractions appear to stimulate various adaptive responses in the muscle. Activation of redox-sensitive transcription factors such as NFκB is one pathway by which these changes occur, but many others are feasible. Activation of these responses is one of the key functions of the ROS generated during contractions and is essential for maintenance of muscle cell homeostasis during repeated episodes of contractile activity.

1.3. Developments in approaches to study ROS in skeletal muscle

Most studies in this area have used indirect approaches to assess ROS activities in contracting skeletal muscle, and traditional biochemical analyses to study the redox couples for glutathione, etc., have utilized whole tissues. In order to understand the manner in which ROS regulate redox-sensitive processes and cellular adaptations in discrete compartments of the cell, it is necessary to develop approaches that permit analyses of ROS in specific body sites and within single cells and cell organelles and to differentiate the sites of ROS activities.

2. MONITORING EXTRACELLULAR ROS USING MICRODIALYSIS TECHNIQUES

The detection of ROS in biological systems is difficult because they exist in very low concentrations and react rapidly close to their site of formation (with either endogenous antioxidants or cellular components), thus having little capacity to accumulate Camus et al. (1994). Consequently, many of the discrepancies in the literature regarding the precise role of ROS in skeletal muscle may be attributable to the difficulties and inaccuracies in the measurement of ROS (Close, AshtonT, Doran, & MacLaren, 2004). Most studies examining ROS generation by contracting skeletal muscle have relied on indirect assays including changes in endogenous antioxidant levels (e.g., Camus et al., 1994; Duthie, Robertson, Maughan, & Morrice, 1990) and the measurement of indicators of total ROS activity, such as products of lipid peroxidation (e.g., Close et al., 2004; Maughan et al., 1989).

Since primary ROS are only found close to their site of synthesis, an assay system that is designed to measure specific primary ROS must have access to this site. We have developed *in vivo* microdialysis approaches as a technique that potentially permits this in tissues (Close, Ashton, McArdle, & Jackson, 2005). For a full review of the general technique and applications of *in vivo* microdialysis, see Benveniste and Huttemeier (1990). Briefly, *in vivo* microdialysis involves inserting a small probe containing a dialysis membrane into the tissue of interest. Once the probe is in place, the membrane is perfused on the inside with a physiological solution, whilst the outside of the membrane is in contact with the extracellular space. Small compounds will perfuse across the dialysis membrane. The physiological solution is perfused at a very slow flow rate allowing equilibrium to be reached between the solution inside the membrane and the solution in contact with the membrane on the outside. The dialysate can then be collected via the outlet port for subsequent analysis.

The technique of *in vivo* microdialysis was first described in 1966 (Bito, Davson, Levin, Murray, & Snider, 1966) to allow the collection of amino acids from remote regions of the brain. The advantage of *in vivo* microdialysis is that it is the only technique that can collect many substances from remote regions without causing major disruption of tissues (Benveniste & Huttemeier, 1990). It has since been used in skeletal muscle to allow the measurement of various substances including pain producing substances (McArdle, Khera, Edwards, & Jackson, 1999), metabolites (Maclean, Bangsbo, & Saltin, 1999), and most recently ROS (Close et al., 2005; McArdle et al., 2001).

Microdialysis analyses of ROS in the extracellular space offer potential advantages over analyses in the circulation since the ROS are removed from potential large molecular weight reactants, and there is a reduced dilution compared with the blood stream (Benveniste & Huttemeier, 1990). Studies have used the technique of microdialysis to measure extracellular superoxide release and extracellular hydroxyl radical activity (Close et al., 2005). These studies involved perfusing cytochrome *c* and monitoring the reduction of cytochrome *c* as a measure of superoxide release and perfusing salicylate and measuring the production of 2,3-dihydroxybenzoic acid (DHB) for assessment of hydroxyl radical activity. The cytochrome *c* assay for detection of superoxide is based upon the reduction of ferricytochrome *c* by superoxide to form ferrocytochrome *c*. This reduction results in an increased absorbance of cytochrome *c* at 550 nm that can be measured spectrophotometrically (McArdle et al., 2001). We have reported an increase in the reduction of cytochrome *c in vivo* during a period of nondamaging isometric contractions using this assay (Close, Kayani, Ashton, McArdle, & Jackson, 2007; McArdle et al., 2001), although the validity of the cytochrome *c* assay has been questioned (Close et al., 2005).

Reaction of hydroxyl radical with the aromatic ring of salicylate at either the 3 or 5 position results in the formation of two stable metabolites, 2,3- and 2,5-DHB. These metabolites can then be separated and measured using HPLC with electrochemical detection (HPLC-EC) as an indicator of hydroxyl radical activity. It is recognized that 2,5-DHB can also be produced enzymatically and hence may not be an accurate assay for hydroxyl radical activity, and therefore, measurement of 2,3-DHB is preferred. This technique is frequently cited as a measure of hydroxyl radical activity (Halliwell & Gutteridge, 1989), although the specificity of this assay has also been questioned. Specifically, it has been suggested that peroxynitrite ($ONOO^-$) may also contribute to the hydroxylation of salicylate and that hydroxyl radical is not a prerequisite for these hydroxylation reactions (Close et al., 2005).

2.1. Use of microdialyis to measure extracellular ROS in skeletal muscle

The increasing availability of microdialysis probes from commercial sources has greatly improved the range and reliability of probes for study of skeletal muscle. Two main types of probe have been used; the "loop" dialysis membrane is effectively a short segment of a continuous tube, whereas "linear" probes have a concentric tube arrangement whereby the perfusion fluid enters through an inner tube, flows to its distal end, exits the tube, and enters the space between the inner tube and an outer dialysis membrane. The perfusion fluid then moves toward the proximal end of the probe, and this is where the dialysis takes place. A wide variety of different probes with different molecular weight cutoffs are available. The "linear" probes are widely used in comparison with other types. Examples of probes can be found at the following Web site http://www.microbiotech.se.

We have used the "linear" probes and have regularly used 10- or 4-mm membrane probes with 0.5 mm diameter and a molecular weight cutoff of 35,000 Da (e.g., from *Metalant*, Sweden). Similar probes are commercially available for use in a wide variety of animal species, and some have also been specifically designed and approved for clinical use in man.

2.2. Experimental procedures

Detailed descriptions of the placement of the probes into suitable muscles of anaesthetized rodents (or humans) have been published (Close et al., 2005; McArdle et al., 2001). In the mouse, the *gastrocnemius* muscle has been more frequently used to obtain the necessary muscle bulk. It is technically feasible to place three or four probes in a single mouse *gastrocnemius* facilitating multiple measurements from the same muscle with the probes remaining in the muscle of the anaesthetized rodent for up to 4 h. We have used this approach to simultaneously monitor superoxide, hydrogen peroxide, NO, and hydroxyl radical activity in mouse muscle extracellular fluid (Vasilaki, Mansouri, et al., 2006). In man, probes can be placed under local anaesthetic. We have placed clinical probes in the *quadriceps* and *tibialis anterior* muscles, although other muscles could undoubtedly be examined. An overview of the approach that we have used is shown in Fig. 17.1.

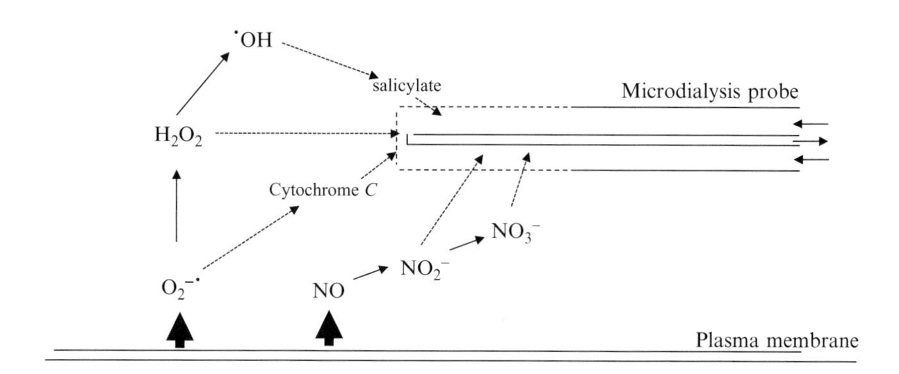

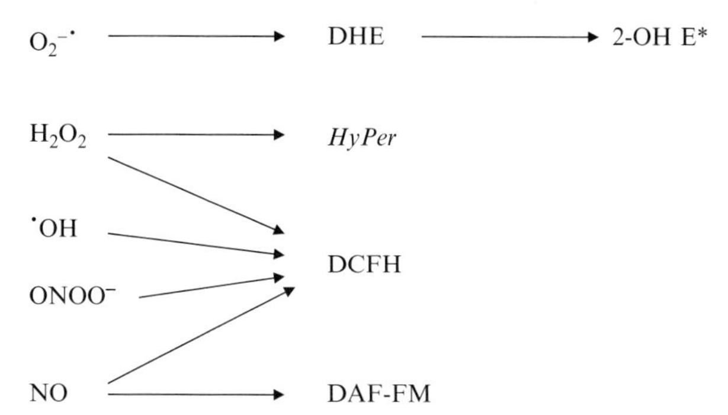

Figure 17.1 Schematic representation of the approaches used to detect ROS in tissue extracellular space using chemical approaches and microdialysis techniques. The lower section of the figure shows the general approaches that have been used to monitor intracellular ROS using fluorescent probes. DCFH is widely used, but the lack of specificity of this probe and the problems with its use are widely described. The other probes shown (DAF-FM, DHE, and *HyPer*) can offer greatly improved specificity but suffer from other complications in use. Thus, the *HyPer* probe is genetically encoded and must, therefore, be transfected into muscle, whereas the 2-hydroxyethidium product should be specifically analyzed to obtain a specific measure of superoxide using DHE as the indicator.

2.2.1 Detection of hydroxyl radical activity in muscle extracellular space in mice

Procedure: Microdialysis probes are perfused with 20 m*M* salicylate in normal saline at a flow rate of 4 μl/min and allowed to stabilize for 30 min. Samples are then collected from the outlet tubes of the probe over sequential 15 min periods. The 2,3-DHB and 2,5-DHB generated from salicylate in the microdialysis fluids are measured as an index of reaction with hydroxyl radicals. 2,3-DHB and 2,5-DHB are measured by HPLC with electrochemical detection. Our HPLC system consisted of a Rheodyne injector, HPLC pump (Gilson Model 303), Spherisorb 5 ODS column (HPLC technology): 25 cm × 4.6 mm with guard column and C-8 cartridge (BDH) and an electrochemical detector (Gilson Model 141). The HPLC eluant consists of 34 m*M* sodium citrate, 27.7 m*M* acetate buffer (pH 4.75) mixed with methanol 97.2:2.8 (v/v). Standard solutions of 2,3-DHB and 2,5-DHB are prepared in HPLC grade water. Twenty microliters of samples or standards are eluted at a flow rate of 0.9 ml/min and monitored at +65 V with the electrochemical detector.

With the conditions and probes described earlier, we have reported the formation of approximately 150 pmol 2,3-DHB/15 min from mouse skeletal muscle at rest (Close et al., 2005; Vasilaki, Mansouri, et al., 2006). These values change if different length probes are used, the molecular weight cutoff differs or if the flow rate is varied. They are also likely to change with probes of different composition, although we have not examined this.

We have examined the effect of electrical stimulation of contraction on 2,3-DHB formation in the extracellular fluid of the mouse *gastrocnemius* muscle, and a rise of approximately 100% was observed (Close et al., 2005; Vasilaki, Mansouri, et al., 2006).

2.2.2 Comment and limitations

There are a number of general drawbacks with the microdialysis to assess ROS in skeletal muscle and some specific points related to the technique used for assessment of extracellular hydroxyl radical activity which should be considered:

1. The cellular source of material detected in the microdialysates cannot be defined since the skeletal muscle ECF will be influenced by multiple cell types, such as endothelial cells and white cells in addition to skeletal muscle cells.

2. Insertion of the probe must cause some local trauma to the tissue, and this might theoretically influence ROS measurements to a greater degree than other substances that are measured by microdialysis.
3. Recoveries of ROS across the microdialysis membrane have not been defined and are much more difficult to quantify than for relatively stable molecules. Attempts to undertake such studies have been relatively unsuccessful (Close et al., 2005).
4. Formation of 2,3-DHB from salicylate in biological systems has been claimed to occur specifically through hydroxyl radical-mediated hydroxylation, although *in vitro* studies have indicated that reaction with peroxynitrite also leads to hydroxylation of salicylate. Inhibitor studies performed by our group are compatible with a major role for hydroxyl radical in forming 2,3-DHB from salicylate in muscle microdialysates (Close et al., 2005).

2.2.3 Detection of superoxide anion in muscle extracellular space in mice

Procedure: Microdialyis probes are perfused with 50 μ*M* cytochrome *c* in normal saline at a flow rate of 4 μl/min and allowed to stabilize for 30 min. Samples are then collected from the outlet tubes of the probe over sequential 15 min periods. Reduction of cytochrome *c* in the microdialysate is used as an index of superoxide radical in the microdialysate. Samples are analyzed using scanning visible spectrometry, and the superoxide content calculated from the absorbance at 550 nm in comparison with the isobestic wavelengths at 542 and 560 nm. A molar extinction coefficient for reduced cytochrome *c* of 21,000 is used for calculation of the superoxide anion concentration.

With the conditions and probes described earlier, we have reported that levels of approximately 0.4 nmol superoxide/15 min can be detected in the ECF of mouse skeletal muscle at rest (Close et al., 2005; McArdle et al., 2001). Again, we have experience that demonstrates that these values change if different length probes are used, the molecular weight cutoff differs, or if the flow rate is varied and are also likely to change with probes of different composition, although we have not examined this. The values for superoxide tend to decrease over the initial sequential 15-min collection periods, but eventually stabilize (Close et al., 2005). We have examined the effect of electrical stimulation of contraction on the reduction of cytochrome *c* in microdialysates from the extracellular fluid of the mouse *gastrocnemius* muscle, and a rise of approximately 60% was observed.

2.2.4 Comment and limitations

1. The effect of trauma to tissues during probe insertion appears to be particularly important for superoxide detection.
2. The lack of specificity of cytochrome *c* reduction as a measure of superoxide is a potentially important drawback with this approach. In our initial study, we observed that levels of cytochrome *c* reduction in mouse microdialysates were reduced by ~50% on addition of purified SOD to the perfusate (McArdle et al., 2001). It is likely that at least a proportion of the reduction of cytochrome *c* occurs outside the microdialysis probe since cytochrome *c* has a molecular weight of ~12kDa, and the dialysis membrane cutoff is 35 kDa. Hence, there will be substantial diffusion of the cytochrome *c* out of the probe and because of the slow flow rate there will also be diffusion back into the probe. In that case, addition of high-molecular-weight purified SODs to the microdialysis fluid could not prevent the reduction of cytochrome *c* by superoxide that occurs outside the probe. Our inhibitor data indicate that NO was unlikely to make a significant contribution to the reduction of cytochrome *c* in this system. It has also been suggested that small molecules such as ascorbate or glutathione can reduce cytochrome *c in vivo*, but in unpublished studies, we have observed that reduction of microdialysate cytochrome *c* stopped immediately on death of the mouse suggesting a dependence of this reduction on metabolic activity that is not compatible with this hypothesis. Overall, therefore, the data are consistent with superoxide playing a substantial role in the reduction of cytochrome *c* in microdialysates, but the lack of complete suppression of the reduction by exogenous SOD means that we cannot define precise levels from the current data (Close et al., 2005).

Other analyses that are feasible with microdialysis approaches include detection of ROS using spin traps with electron spin resonance detection, direct chemical analyses of hydrogen peroxide, or measurement of nitrate and nitrite as an index of NO release to the extracellular fluid (Vasilaki, Mansouri, et al., 2006).

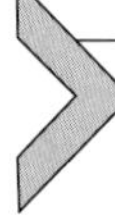

3. ASSESSMENT OF INTRACELLULAR ROS ACTIVITIES

3.1. Choice of model system

Skeletal muscle tissues contain numerous cell types including endothelial cells, vascular smooth muscle cells, and inflammatory cells that can generate ROS in addition to skeletal muscle fibers and hence evaluation of intracellular ROS activities specifically in muscle fibers requires some system to allow direct access to the fibers. Several authors have examined the superficial fibers of limb muscles and diaphragm as a means of undertaking single fiber measurements, and others have laboriously dissected single fibers from small rodent muscles. We have focused on approaches to monitor ROS activities in single intact muscle fibers that are readily isolated by collagenase digestion (Palomero et al., 2008; Pye et al., 2007) and in single myotubes. These studies have used both general (McArdle, Pattwell, Vasilaki, McArdle, & Jackson, 2005; Palomero et al., 2008; Vasilaki, Csete, et al., 2006) and more specific (Sakellariou et al., 2011, 2013) probes for ROS and NO (Pye et al., 2007). The single intact mature fiber preparation has a variety of advantages over other approaches since these fibers are mature in comparison with cultured myotubes, and the preparation is relatively uncontaminated with nonmuscle cells. In our studies, fibers have been isolated from the mouse *flexor digitorum brevis* (FDB) and used it to monitor in real-time the changes in ROS activities in skeletal muscle cells, but isolation from other small rodent muscles is also feasible. We have demonstrated the use of this preparation with DAF-FM (to monitor NO) and dichlorodihydrofluorescein (DCFH or CM-DCFH), an indicator that has been used for a number of years and shown to be sensitive to a number of ROS and RNS including hydrogen peroxide, superoxide, NO, and peroxynitrite and with DHE (to monitor superoxide).

3.1.1 Specific considerations for monitoring of superoxide and H_2O_2 in single fibers

3.1.1.1 Cytosolic superoxide activity

Ethidium fluorescence following dihydroethidium (DHE) loading of isolated cells has been used as an assay for intracellular superoxide (Zuo et al., 2000). In the standard technique, oxidation of DHE within all parts of the cell leads to the formation of ethidium that intercalates into nuclear DNA, and fluorescence measurements have previously been made from either the cytosol or nuclei of muscle cells. We have used this technique to monitor the contraction-induced generation of superoxide in single mature muscle fibers (Sakellariou et al., 2011). A development of this assay to improve the specificity is analysis of the specific product of the reaction of superoxide and DHE, 2-hydroxyethidium (Zhao et al., 2005). We have developed a HPLC approach to measure 2-hydroxyethidium (Sakellariou et al., 2011), but in addition, an alternative excitation wavelength (396 nm) for fluorescence microscopy has been proposed that allows

2-hydroxyethidium to be measured using fluorescence microscopy (Robinson, Janes, & Beckman, 2008), although the validity of this approach has been questioned.

3.1.1.2 Hydrogen peroxide

Although many previous studies have used 2′,7′-DCFH as an intracellular fluorescent probe for H_2O_2 in cells (e.g., McArdle et al., 2005; Vasilaki, Csete, et al., 2006), the lack of specificity of this compound actually precludes its use other than as a general probe for ROS (Murrant, Andrade, & Reid, 1999). Alternative and specific assays for H_2O_2 have been developed, such as the genetically encoded probe *HyPer,* in which yellow fluorescent protein is inserted into the regulatory domain of the prokaryotic H_2O_2 sensing protein, OxyR (Belousov et al., 2006).

3.1.1.3 Mitochondrial superoxide and hydrogen peroxide activities

Modifications of the above approaches permit the analyses of superoxide activity and hydrogen peroxide in mitochondria. DHE has been linked through a hexyl carbon chain to a triphenylphosphonium group to produce a compound that accumulates in mitochondria in response to the negative membrane potential. This compound is known as MitoSOX or Mito-HE (Robinson et al., 2008), and we have used this in conjunction with confocal microscopy to monitor mitochondrial superoxide generation (Sakellariou et al., 2013).

To facilitate specific measurements of hydrogen peroxide activity in mitochondria, a modified version of the genetically encoded probe (*HyPer-M*) has been targeted to mitochondria by linkage to a mitochondrial target sequence derived from subunit VIII precursor of cytochrome *c* oxidase (*Evrogen*, Moscow) (Belousov et al., 2006). An overview of the approaches that we have used to monitor intracellular ROS in muscle fibers is shown in Fig. 17.1, and typical images of single isolated fibers loaded with different probes are shown in Fig. 17.2.

3.1.2 Procedure: Use of DCFH and DHE as probes to monitor ROS in isolated single muscle fibers

3.1.2.1 Isolation of single skeletal muscle fibers

The FDB muscles are removed from mice following humane killing and placed into 0.4% type H collagenase (EC 3.4.24.3) (Sigma–Aldrich Co. St. Louis, MO) solution in culture medium composed of minimum essential medium (MEM) eagle (Sigma–Aldrich Co. St. Louis, MO) supplemented

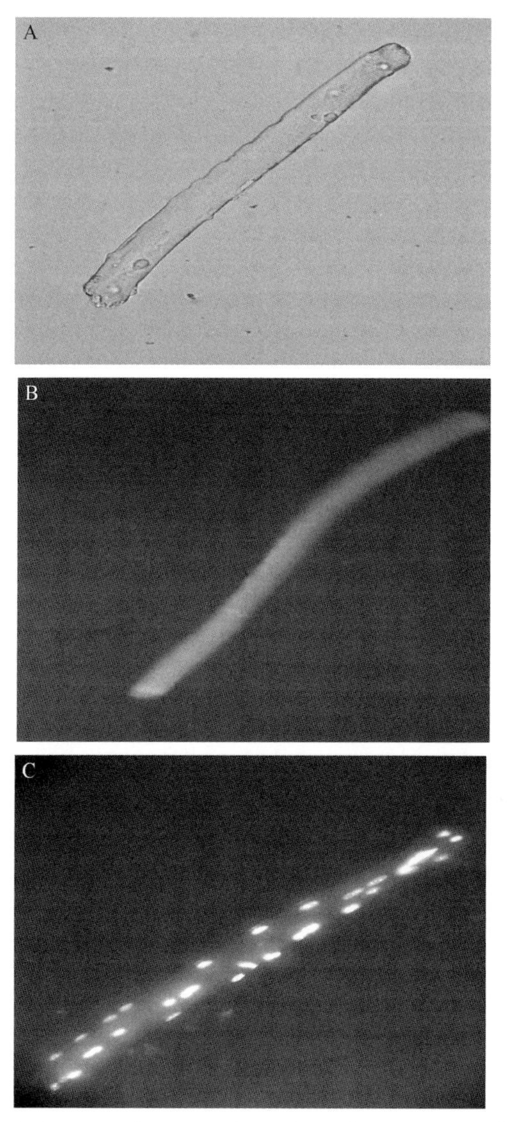

Figure 17.2 Example images of single isolated fibers from mouse FDB muscles loaded with ROS sensitive probes. (A) Bright-field image; (B) fiber loaded with DCFH; (C) Fiber loaded with DHE showing nuclear localization of the ethidium product. (See Color Insert.)

with 10% fetal bovine serum (Invitrogen Ltd., Paisley, UK) containing 2 m*M* glutamine, 50 i.u. penicillin, and 50 μg ml^{-1} streptomycin.

The FDB muscles are incubated in collagenase solution at 37 °C for 2 h, and the mixture is manually shaken every 30 min to improve digestion of the connective tissue. Fiber bundles that have not been separated during the incubation are gently triturated by a wide-bore plastic pipette to separate the fibers.

Free single muscle fibers are separated from damaged fibers and contaminating cells by centrifugation at low speed (600 g for 30 s). The fibers are washed four times in fresh culture medium. Cleaned fibers are plated onto 35-mm dishes precoated with 120 μl of a 1:6 mixture of *Vitrogen* collagen (Cohesion Technologies Inc., Palo Alto, CA) in 7 × Dulbecco's modified Eagle's medium (Invitrogen Ltd., Paisley, UK). Fibers are allowed to attach for 30 min at 37 °C, 1 ml of fresh culture medium was added to each plate, and fibers are cultured for 18 h at 37 °C in a 5% CO_2 atmosphere.

3.1.2.2 Viability of single skeletal muscle fibers

The viability of the isolated muscle fibers in culture is determined by examination of their morphology by bright-field microscopy and by their ability to exclude the vital stain, trypan blue. Fibers are incubated in a 0.04 mg/ml solution of trypan blue (Sigma–Aldrich Co. St. Louis, MO) in culture medium and examined under bright-field microscopy.

3.1.3 Loading of single skeletal muscle fibers with CM-DCFH DA

A solution of 5-(and-6)-chloromethyl-2′,7′-DCFH diacetate (CM-DCFH DA) (Molecular Probes™, Invitrogen, Eugene, OR) is prepared daily in absolute ethanol and kept at 4 °C. Culture plates containing isolated fibers are washed with Dulbecco's phosphate-buffered saline (D-PBS) (Sigma–Aldrich Co. St. Louis, MO), and fibers are preincubated in D-PBS at 37 °C for 30 min. The medium is replaced by D-PBS-containing CM-DCFH DA (17.5 μ*M*) and incubated for 30 min at 37 °C. The fibers are then washed with D-PBS and covered with MEM without Phenol Red (Sigma–Aldrich Co. St. Louis, MO) for fluorescence microscopy. 5-(and-6)-CM-2′, 7′-DCFH is a derivative of DCFH modified to improve retention by cells, the diacetate form (CM-DCFH DA) is a nonpolar molecule that diffuses into the fibers and within the cytosol the acetate group is cleaved by cytosolic esterases to yield CM-DCFH, a hydrophilic nonfluorescent molecule that is retained by the cell.

3.1.4 Loading of fibers with DHE

Fibers are loaded by incubation in 2 ml D-PBS containing 5 μ*M* DHE for 30 min at 37 °C. Cells are then washed twice with D-PBS, and the fibers are maintained in MEM without Phenol Red during the experimental protocol.

3.1.5 Fluorescence microscopy and image analysis of CM-DCF

We have used a Zeiss Axiovert 200 *M* epifluorescence microscope equipped with a ×10 and ×20 objectives and a 450–490 nm excitation/515–565 nm emission filter set for CM-DCF fluorescence (Carl Zeiss Microimaging GmbH, Jena, Germany). Images were acquired and analyzed using a computer-controlled Zeiss HRc charged-coupled device (CCD) camera (Carl Zeiss Microimaging GmbH, Jena, Germany) and by AxioVision 4.4 image capture and analysis software (Carl Zeiss Microimaging GmbH, Jena, Germany) for quantification of changes in emission fluorescence. This software allows measurements to be made from user-defined areas of the microscope field; in this case, the fluorescence measurements were localized to selected areas of the muscle fibers.

CM-DCF fluorescence from fibers is recorded at 15-min intervals over 45 min. Exposure of fibers to ultraviolet light is minimized by the use of a 500-ms exposure time.

3.1.6 Monitoring of DHE oxidation to assess intracellular superoxide activity in isolated fibers

Fibers were loaded by incubation in 2 ml D-PBS containing 5 μ*M* DHE for 30 min at 37 °C. Cells were then washed twice with D-PBS, and the fibers were maintained in MEM without Phenol Red during the experimental protocol.

3.1.7 Fluorescence microscopy and image analysis of DHE

The same image capture system and microscope equipped with both 500/20 excitation, 535/30 emission, and 510/60 excitation, 590 emission filter sets (Carl Zeiss GmbH, Germany) was used. Using a ×20 objective, fluorescence images were captured with a computer-controlled Zeiss MRm CCD camera (Carl Zeiss GmbH, Germany) and analyzed with the Axiovision 4.0 image capture and analysis software (Carl Zeiss Vision, GmbH, Germany).

3.1.8 Analysis of 2-hydroxyethidium by HPLC

In order to determine the specific product of DHE reaction with superoxide (2-hydroxyethidium), the content was analyzed by HPLC essentially as described by Zhao et al. (2005). All isolated fibers in a 35-mm dish were loaded with 5 μ*M* DHE for 30 min at 37 °C as described earlier and washed twice with D-PBS. At 15 min following loading 2-hydroxyethidium was extracted by addition of 0.5 ml *n*-butanol to the sample. Following vortex mixing, the butanol phase was separated and dried under nitrogen. The sample was reconstituted in 0.1 ml HPLC grade water prior to injection onto the HPLC system. We had undertaken separation of 2-hydroxyethidium from ethidium using a Ginkotech HPLC system with a C18 reverse phase column (Partisil ODS-3, 250 × 4.5 mm). The solvent was initially 10% acetonitrile in 0.1% trifluoroactetic acid and increased linearly to 70% acetonitrile over 46 min with a flow rate of 0.5 ml/min. Detection was by fluorescence with excitation at 510 nm and emission at 595 nm (Sakellariou et al., 2011).

3.1.9 Comments and limitations

Although the use of isolated fibers facilitates the monitoring of ROS specifically in muscle tissue, there are clear limitations of removing the fibers from their *in vivo* environment. Thus, the fibers have no blood or nerve supply and have been subjected to mechanical and chemical stress during their isolation. We have examined the viability of isolated FDB in culture, and they retain viability for 6–7 days, but significant changes in morphology occur over that time.

The limitations of DCFH as a probe for ROS have been widely discussed, and it can only be thought of as a general probe. Readers are referred to the recent review by Murphy et al. (2011) to be informed of this. DHE is accepted as a specific probe when 2-hydroxyethidium is measured as the specific product of the reaction, and this should be undertaken to validate the fluorescence microscopy. The specificity of confocal approaches may be improved with the new probes such as *HyPer* which are becoming used in this area.

4. CONCLUDING REMARKS

A great deal has been learned about the generation of hydrogen peroxide and other ROS in skeletal muscle and on the effects of these reactive substances on tissue function. The approaches described here provide a basis for further studies of the nature and functions of these species in muscle and in the interactions of muscle with other tissues.

ACKNOWLEDGMENTS

Generous funding for the author's work in this area has been obtained from various agencies including the Medical Research Council, Biotechnology and Biological Sciences Research Council, Wellcome Trust, US National Institute on Aging, *Research into Ageing* and *Arthritis Research UK*.

REFERENCES

Arora, S., Vaishya, R., Dabla, P. K., & Singh, B. (2010). NAD(P)H oxidases in coronary artery disease. *Advances in Clinical Chemistry*, *50*, 65–86.

Bakkar, N., Wang, J., Ladner, K. J., Wang, H., Dahlman, J. M., Carathers, M., et al. (2008). IKK/NF-kappaB regulates skeletal myogenesis via a signaling switch to inhibit differentiation and promote mitochondrial biogenesis. *The Journal of Cell Biology*, *180*, 787–802.

Bar-Shai, M., Carmeli, E., & Reznick, A. Z. (2005). The role of NF-kappaB in protein breakdown in immobilization, aging, and exercise: From basic processes to promotion of health. *Annals of the New York Academy of Sciences*, *1057*, 431–447.

Belousov, V. V., Fradkov, A. F., Lukyanov, K. A., Staroverov, D. B., Shakhbazov, K. S., Terskikh, A. V., et al. (2006). Genetically encoded fluorescent indicator for intracellular hydrogen peroxide. *Nature Methods*, *3*, 281–286.

Benveniste, H., & Huttemeier, P. C. (1990). Microdialysis—Theory and application. *Progress in Neurobiology*, *35*, 195–215.

Bito, L., Davson, H., Levin, E., Murray, M., & Snider, N. (1966). The concentration of free amino acids and other electrolytes in cerebrospinal fluid, in vivo dialysate of brain, and blood plasma of the dog. *Journal of Neurochemistry*, *13*, 1057–1067.

Boveris, A., & Chance, B. (1973). The mitochondrial generation of hydrogen peroxide. General properties and effect of hyperbaric oxygen. *The Biochemical Journal*, *134*, 707–716.

Camus, G., Felekidis, A., Pincemail, J., Derby-Dupont, G., Deby, C., Juchmes-Ferer, A., et al. (1994). Blood levels of reduced/oxidized glutathione and plasma concentration of ascorbic acid during eccentric and concentric exercises of similar energy cost. *Archives Internationales De Physiologie, De Biochimie Et De Biophysique*, *102*, 67–70.

Close, G. C., Ashton, T., McArdle, A., & Jackson, M. J. (2005). Microdialysis studies of extracellular reactive oxygen species in skeletal muscle: Factors influencing the reduction of cytochrome c and hydroxylation of salicylate. *Free Radical Biology and Medicine*, *39*, 1460–1467.

Close, G. L., Ashton, T., Cable, T., Doran, D., & MacLaren, D. P. (2004). Eccentric exercise, isokinetic muscle torque and delayed onset muscle soreness: The role of reactive oxygen species. *European Journal of Applied Physiology*, *91*, 615–621.

Close, G. L., Kayani, A. C., Ashton, T., McArdle, A., & Jackson, M. J. (2007). Release of superoxide from skeletal muscle of adult and old mice: An experimental test of the reductive hotspot hypothesis. *Aging Cell*, *6*, 189–195.

Cotto, J. J., & Morimoto, R. I. (1999). Stress-induced activation of the heat-shock response: Cell and molecular biology of heat-shock factors. *Biochemical Society Symposium*, *64*, 105–118.

Dahlman, J. M., Wang, J., Bakkar, N., & Guttridge, D. C. (2009). The RelA/p65 subunit of NF-kappaB specifically regulates cyclin D1 protein stability: Implications for cell cycle withdrawal and skeletal myogenesis. *Journal of Cellular Biochemistry*, *106*, 42–51.

Davies, K. J., Quintanilha, A. T., Brooks, G. A., & Packer, L. (1982). Free radicals and tissue damage produced by exercise. *Biochemical and Biophysical Research Communications*, *107*, 1198–1205.

Dröge, W. (2002). Free radicals in the physiological control of cell function. *Physiological Reviews*, *82*, 47–95.

Duthie, G. G., Robertson, J. D., Maughan, R., & Morrice, J. C. (1990). Blood antioxidant status and erythrocyte lipid peroxidation following distance running. *Archives of Biochemistry and Biophysics*, *282*, 78–83.

Espinosa, A., Leiva, A., Pena, M., Muller, M., Debandi, A., Hidalgo, C., et al. (2006). Myotube depolarization generates reactive oxygen species through NAD(P)H oxidase; ROS-elicited Ca2+ stimulates ERK, CREB, early genes. *Journal of Cellular Physiology*, *209*, 379–388.

Febbraio, M. A., & Pedersen, B. K. (2005). Contraction-induced myokine production and release: Is skeletal muscle an endocrine organ? *Exercise and Sport Sciences Reviews*, *33*, 114–119.

Gomez-Cabrera, M. C., Borras, C., Pallardo, F. V., Sastre, J., Ji, L. L., & Vina, J. (2005). Decreasing xanthine oxidase-mediated oxidative stress prevents useful cellular adaptations to exercise in rats. *The Journal of Physiology*, *567*, 113–120.

Gomez-Cabrera, M. C., Close, G. L., Kayani, A., McArdle, A., & Jackson, M. J. (2010). Effect of xanthine oxidase-generated extracellular superoxide on skeletal muscle force generation. *American Journal of Physiology. Regulatory, Integrative and Comparitive Physiology*, *298*, R2–R8.

Haddad, J. J. (2002). Antioxidant and pro-oxidant mechanisms in the regulation of redox(y)-sensitive transcription factors. *Cellular Signalling*, *14*, 879–897.

Halliwell, B., & Gutteridge, J. M. C. (1989). *Free radical biology and medicine*. U.K.: Oxford University Press.

Jackson, M. J. (2008). Free radicals generated by contracting muscle: By-products of metabolism or key regulators of muscle function? *Free Radical Biology and Medicine*, *44*, 132–141.

Jackson, M. J. (2011). Control of reactive oxygen species production in contracting skeletal muscle. *Antioxidants & Redox Signaling*, *15*, 2477–2486.

Jackson, M. J., Edwards, R. H., & Symons, M. C. (1985). Electron spin resonance studies of intact mammalian skeletal muscle. *Biochimica et Biophysica Acta*, *847*, 185–190.

Jackson, M. J., Papa, S., Bolanos, J., Bruckdorfer, R., Carlsen, H., Elliott, R. M., et al. (2002). Antioxidants, reactive oxygen and nitrogen species, gene induction and mitochondrial function. *Molecular Aspects of Medicine*, *23*, 209–285.

Jackson, M. J., Pye, D., & Palomero, J. (2006). The production of reactive oxygen and nitrogen species by skeletal muscle. *Journal of Applied Physiology*, *102*, 1664–1670.

Janssen-Heininger, Y. M., Mossman, B. T., Heintz, N. H., Forman, H. J., Kalyanaraman, B., Finkel, T., et al. (2008). Redox-based regulation of signal transduction: Principles, pitfalls, and promises. *Free Radical Biology & Medicine*, *45*, 1–17.

Javesghani, D., Magder, S. A., Barreiro, E., Quinn, M. T., & Hussain, S. N. (2002). Molecular characterization of a superoxide-generating NAD(P)H oxidase in the ventilatory muscles. *American Journal of Respiratory and Critical Care Medicine*, *165*, 412–418.

Kanter, M. M. (1994). Free radicals, exercise, and antioxidant supplementation. *International Journal of Sport Nutrition*, *4*, 205–220.

Loschen, G., Azzi, A., Richter, C., & Flohe, L. (1974). Superoxide radicals as precursors of mitochondrial hydrogen peroxide. *FEBS Letters*, *42*, 68–72.

Maclean, D. A., Bangsbo, J., & Saltin, B. (1999). Muscle interstitial glucose and lactate levels during dynamic exercise in humans determined by microdialysis. *Journal of Applied Physiology*, *87*, 1483–1490.

Maughan, R. G., Donnelly, A. E., Gleeson, M., Whiting, P. H., Walker, K. A., & Clough, P. J. (1989). Delayed-onset muscle damage and lipid peroxidation in man after a downhill run. *Muscle & Nerve*, *12*, 332–336.

McArdle, A., Khera, G., Edwards, R. H., & Jackson, M. J. (1999). In vivo microdialysis—A technique for analysis of chemical activators of muscle pain. *Muscle & Nerve*, *22*, 1047–1052.
McArdle, A., Pattwell, D., Vasilaki, A., Griffiths, R. D., & Jackson, M. J. (2001). Contractile activity-induced oxidative stress: Cellular origin and adaptive responses. *American Journal of Physiology. Cell Physiology*, *280*, C621–C627.
McArdle, F., Pattwell, D. M., Vasilaki, A., McArdle, A., & Jackson, M. J. (2005). Intracellular generation of reactive oxygen species by contracting skeletal muscle cells. *Free Radical Biology & Medicine*, *39*, 651–657.
Michaelson, L. P., Shi, G., Ward, C. W., & Rodney, G. G. (2010). Mitochondrial redox potential during contraction in single intact muscle fibers. *Muscle & Nerve*, *42*, 522–529.
Moran, L. K., Gutteridge, J. M., & Quinlan, G. J. (2001). Thiols in cellular redox signalling and control. *Current Medicinal Chemistry*, *8*, 763–772.
Murphy, M. P., Holmgren, A., Larsson, N. G., Halliwell, B., Chang, C. J., Kalyanaraman, B., et al. (2011). Unraveling the biological roles of reactive oxygen species. *Cell Metabolism*, *13*, 361–366.
Murrant, C. L., Andrade, F. H., & Reid, M. B. (1999). Exogenous reactive oxygen and nitric oxide alter intracellular oxidant status of skeletal muscle fibres. *Acta Physiologica Scandinavica*, *166*, 111–121.
Nishino, T., Okamoto, K., Eger, B. T., Pai, E. F., & Nishino, T. (2008). Mammalian xanthine oxidoreductase—Mechanism of transition from xanthine dehydrogenase to xanthine oxidase. *The FEBS Journal*, *275*, 3278–3289.
Pacher, P., Nivorozhkin, A., & Szabó, C. (2006). Therapeutic effects of xanthine oxidase inhibitors: Renaissance half a century after the discovery of allopurinol. *Pharmacological Reviews*, *58*, 87–114.
Palomero, J., Pye, D., Kabayo, T., Spiller, D. G., & Jackson, M. J. (2008). In situ detection and measurement of intracellular reactive oxygen species in single isolated mature skeletal muscle fibres by real-time fluorescence microscopy. *Antioxidants & Redox Signalling*, *10*, 1463–1474.
Pattwell, D. M., McArdle, A., Morgan, J. E., Patridge, T. A., & Jackson, M. J. (2004). Release of reactive oxygen and nitrogen species from contracting skeletal muscle cells. *Free Radical Biology & Medicine*, *37*, 1064–1072.
Peterson, J. M., & Guttridge, D. C. (2008). Skeletal muscle diseases, inflammation, and NF-kappaB signaling: Insights and opportunities for therapeutic intervention. *International Reviews of Immunology*, *27*, 375–387.
Powers, S. K., & Jackson, M. J. (2008). Exercise-induced oxidative stress: Cellular mechanisms and impact on muscle force production. *Physiological Reviews*, *88*, 1243–1276.
Pye, D., Kabayo, T., Palmero, J., & Jackson, M. J. (2007). Real-time measurements of nitric oxide in mature skeletal muscle fibres during contractions. *The Journal of Physiology*, *581*, 309–318.
Reid, M. B., Shoji, T., Moody, M. R., & Entman, M. L. (1992). Reactive oxygen in skeletal muscle. II. Extracellular release of free radicals. *Journal of Applied Physiology*, *73*, 1805–1809.
Robinson, K. M., Janes, M. S., & Beckman, J. S. (2008). The selective detection of mitochondrial superoxide by live cell imaging. *Nature Protocols*, *3*, 941–947.
Sakellariou, G. K., Pye, D., Vasilaki, A., Zibrik, L., Palomero, J., Kabayo, T., et al. (2011). Role of superoxide-nitric oxide interactions in the accelerated age-related loss of muscle mass in mice lacking Cu, Zn superoxide dismutase. *Aging Cell*, *10*, 749–760.
Sakellariou, G. K., Vasilaki, A., Palomero, J., Kayani, A., Zibrik, L., McArdle, A., et al. (2013). Studies of mitochondrial and nonmitochondrial sources implicate nicotinamide adenine dinucleotide phosphate oxidase(s) in the increased skeletal muscle superoxide

generation that occurs during contractile activity. *Antioxidants & Redox Signaling, 18*, 603–621.
St-Pierre, J., Buckingham, J. A., Roebuck, S. J., & Brand, M. D. (2002). Topology of superoxide production from different sites in the mitochondrial electron transport chain. *The Journal of Biological Chemistry, 277*, 44784–44790.
Urso, M. L., & Clarkson, P. M. (2003). Oxidative stress, exercise, and antioxidant supplementation. *Toxicology, 189*, 41–54.
Van Gammeren, D., Damrauer, J. S., Jackman, R. W., & Kandarian, S. C. (2009). The IkappaB kinases IKKalpha and IKKbeta are necessary and sufficient for skeletal muscle atrophy. *The FASEB Journal, 23*, 362–370.
Vasilaki, A., Csete, M., Pye, D., Lee, S., Palomero, J., McArdle, F., et al. (2006). Genetic modification of the MnSOD/GPx1 pathway influences intracellular ROS generation in quiescent, but not contracting myotubes. *Free Radical Biology & Medicine, 41*, 1719–1725.
Vasilaki, A., Mansouri, A., Remmen, H., van der Meulen, J. H., Larkin, L., Richardson, A. G., et al. (2006). Free radical generation by skeletal muscle of adult and old mice: Effect of contractile activity. *Aging Cell, 5*, 109–117.
Vasilaki, A., McArdle, F., Iwanejko, L. M., & McArdle, A. (2006). Adaptive responses of mouse skeletal muscle to contractile activity: The effect of age. *Mechanisms of Ageing and Development, 127*, 830–839.
Viña, J., Gimeno, A., Sastre, J., Desco, C., Asensi, M., Pallardó, F. V., et al. (2000). Mechanism of free radical production in exhaustive exercise in humans and rats; role of xanthine oxidase and protection by allopurinol. *IUBMB Life, 49*, 539–544.
Xia, R., Webb, J. A., Gnall, L. L., Cutler, K., & Abramson, J. J. (2003). Skeletal muscle sarcoplasmic reticulum contains a NADH-dependent oxidase that generates superoxide. *The American Journal of Physiology, 285*, C215–C221.
Zhao, H., Joseph, J., Fales, H. M., Sokoloski, E. A., Levine, R. L., Vasquez-Vivar, J., et al. (2005). Detection and characterization of the product of hydroethidine and intracellular superoxide by HPLC and limitations of fluorescence. *Proceedings of the National Academy of Sciences of the United States of America, 102*, 5727–5732.
Zuo, L., Christofi, F. L., Wright, V. P., Liu, C. Y., Merola, A. J., Berliner, L. J., et al. (2000). Intra- and extracellular measurement of reactive oxygen species produced during heat stress in diaphragm muscle. *American Journal of Physiology. Cell Physiology, 279*, C1058–C1066.

AUTHOR INDEX

Note: Page numbers followed by "*f*" indicate figures, and "*t*" indicate tables, and "*np*" indicate footnotes.

A

Abdalla, F. C., 218, 225, 226, 227–228
Abeliovich, H., 218, 225, 226, 227–228
Abraham, R. T., 218, 225, 226, 227–228
Abrams, J. M., 207
Abramson, J. J., 280–281
Acevedo-Arozena, A., 218, 225, 226, 227–228
Acs, G., 81
Adeli, K., 218, 225, 226, 227–228
Adjei, A. A., 35–36, 42–43
Affourtit, C., 258, 259–260, 261–264, 263*f*
Agapite, J., 207
Agnoletto, C., 82
Agostinis, P., 227, 228
Agrawal, D. K., 227, 228
Agudo, M., 199–200
Akers, M., 50
Albers, A. E., 20
Ale-Agha, N., 112–114
Alexander, R. W., 50
Alfano, I., 130–131
Ali, S. M., 221–222
Aliev, G., 227, 228
Alleva, E., 101–102
Allister, E. M., 258
Alvarez, B., 6
Amadio, L., 101–102
Amelotti, M., 225
Ammendola, A., 119–120
Amoscato, A. A., 10
Andersen, J. N., 6, 15
Anderson, E. J., 271, 273
Anderson, W. B., 82, 85, 86–87
Andrade, F. H., 292
Andrae, J., 189–190
Angerman-Stewart, J., 131–132
Anton, R., 225
Antunes, F., 17–18, 160–161, 163, 164, 167–168, 174–175, 177–178, 179, 180–181, 182, 183, 184–185, 186, 186*f*
Aoki, K., 65–66, 76
Apperson-Hansen, C., 117
Archuleta, T. L., 273, 275
Areces, L. B., 81
Arganda-Carreras, I., 245
Armstead, B., 112–114
Arnold, R. S., 39–40
Arora, S., 282
Artieri, C. G., 200–201
Asabere, A., 174–175
Asara, J. M., 270
Asensi, M., 282
Ashman, K., 119–120
Ashton, T., 284, 285–287, 288, 289, 290
Askew, D. S., 227, 228
Atesoh Awuh, J., 224
Augusto, O., 7
Aumann, K. D., 11
Authier, H., 222
Avruch, J., 28–29
Avshalumov, M. V., 100
Azad, M. B., 218–221, 224, 227–228, 230
Azam, M. A., 117
Azar, Z. M., 42–43
Azzi, A., 281–282

B

Bacher, F., 204
Bae, E., 204–205
Bae, S. H., 11–12, 12*f*, 13, 15
Bae, Y. S., 15, 50, 80, 198
Baffy, G., 258
Bai, Y., 258
Baines, I. C., 199–200
Bajrami, B., 240
Baker, B. S., 200–201
Baker, L. M., 12–13
Baker, S. J., 190
Bakhmutova-Albert, E. V., 7
Bakkar, N., 283–284

Baldari, C. T., 101–102
Baldini, C., 82
Balligand, J. L., 62
Ballou, L. M., 218–220
Bangsbo, J., 285
Bao, L., 100
Barancík, M., 28–29
Bardswell, S. C., 112–114, 117–119
Barford, D., 15
Barinova, Y., 205–206
Barja, G., 102
Barlow, J., 258
Barnard, M. L., 238
Barnes, S., 6
Barr, A. J., 130–131
Barrault, M. B., 13
Barreiro, E., 280–281
Barrett, C. W., 38–39
Bar-Shai, M., 283–284
Bartrons, R., 223–224
Bashan, N., 272, 273, 275
Bauer, M., 76
Bauvy, C., 225
Beaton, A., 204–205
Beck Previs, S., 117
Becker, D., 10
Beckman, J. S., 291–292
Bedard, K., 238
Begum, S., 117–119
Beinert, N., 204–205
Bell, E. L., 100
Bell, J. R., 117–119
Bellen, H. J., 204–205
Belousov, V. V., 19, 68–70, 239, 292
Beltrami, E., 101
Belvin, M., 201, 205
Bennett, A. M., 37–38
Bensaad, K., 223–224
Benveniste, H., 284–285
Berger, Z., 221–222
Berk, B. C., 40–41
Berliner, L. J., 291–292
Bernardi, P., 104, 105, 107, 108
Berniakovich, I., 101, 107, 108
Berry, A., 101–102, 108
Bers, D. M., 62, 116–117
Besancon, F., 225
Betsholtz, C., 189–190
Betz, A., 8
Beyerle, A., 42–43
Bhatnagar, A., 112–114
Bhattacharjee, A., 258
Bhattacharya, S., 130–132
Biard, D., 112–114
Biard-Piechaczyk, M., 218–220
Bienert, G. P., 17–18, 238, 239–240
Binari, R., 205–206
Birben, E., 28
Birket, M. J., 261
Birrer, M. J., 38–39
Bissonnette, R. P., 178*f*, 184–185
Biteau, B., 199–200, 202–203
Bito, L., 285
Blair, A. S., 271–272, 275
Blanz, J., 227–228
Blommaart, E. F., 222
Blumberg, P. M., 81
Blumenstiel, J. P., 208–209
Blumenthal, D. K., 117–119
Blumer, K. J., 119–120
Boguslowicz, C., 29
Böhmer, F.-D., 130–131
Bokel, C., 204, 208–209
Boland, B., 221–222
Bolanos, J., 282–283
Bonnard, C., 271–272, 273–274
Bononi, A., 82
Bonora, M., 82
Boonstra, J., 35–36
Borbouse, L., 119–120
Borman, M. A., 119–120
Borras, C., 281
Botti, H., 8, 12–13
Boulden, B. M., 55–56
Bourey, R. E., 273
Boutros, T., 28–29, 37–38
Boveris, A., 281–282
Bowitz Larsen, K., 224
Bowles, H., 35–36
Bowman, T., 242–243
Bowtell, D. D., 198, 200–201, 204, 206, 211–212
Boyle, C. E., 271, 273
Bozonet, S. M., 13
Bradford, M. M., 135–136, 162
Brady, A. J., 62
Brand, M. D., 259–260, 261–264, 263*f*, 281–282

Brandes, R. P., 43
Brennan, J. P., 112–114, 117–120
Brenowitz, M., 132–133
Breton-Romero, R., 50
Brewer, A. C., 112–114
Brigelius-Flohé, R., 11, 174–175
Bright, R., 94
Britton, F. C., 119–120
Broihier, H. T., 207, 208
Brookes, P. S., 100
Brooks, A. N., 209
Brooks, G. A., 280–281
Brown, D. I., 39–40
Brown, K. L., 30–32
Bruckdorfer, R., 282–283
Brummer, T., 37–38
Brunk, U. T., 10, 19–20, 174–175
Bryan, N. S., 55–56
Buckingham, J. A., 281–282
Buckley, D. A., 131–132
Budanov, A. V., 223–224
Bulleid, N. J., 11–12
Bunemann, M., 116–117
Bupha-Intr, T., 273, 274–275
Burchiel, S. W., 35–36
Burgoyne, J. R., 112–114, 117–120
Burke, C., 242–243
Burkhoff, D., 116–117
Burkitt, M. J., 19–20
Bushkin, I., 130–131, 271–272
Bussolari, S. R., 52
Bylund, J., 30–32
Byun, Y. J., 220–221, 225

C

Cable, T., 284
Cadenas, E., 17–18, 160–161, 163, 164, 174–175, 179, 180–181
Cadenas, S., 102
Cai, H., 62–63
Campbell, D. S., 247
Campos-Ortega, J. A., 208
Camus, G., 284
Cantley, L. C., 222
Cao, J., 13, 15, 221–222
Capone, F., 101–102
Carathers, M., 283–284
Carballal, S., 6
Carlsen, H., 282–283
Carlson, J. W., 204–205, 209
Carmeli, E., 283–284
Carpenter, A. T., 200–201
Carpenter, L. R., 240–241
Carpi, A., 101–102
Carroll, K. S., 6, 15, 112–114, 130–131
Carson, B. B., 206
Cartee, G. D., 270–271
Carter, G., 270
Catravas, J. D., 119–120
Cave, A. C., 112–114
Ceol, C., 242–243
Cerchiaro, G., 7
Chae, G. T., 220–221
Chae, H. Z., 199–200
Chance, B., 281–282
Chandel, N. S., 220–221
Chang, C. J., 19, 20, 100, 130–131, 296
Chang, H. C., 200–201, 203, 204–205
Chang, L., 43
Chang, T.-S., 32–33, 50, 80, 273–274
Chang, Y. C., 30
Chanseaume, E., 271–272, 273–274
Charles, R. L., 112–114, 119–120
Chattopadhyay, M., 218–220
Chau, J. F., 223–224
Chaudhary, P., 101–102
Chauvin, M.-A., 271–272, 273–274
Cheh, C. W., 223
Chen, C., 158
Chen, D., 205–206
Chen, E., 241
Chen, K., 131–132
Chen, L. F., 117–119, 174–175
Chen, M., 117
Chen, Q. M., 159, 167
Chen, S. J., 86, 221–222
Chen, Y. W., 205, 218–221, 223, 224, 226, 227–228, 229–230, 231
Chen, Z. H., 85, 86–87, 90, 93
Cheng, C. M., 42–43
Chevet, E., 28–29, 37–38
Chevion, M., 9
Chhabra, D., 85
Chiba, T., 158
Chien, S., 50
Chikuma, S., 17–18
Chinopoulos, C., 261

Cho, D. Y., 200–201
Choi, J., 87
Chou, I. N., 238–239
Chouchani, E. T., 198
Chowdhury, G., 131–132
Christensen, S. J., 201, 205
Christofi, F. L., 291–292
Christou, D. D., 32–33
Chuang, L. M., 30
Chung, H. R., 204–205
Chung, T., 224
Cipolat, S., 94
Cirino, G., 55–56
Claiborne, A., 6
Clarkson, P. M., 281–282
Close, G. C., 284–287, 288, 289, 290
Close, G. L., 281, 282, 284, 285
Clough, P. J., 284
Coburn, R. A., 201, 205
Codogno, P., 218–220, 222, 225
Codreanu, S. G., 6
Cohen, S. M., 205
Cole, P. A., 130–131
Coleman, K., 112–114
Collins, T., 42–43
Colonna, R., 105
Comb, W., 112–114
Confalonieri, S., 101
Conrad, M., 131–132
Contursi, C., 102, 105, 106, 107
Cook, K. R., 201, 205
Cook, R. K., 201, 205
Cooley, L., 158
Cooper, D. R., 270
Corcelle-Termeau, E., 218–220
Corrales, R. M., 162
Costantini, P., 105
Cotto, J. J., 283–284
Covas, G., 181, 182
Cox, A. G., 12–13, 14–15
Cozza, G., 11–12, 12*f*
Criollo, A., 222
Cronin, M. T. D., 30
Crosby, D., 80–81
Crosier, K., 242–243, 247
Crosier, P., 242–243, 247
Cross, J. V., 158
Csete, M., 280, 290–291, 292
Cuadrado, A., 36
Cuello, F., 112–114, 119–120
Cuervo, A. M., 229–230
Culotta, V. C., 112–114
Cummings, A. H., 130–131
Curado, S., 204–205
Cutler, K., 280–281
Cyrne, L., 160–161, 163, 164, 167–168, 174–175, 177–178, 179, 180–181, 182, 183, 184–185, 186, 186*f*
Czapski, G., 9
Czech, B., 205–206

D

Dabla, P. K., 282
Dabrowska, M., 29
Dabrowski, A., 29
Dagnell, M., 131–132
Dahlgren, C., 30–32
Dahlman, J. M., 283–284
Dale, R. K., 200–201
Daly, R. J., 37–38
Damrauer, J. S., 283–284
Damron, D. S., 117
Daniel, K., 258
Darley-Usmar, V., 19
Datta, S. R., 222
D'Autreaux, B., 15, 50, 62–63
Dave, K. R., 82
Davies, C., 36–37
Davies, J. E., 218–220, 221–222
Davies, K. J., 19, 280–281
Davies, M. J., 10
Davies, P. F., 52
Davis, R. J., 38–39
Davis, R. W., 200–201
Davson, H., 285
Day, A. M., 13
de Alvaro, C., 50
de Cingolani, G. C., 117
de Kloet, E. R., 101–102
de la Harpe, J., 7
de Nigris, F., 101–102
Deal, J. A., 201, 205
Deal, M. E., 201, 205
Debandi, A., 280–281, 282
DeBoer, C. J., 7
Deby, C., 284

Dedek, M., 159, 167
DeGennaro, M., 199–200, 202–203
Delaunay, A., 13
den Hertog, J., 112–114, 130–131
Denell, R. E., 200–201
Denevan, D. E., 7
Deng, Q., 239–240
Denicola, A., 7, 8
Dennery, P. A., 19
Denu, J. M., 130–132, 141
Derby-Dupont, G., 284
Desco, C., 282
Desnoyer, R. W., 117
Despa, S., 116–117
Dessy, C., 62
Deutschbauer, A. M., 200–201
Devreotes, P. N., 239–240
Dewey, C. F. Jr., 52
Di Fiore, P. P., 101
Di Lisa, F., 101–102, 104
Diamond-Stanic, M. K., 270, 271, 273, 275
Dick, T. P., 19, 246
Dickenson, J. M., 29
Dickinson, B. C., 130–131
Diebold, B., 130–131
Dietzl, G., 205–206
Dikalov, S. I., 55–56
Dimitrie, I. P., 231
Ding, X., 221–222
Djapri, C., 206
Djavaheri-Mergny, M., 225
Dodson, G. S., 198, 200–201, 204, 206, 211–212
Dokken, B. B., 273, 275
Dompe, N. A., 201, 202, 204–205
Dong, Z., 218–220
Donko, A., 239, 246
Donnelly, A. E., 284
Dora, K. A., 119–120
Doran, D., 284
Doshi, A., 162
Dou, Z., 218–220
Dowe, G., 204–205
Dranka, B. P., 20
Drenan, R. M., 119–120
Dröge, W., 282–283
Dror, R., 130–131, 271–272
Dudek, H., 222
Dudley, S. C. Jr., 55–56
Dudzinski, D. M., 62
Duff, M. O., 209
Dunn, M. J., 117–119
Durand, A., 271–272, 273–274
Durazo, D. E., 273, 275
Duthie, G. G., 284

E

Eaton, J. W., 10
Eaton, P., 112–114, 117–120
Edwards, R. H., 280–281, 285
Eger, B. T., 282
Ekas, L. A., 206
Ekker, S. C., 241
Elazar, Z., 220–221
Elhiani, A. A., 82–83, 90
Ellett, F., 246–247
Elliott, R. M., 282–283
Elouardighi, H., 82
Elworthy, S., 242–243
Engel, J. D., 158
Ennis, T., 112–114
Entman, M. L., 280–281
Enyedi, B., 239, 246
Enyedi, P., 238
Erdjument-Bromage, H., 221–222
Erdodi, F., 119–120
Erzurum, S., 28
Eskelinen, E. L., 227–228
Espenson, J. H., 150
Espert, L., 218–220
Espinosa, A., 280–281, 282
Esteves, T. C., 261
Eulenberg, K., 204–205
Eulitt, P., 231
Evans, J. L., 271, 272, 273, 275
Evans, J. M., 13
Evans, S., 206
Evans-Holm, M., 204–205
Everhart, A. L., 131–132
Exton, J. H., 88–89
Eyries, M., 42–43

F

Fahimi, H. D., 11, 17–18
Fales, H. M., 291–292, 295–296
Fan, R., 55–56

Fan, S., 223
Fang, J., 42–43
Farkas, A., 223
Farkas, T., 218–220
Farley, W., 162
Fassina, A., 101
Fawcett, R., 201, 205
Fearnley, I. M., 198
Fearon, E. R., 190
Febbraio, M. A., 283–284
Feelisch, M., 55–56
Felekidis, A., 284
Fellner, M., 205–206
Feng, H. Z., 117
Feng, Y., 238
Feng, Z., 223–224
Feron, O., 62
Ferrans, V. J., 192–193
Ferrer-Sueta, G., 8, 11–12
Fetrow, J. S., 11–12
Field, D., 184
Filippakopoulos, P., 130–131
Findlay, V. J., 13
Fink, M. A., 117
Finkel, T., 192–193, 198, 220–221, 282–283
Fischer, F., 107
Fisher, A. B., 11–12
Flaherty, D. M., 38–39
Floering, L. M., 258
Flohé, L., 11–13, 12*f*, 174–175, 198, 281–282
Flores, M. V., 242–243, 247
Floyd, R. A., 29–30
Flynn, C. R., 117
Forman, H. J., 15, 19, 29, 87, 131–132, 198, 282–283
Förster, H., 131–132
Forte, M., 104
Foster, D. B., 112–114
Fourquet, S., 112–114
Foyer, C. H., 11, 13
Fradkov, A. F., 19, 68–70, 239, 292
Franch, J., 271, 273–274
Frank, J., 43–44
Franklin, R. A., 29
Freeman, B. A., 6, 62–63
Freisinger, C. M., 246–247
Frennesson, C. I., 19–20
Frezza, C., 94
Friday, B. B., 35–36, 42–43
Frijhoff, J., 130–132
Frise, E., 245
Fu, H., 222
Fujii, M., 15–16
Fujiki, H., 92
Fujimoto, E., 247
Fujimura, Y., 90
Fujino, G., 40–42
Fujisawa, T., 40–41
Fukao, M., 119–120
Fuller, W., 112–114, 117–119
Funato, Y., 33
Furnish, E. J., 117
Futter, M., 218–220, 221

G

Gabbita, S. P., 29–30
Gabryelewicz, A., 29
Galeotti, T., 223–224
Galic, S., 131–132
Gallagher, E., 36–37
Gallini, R., 189–190
Gallogly, M. M., 15–16
Gamaley, I. A., 238–239
Ganley, I. G., 221–222
Gao, T., 116–117
Garabatos, M. N., 131–132
Garcia, F. J., 6
Garcia-Arencibia, M., 218–220, 221
Garcia-Cardena, G., 55–56
Garland, C. J., 119–120
Garuti, R., 225
Gates, K. S., 130–132
Gebremedhin, D., 119–120
Geiszt, M., 238, 239, 246
Geng, S., 223–224
Gerencser, A. A., 261
Gerhardstein, B. L., 116–117
Germino, F. W., 119–120
Gertler, F. B., 76
Gertz, M., 107
Gestwicki, J. E., 221–222
Ghansah, T., 270
Ghosh, S., 174–175, 238

Gibson, S. B., 218–221, 223, 224, 226, 227–228, 229–230, 231
Giffard, R. G., 94
Gilla, S. C., 135–136
Gilliland, W. D., 208–209
Gimbrone, M. A. Jr., 52
Gimeno, A., 282
Giorgi, C., 82
Giorgio, M., 101–102, 105, 106, 107, 108
Glascow, W. C., 131–132
Gleeson, M., 284
Glogauer, M., 240
Gnall, L. L., 280–281
Go, Y. M., 15–16
Godlewski, J., 38–39
Gogvadze, V., 30–32
Gohda, J., 41–42
Goldfine, A. L., 272, 273, 275
Goldfine, I. D., 271, 272, 273, 275
Goldstein, B. J., 131–132
Goldstein, I. M., 240–241
Goldstein, S., 9
Gomez-Cabrera, M. C., 281, 282
Gonsalves, F. C., 206
Gonzalez, A., 86
Gonzalez de Orduna, C., 50, 55–56
Goodyear, L. J., 270–271
Gopalakrishna, R., 80, 81, 82–83, 85, 86–87, 90, 93, 94–95
Gotoh, Y., 222
Grabher, C., 238, 239, 241–242, 246–247
Graham, E. T., 132–133
Gralla, E. B., 225
Grant, G. E., 240–241
Gravel, S., 240–241
Graveley, B. R., 209
Greaser, M. L., 117
Green, D. R., 104
Greene, W. C., 174–175
Green-Thompson, Z. W., 218–220, 221
Gregory, S. L., 204–205
Greif, D., 62
Gresens, J. M., 201, 205
Grether, M. E., 207
Grider, J. R., 117
Griendling, K. K., 39–40, 50
Grieve, D. J., 112–114
Griffin, M., 29
Griffiths, J. A., 208–209
Griffiths, R. D., 280–281, 283, 285, 286–287, 289, 290
Griner, E. M., 80–81, 82
Grisham, M. B., 19
Grodsky, G. M., 271
Groen, A., 130–131
Gross, T. J., 38–39
Grubert, F., 174–175
Grujic, D., 258
Guan, K. L., 270
Guay, J., 240–241
Guerois, R., 112–114
Guerriero, J. L., 218–220
Gulick, J., 117
Gulve, E. A., 274
Gundimeda, U., 82–83, 85, 86–87, 90, 93, 94–95
Guo, J., 82
Gurney, M., 10
Gurovich, A. N., 32–33
Gustafsson, B., 10
Gutteridge, J. M. C., 140, 281, 283–284, 285–286
Guttridge, D. C., 280, 281–282, 283–284

H

Hably, C., 238
Hadari, Y., 271–272
Haddad, J. J., 282–283
Hadi, R., 206
Haga, J. H., 50
Hagopian, K., 108
Hahn, H., 225
Hajduch, E., 271–272, 275
Hall, A., 8, 11–12
Hall, C., 242–243, 247
Hall, J. C., 200–201
Hall, J. P., 38–39
Halliwell, B., 19, 140, 281, 285–286, 296
Hamanaka, R. B., 220–221
Hamed, H. A., 231
Hamilton, S. R., 190
Hampton, M. B., 6*np*, 8, 12–13, 14–15, 17–18
Han, B.-B., 131–132
Hannink, M., 158
Hansen, P. A., 270, 274

Haque, A., 15
Hara, T., 221–222
Hara-Chikuma, M., 17–18
Hardie, D. G., 270–271
Hardy, M. E., 20, 247
Hariharan, M., 174–175
Harris, T. J., 199–200
Harrison, D. G., 50
Hart, C. M., 55–56
Hartenstein, V., 208
Hartshorne, D. J., 119–120
Hawkins, C. L., 10
Hayden, M. S., 238
Hayes, G. R., 271–272
Hayes, J. D., 158
He, C., 223–224
He, Y., 151–152, 204–205
Hebbar, V., 158
Hecht, D., 130–131
Heffelfinger, C., 174–175
Heffetz, D., 130–131, 271–272
Heintz, N. H., 198, 282–283
Helmcke, I., 43
Hemi, R., 272, 275
Henriksen, E. J., 270, 271, 273, 274–275
Hensley, K., 29–30
Heo, J., 229–230
Hernandez-Hernandez, A., 131–132
Hers, I., 80–81
Herscovitch, M., 112–114
Hey-Mogensen, M., 271, 273–274
Hidalgo, C., 280–281, 282
Higueras, M. A., 55–56
Hikiba, Y., 40–41
Hiley, C. R., 119–120
Hill, B. G., 112–114
Hillenmeyer, M. E., 200–201
Hinks, J. A., 6
Hinton, D. R., 82–83, 90
Hirata, H., 43
Hirschberg, V., 258
Hisamatsu, Y., 116–117
Hoffman, D. L., 100
Hoffman, T. A., 107
Holderbaum, L., 205–206
Holloszy, J. O., 270, 273, 274
Holmgren, A., 19, 296
Honda, S., 43
Hong, H. J., 42–43
Horowitz, B., 119–120
Hosaka, M., 43
Hosey, M. M., 116–117
Hosokawa, N., 221–222
Hou, N., 43
Hoyer-Hansen, M., 218–220
Hu, R., 158
Hu, W., 223–224
Hu, X., 223–224
Huang, B. W., 50
Huang, D., 223
Huang, H. C., 158
Huang, X., 223–224
Huber, A., 119–120
Hugo, M., 12–13
Hui, R., 131–132
Hundal, H. S., 271–272, 275
Hunninghake, G. W., 38–39
Hunter, T., 130–131
Huppert, K., 201, 205
Hurd, T. R., 199–200, 202–203
Hurlstone, A., 238
Hurt, D., 18–19
Hussain, S. N., 280–281
Huttemeier, P. C., 284–285
Huttenlocher, A., 239–240, 246–247
Hwang, Y. S., 11–12
Hylemon, P. B., 231

I

Ichijo, H., 28, 36–37, 40–41
Igarashi, J., 62
Igarashi, K., 158
Iglesias, P. A., 239–240
Iijima, K., 199–200
Iijima-Ando, K., 199–200
Ikejima, T., 225
Ikenoue, T., 270
Illert, A. L., 227–228
Imlay, J. A., 8–9, 10
Ingham, P. W., 242–243
Inoki, K., 221–222
Inoue, S., 17–18
Irani, K., 192–193
Isacoff, E. Y., 20
Ishii, T., 158

Israels, S. J., 220–221, 223, 226, 227–228, 229–230
Itoh, K., 158
Iwanejko, L. M., 280, 283–284

J

Jackle, H., 204–205
Jackman, R. W., 283–284
Jackson, M. J., 280–283, 285–287, 288, 289, 290–291
Jacob, S., 274
Jahn, T. P., 17–18, 238, 239–240
Jain, A., 224
Jain, M., 63
Jaken, S., 80, 81, 82
Jakob, U., 198
James, A. M., 198
Janakidevi, K., 161–162
Janes, M. S., 291–292
Jang, J., 206
Jang, S., 10
Janssen-Heininger, Y. M., 198, 282–283
Jaquet, V., 19, 39–40
Jaramillo, D. F., 200–201
Jasper, H., 199–200, 202–203
Jastroch, M., 261, 263–264, 263*f*
Javesghani, D., 280–281
Jayaraman, T., 116–117
Jayawickreme, S. P., 131–132
Jeong, S. W., 220–221
Jeong, W., 32–33, 240, 273–274
Jeppesen, M. J., 271, 273–274
Jessen, N., 270–271
Jessup, J. M., 190
Ji, L. L., 281
Jia, D., 223–224
Jiang, B. H., 42–43
Jiang, J., 10
Jiang, L., 200–201
Jiang, M., 218–220
Jiang, X., 221–222
Jiang, Y. P., 218–220
Jiang, Z.-H., 151–152
Jin, B. Y., 76
Jin, J. P., 117
Jin, L., 223–224
Johansen, T., 224
Johnson, J., 159, 167
Jones, D. P., 15–16
Jones, L., 204–205
Jones, M. L., 80–81
Jorgensen, E. M., 198, 200, 203
Joseph, J. A., 10, 20, 90–91, 291–292, 295–296
Jourd'Heuil, D., 55–56
Juan, S. H., 42–43
Juchmes-Ferer, A., 284
Jun, C. B., 221–222
Jung, C. H., 221–222

K

Kabashima, K., 17–18
Kabayo, T., 280–281, 290–292, 295–296
Kabeya, Y., 218–220
Kagan, H. M., 238–239
Kagan, V. E., 10
Kahn, B. B., 270
Kai, Y., 92
Kaizuka, T., 221–222
Kalayci, O., 28
Kallunki, T., 218–220
Kaludercic, N., 101–102
Kalwa, H., 62–63, 65–66, 69*f*, 76, 77*f*
Kalyanaraman, B., 10, 19, 20, 198, 282–283, 296
Kamata, H., 43
Kanda, Y., 50
Kandarian, S. C., 283–284
Kane, D. A., 271, 273
Kane, D. J., 225
Kane, S., 270
Kanety, H., 272, 275
Kang, D., 131–132, 273–274
Kang, M. I., 158
Kang, R., 223
Kang, S. W., 32–33, 50, 80, 199–200
Kanter, M. M., 281–282
Kao, A. W., 101–102
Karim, F. D., 200–201, 203, 204–205
Karin, M., 36–37, 43, 223–224
Karlsson, A., 30–32
Karlsson, M., 19–20
Karplus, P. A., 8, 11–12, 16–17
Karton, A., 8
Kasorn, A., 240
Kasowski, M., 174–175

Kass, D. A., 112–114
Katagiri, K., 36–37
Katoh, Y., 158
Katsuyama, M., 30–32
Katz, S., 206
Kaul, N., 87
Kawahara, T., 130–131
Kayani, A. C., 280–282, 285, 290–291, 292
Kaynig, V., 245
Kazanietz, M. G., 80–81, 82
Kazlauskas, A., 190–191, 192–193, 194
Keaney, J. F. J., 131–132
Keaney, J. F. Jr., 62–63
Keef, K. D., 119–120
Keerthi, K., 130–131
Kemp, M., 15–16
Kennedy, B. P., 134, 135
Kensler, T. W., 158
Kenyon, J. L., 119–120
Keshet, Y., 28–29
Kettle, A. J., 17–18
Khachigian, L. M., 42–43
Khamaisi, M., 272, 275
Khera, G., 285
Kiens, B., 271, 273–274
Kihara, A., 218–220
Kikkawa, U., 82
Kil, I. S., 11–12, 12*f*, 13, 15
Kim, C. S., 107, 117–119
Kim, D. H., 221–222
Kim, D. J., 225
Kim, E. G., 88–89
Kim, H. J., 11–12, 229–230
Kim, J. S., 271, 273, 274–275
Kim, K. M., 11–12, 108, 199–200
Kim, P. K., 231
Kim, S. K., 220–221
Kim, S. R., 130–132
Kim, Y. J., 11–12
Kim, Y. M., 221–222
Kim, Y. R., 107
Kinderman, F. S., 117–119
King, J. E., 221–222
King, O. N., 130–131
Kinkade, P., 132–133
Kirber, M. T., 131–132
Kirk, M. C., 6
Kirpichnikova, K. M., 238–239
Kirschner, M. W., 200–201
Kishi, C., 221–222
Kitz, R., 146–147, 151–152
Klann, E., 82, 86
Kleiman, E., 270
Klein, K., 206
Kleinhenz, D. J., 55–56
Klionsky, D. J., 218–220, 221–222, 225, 226, 227–228, 229–230
Klomsiri, C., 6, 11–12
Klyubin, I. V., 238–239
Knapp, L. T., 82
Knutson, S. T., 11–12
Kobayashi, A., 158
Koch, W. J., 116
Koga, K., 90
Koh, M. S., 36
Kohen, R., 32, 91–92
Koike, K., 119–120
Komalavilas, P., 117
Kondoh, K., 37–38
Kong, A. N., 158
Kong, J., 220–221, 223, 226, 227–228, 229–230
Konishi, H., 82
Kopp, E. B., 174–175
Koranyi, L., 273
Koshkin, V., 258
Kozlovsky, N., 272, 273, 275
Kranias, E. G., 116–117
Krause, K. H., 19, 39–40, 238
Krauss, S., 258
Kristiansen, K. A., 17–18
Kroemer, G., 104, 218
Krook, A., 270
Kugler, J. M., 205
Kuiper, J. W., 240
Kumar, A., 107, 221–222
Kumar, S., 107, 151–152
Kumm, J., 200–201
Kuroki, Y., 50
Kurz, T., 10, 19–20
Kwan, K. M., 247
Kwiatkowski, D., 184
Kwon, J., 15, 50, 198, 240
Kwon, K. S., 11–12, 130–132
Kyriakis, J. M., 28–29

L

LaButti, J. N., 131–132
Ladner, K. J., 283–284
Laffranchi, R., 30–32
Lahair, M. M., 29
Lai, K. M., 101–102
Lai, K. P., 223–224
Lakatos, G., 238
Lakshminarayanan, V., 161–162, 240–241
Lam, A., 108
Lam du, H., 221–222
Lam, H. A., 206
Lamark, T., 224
Lambeth, J. D., 18–19, 130–131
Lancaster, J. R. Jr., 17*np*
Landolin, J. M., 209
Lane, W. S., 270
Lanks, K., 7
Larkin, L., 280–281, 286–287, 288, 290
Larsson, N. G., 19, 296
Lassegue, B., 39–40
Latek, R. R., 221–222
Lauer, F. T., 35–36
Laverty, T. R., 198, 200–201, 203, 204–205, 206, 211–212
Lawal, H. O., 206
Lawrence, J. C. Jr., 272, 273, 275
Lawson, N. D., 242–243
Layton, J. E., 246–247
Lazarovici, P., 91–92
Leaner, V., 38–39
LeBert, D. C., 246–247
LeBlanc, J., 242–243
Lee, C. R., 100, 240
Lee, H. J., 229–230
Lee, H. O., 225
Lee, J. H., 199–200, 223–224
Lee, K. S., 199–200
Lee, S. B., 221–222, 225, 280, 290–291, 292
Lee, S. P., 11–12
Lee, S.-R., 15, 50, 80, 130–132, 198, 240
Lee, W. H., 130–131
Lee, W. J., 199–200
Leeuwenburgh, C., 32–33
Lefkowitz, R. J., 116
Lehmann, R., 199–200, 202–203, 207–208
Lehoux, S., 50
Lei, H., 190–191, 192–193, 194
Leibfritz, D., 30
Leipelt, M., 107
Leitch, J. M., 112–114
Leiva, A., 280–281, 282
Lemieux, A. M., 273, 275
Lemmon, M. A., 130–131
Lenardo, M. J., 224
Lengyel, J. A., 211–212
León-Buitimea, A., 35–36
Leong, W. F., 223–224
LeRoith, D., 271–272
Leto, T. L., 18–19
Levin, E., 221, 285
Levine, A., 223–224
Levine, B., 218–220, 221, 225, 226, 229–230
Levine, R. L., 291–292, 295–296
Levis, R. W., 204–205
Lewin, N. E., 81
Lewis, E. B., 204
Lewis, S. M., 130–132
Ley, H. G., 273
Li, D. Q., 162
Li, L., 231
Li, R., 119–120
Li, W., 238–239
Li, Y. S., 50, 240
Lian, H., 223
Liang, X. H., 225
Liao, G., 204–205
Liao, R., 63, 76
Liebler, D. C., 6
Lieschke, G. J., 65–66, 68–70, 74, 76, 241
Lilley, K. S., 198
Lim, S. F., 205
Lin, C.-T., 271, 273
Lin, S. F., 189–190
Lin, Y., 159, 167
Lindborg, K. A., 273, 275
Lindsley, D. L., 200–201
Lipinski, M. M., 225
Lippard, S. J., 62–63, 65–68, 69*f*, 76, 77*f*
Litherland, G. J., 271–272, 275
Liu, C. Y., 291–292
Liu, G., 238–239
Liu, H. L., 81, 223–224
Liu, J. C., 42–43
Liu, L. P., 205–206

Liu, S., 151–152
Liu, X., 174–175
Liu, Y., 43
Liu, Z., 224
Livesey, J. H., 17–18
Livesey, K. M., 223
Lo, S. C., 158
Lockwood, D. H., 271–272
Loh, S. H., 42–43
Loison, F., 240
Longair, M., 245
Look, A. T., 238, 239, 241–242, 246–247
Lopes, L. B., 117
Lopez-Ferrer, D., 55–56
Lopez-Revuelta, A., 131–132
López-Torres, M., 102
Lornejad-Schäfer, M. R., 43–44
Loschen, G., 281–282
Lotze, M. T., 223
Loughran, P., 223
Low, F. M., 8, 12–13
Loynes, C. A., 242–243
Loze, M. T., 223
Lu, T., 225
Lubbe, K., 132–133
Lukyanov, K. A., 19, 68–70, 239, 292
Lunsford, L. B., 207, 208
Luo, H. R., 240
Luo, L., 162
Luo, S., 221–222, 225
Lustig, M. E., 271, 273
Luzi, L., 101

M

Ma, C., 119–120
MacDonald, J. A., 119–120
MacDonald, M. J., 258
Macdonald, P. M., 210*f*
Mackey, J. A., 117
MacLaren, D. P., 284
Maclean, D. A., 285
MacLennan, D. H., 116–117
Maddux, B. A., 271, 272, 273, 275
Madhani, M., 112–114, 119–120
Madsen, K., 271, 273–274
Maeda, S., 40–41, 43
Magalhaes, M. A., 240
Magder, S. A., 280–281
Maghzal, G. J., 8, 12–13, 19, 39–40
Mahboubi, A., 178*f*, 184–185
Mahedev, K., 131–132
Maiorino, M., 11, 12–13, 15
Maiuri, M. C., 222
Makhlouf, G. M., 117
Malik, A. B., 161–162
Malone, J. H., 200–201
Mandel, R. J., 70
Manevich, Y., 13, 15
Mango, S. E., 198, 200, 203
Mangum, B. D., 247
Mansouri, A., 280–281, 286–287, 288, 290
Mansueto, G., 101–102
Manta, B., 8, 12–13
Marchi, S., 82, 107
Marchionne, E. M., 270, 271, 273, 275
Marinho, H. S., 160–161, 163, 164, 167–168, 174–175, 177–179, 180–181, 182, 183, 184–185, 186, 186*f*
Markstein, M., 205–206
Maron, B. A., 62
Marracci, G., 101–102
Martin, A. P., 231
Martin, P., 238
Martin, S. J., 178*f*, 184–185
Martinez-Ruiz, A., 55–56
Martin-Padura, I., 101–102
Marx, S. O., 116–117
Mason, H. S., 119–120
Massberg, S., 240
Masters, S., 222
Mata, J., 204–205
Matalon, S., 238
Mathieu, J., 204–205, 225
Matsuda, M., 65–66, 76
Matsuno, K., 30–32
Matsuzaki, H., 82
Matsuzawa, A., 28, 36–37, 40–42
Mattiazzi, A., 117
Mattiuzzo, N. R., 200–201
Maughan, R. G., 284
May, M. J., 174–175
Mayama, T., 223–224
Mazer, B. D., 240–241
Mazur, M., 30
McArdle, A., 280–282, 283–284, 285–287, 288, 289, 290–291, 292

McArdle, F., 280, 283–284, 290–291, 292
McCubrey, J. A., 29
McDonald, J. A., 207–208
McGahon, A. J., 178*f*, 184–185
McMahon, M., 158
McMillan-Ward, E., 220–221, 223, 226, 227–228, 229–230
McNeill, T. H., 82–83, 90, 94–95
McQuade, L. E., 66–68
McQuibban, G. A., 231
Meczner, A., 238
Mehdi, M. Z., 42–43
Meijer, A. J., 222
Mele, S., 101–102
Menabó, R., 101–102
Mendoza, S. A., 119–120
Meng, D., 42–43
Meng, Q., 223
Meng, T.-C., 6, 131–132
Menini, S., 101–102
Merkler, D. J., 132–133
Merola, A. J., 291–292
Metodiewa, D., 5–6
Metrakos, P., 28–29, 37–38
Meves, A., 42–43
Meyer, A. J., 19
Meyerstein, D., 9
Michaelson, L. P., 281–282
Michaud, N. R., 200–201
Michel, T., 62–63, 65–66, 76, 77*f*
Mieyal, J. J., 15–16
Migliaccio, E., 101–102, 105, 106, 107, 108
Miinea, C. P., 270
Miki, H., 33
Milano, C. A., 116
Milash, B., 202, 204–205
Miller, E. W., 20, 100
Miller, H., 6
Milne, G. W., 81
Min, D. S., 88–89
Minucci, S., 107, 108
Mione, M., 238
Misra, H. P., 135
Mitchell, C., 231
Mitchison, T. J., 238, 239, 241–242, 246–247
Miura, Y., 221–222
Miyazaki, W. Y., 202, 204–205
Mizuno, K., 50
Mizushima, N., 218–220, 225, 226, 227–230
Mnaimneh, S., 55–56
Mochly-Rosen, D., 80–81, 94
Mogil, R. J., 178*f*, 184–185
Molitor, A., 204–205
Moller, A. L., 17–18
Moller, I. M., 17–18
Moncol, J., 30
Mondal, S., 240
Mongue-Din, H., 112–114
Monick, M. M., 38–39
Montalibet, J., 134, 135
Monteiro, G., 12–13
Montell, D. J., 207–208
Moody, M. R., 280–281
Moon, A., 36
Moore, L. A., 207, 208
Moran, L. K., 283–284
Morand, S., 18–19
Morelock, M. M., 132–133
Morgan, B., 246
Morgan, J. E., 280–281
Morimoto, R. I., 283–284
Morio, B., 271–272, 273–274
Morita, K., 41–42
Moroni, M., 102, 105, 106, 107
Morrice, J. C., 284
Morris, B., 101–102
Morrison, D. K., 200–201
Morrissy, S., 159, 167
Morselli, E., 222
Moss, R. L., 117
Mossman, B. T., 198, 282–283
Mott, R., 184
Movitz, C., 30–32
Moy, J. N., 240–241
Moynagh, P. N., 174–175
Mozden, N., 204–205
Mukai, S., 190, 194
Muller, M., 218–220, 280–281, 282
Muller-Delp, J. M., 32–33
Mundina-Weilenmann, C., 117
Murphy, L. O., 221–222
Murphy, M. P., 19, 100, 198, 238, 296
Murrant, C. L., 292
Murray, M. J., 204–205, 285

Murray-Rust, J., 80–81
Murthy, K. S., 117
Myers, M. P., 6

N

Nadiv, O., 271–272
Nagano, H., 223–224
Nagy, D., 70
Nagy, P., 6*np*, 8
Nah, J., 229–230
Nakagawa, H., 40–41
Nakano, H., 41–42
Nakano, K., 223–224
Naldini, L., 70
Napoli, C., 101–102
Naqvi, A., 107
Narayanapanicker, A., 112–114
Nartiss, Y., 231
Natarajan, V., 88–89
Nathan, C. F., 7, 239–240
Nebreda, A. R., 36
Nelson, K. J., 6, 11–12
Netto, L. E., 12–13
Neubauer, G., 119–120
Neumann, C. A., 13, 15
Newton, A. C., 80–81
Ng, A., 225
Nguyen, T., 158
Ni, J. Q., 205–206
Nicholls, D. G., 261
Niethammer, P., 238, 239, 241–242, 246–247
Nigro, J. M., 190
Nilsson, S. E., 19–20
Nioi, P., 158
Nishida, E., 37–38
Nishino, T., 282
Nishio, E., 50
Nishitoh, H., 15–16, 41–42
Nishizuka, Y., 80–81
Nivorozhkin, A., 282
Nobunaga, M., 240–241
Noctor, G., 11, 13
Noda, T., 221–222
Noguchi, T., 40–42
Noll, A. C., 208–209
Nowell, C. J., 65–66, 68–70, 74, 76, 246–247
Nylandsted, J., 218–220
Nyska, A., 32

O

O'Brien, J., 184–185
O'Brien, P. M., 37–38
O'Connell, T. D., 63
Oddi, G., 101–102
O'Donnell-Tormey, J., 7
Ogier-Denis, E., 222
Oh, E., 220–221
Ohkubo, S., 223–224
Ohkura, S., 117
Ohmae, T., 40–41
Ohsumi, Y., 218–220, 221–222
Ohtsuji, M., 158
Oka, S. I., 112–114
Okabe, S., 92
Okada, S., 101–102
Okamoto, K., 282
Okawa, H., 158
O'Keefe, L., 204–205
Oliveira-Marques, V., 167–168, 174–175, 177–178, 179–180, 183, 184–185, 186, 186*f*
Olson, L. E., 189–190
Onodera, S., 225
O'Reilly, R. J., 8
O'Rourke, B., 112–114
Oroz, L. G., 221–222
Orsini, F., 102, 105, 106, 107
Orton, T., 184–185
Osburn, W. O., 158
Osei-Owusu, P., 119–120
Östman, A., 130–131
Otto, N. M., 221–222
Overvatn, A., 224
Overvoorde, J., 112–114
Owuor, E., 158

P

Pace, P., 8
Pacher, P., 282
Packer, L., 33, 280–281
Page, S. M., 240–241
Pai, E. F., 282
Pallardó, F. V., 281, 282
Palmero, J., 280–281, 290–291

Palomero, J., 280–282, 290–292, 295–296
Pan, J. A., 218–220
Pandolfi, P. P., 101–102
Pani, G., 223–224
Paolucci, D., 105, 106, 107
Papa, S., 282–283
Papapetropoulos, A., 55–56
Pargellis, C. A., 132–133
Park, M. A., 231
Parker, P. J., 80–81
Parks, A. L., 201, 205
Parsonage, D., 6, 8
Parsons, Z. D., 130–132
Pase, L., 65–66, 68–70, 74, 76, 246–247
Passamonti, S., 105
Patel, J. C., 100
Patel, N. A., 270
Patel, P., 240–241
Paton, L. N., 8, 12–13, 14–15
Patridge, T. A., 280–281
Patterson, K. I., 37–38
Pattingre, S., 218–220, 225
Pattison, D. I., 10
Pattwell, D. M., 280–281, 283, 285, 286–287, 289, 290–291
Pav, S., 132–133
Pedersen, B. K., 283–284
Pelicci, G., 101–102
Pelicci, P. G., 101–102, 107
Pena, M., 280–281, 282
Penn, R. B., 117
Perera, A. G., 208–209
Perez, H. D., 240–241
Perez-Campo, R., 102
Perez-Pinzon, M. A., 82
Periyasamy-Thandavan, S., 218–220
Permutt, M. A., 273
Peroni, O., 258
Perret, P., 258
Persson, H. L., 10
Pesce, C., 101–102
Peskin, A. V., 8, 12–13, 14–15
Pessin, J. E., 101–102
Peterson, J. M., 283–284
Petiot, A., 222
Petronilli, V., 104, 105
Peus, D., 42–43
Peyrol, S., 271–272, 273–274
Pfeiffer, B., 205–206
Pflieger, D., 13
Pflugfelder, S. C., 162
Pi, J., 258
Piccini, D., 102
Pickett, C. B., 158
Pietzsch, T., 245
Pincemail, J., 284
Pinton, P., 107
Pittelkow, M. R., 42–43
Platt, F. M., 221–222
Plotnikov, A., 33–36, 42–43
Pluth, M. D., 62–63, 65–68, 69*f*, 76, 77*f*
Pognan, F., 184–185
Poletaeva, I., 108
Poole, A. W., 80–81
Poole, L. B., 8, 11–13, 16–17
Porras, A., 50
Post, J. A., 35–36
Potashnik, R., 272, 273, 275
Powers, L. S., 38–39
Powers, P. A., 117
Powers, S. K., 280, 281–282, 283
Poyurovsky, M. V., 223–224
Pralle, A., 20
Prasannarong, M., 273, 274–275
Pregel, M. J., 135–136
Preisinger, A. C., 190
Previs, M. J., 117
Procaccia, S., 33–36, 42–43
Proctor, M., 200–201
Prysyazhna, O., 119–120
Pugieux, C., 204–205
Pugliese, F., 101–102
Pula, G., 80–81
Purdom-Dickinson, S. E., 159, 167
Py, B. F., 225
Pye, D., 280–281, 290–292, 295–296
Pyo, J. O., 229–230

Q

Qu, X., 225
Quinlan, G. J., 283–284
Quinn, J., 13
Quinn, M. T., 280–281
Quintanilha, A. T., 280–281

R

Rada, B., 238
Radi, R., 6, 7, 8, 11–12
Rahman, A., 161–162
Raker, V. A., 102
Ramsey, J. J., 108
Rasband, W. S., 166, 183–184
Rassaf, T., 55–56
Raval, A. P., 82
Ravikumar, B., 218–220, 221–222
Ravingerová, T., 28–29
Ray, P. D., 50
Ray, R., 112–114
Reboldi, P., 101–102
Reddie, K. G., 112–114
Rees, M. D., 10
Reichert, A. S., 218–220
Reid, M. B., 280–281, 292
Reiken, S., 116–117
Reilly, T. J., 131–132
Remmen, H., 280–281, 286–287, 288, 290
Renshaw, S. A., 242–243
Reyes-Aldasoro, C. C., 246–247
Reznick, A. Z., 283–284
Rheaume, M. A., 190, 194
Rhee, S. G., 11–12, 12*f*, 13, 15, 32–33, 50, 80, 130–132, 198, 199–200, 240, 273–274
Rhem, E. J., 204–205
Ricci, C., 101–102
Richardson, A. G., 280–281, 286–287, 288, 290
Richardson, B. E., 199, 207–208
Richardson, D. E., 7
Richter, C., 30–32, 281–282
Rimessi, A., 107
Rinaldi, G., 117
Rinna, A., 131–132
Ritov, V. B., 10
Ro, S. H., 221–222
Robbins, J., 117
Roberson, E. D., 82
Roberts, C. T. Jr., 271–272
Robertson, J. D., 284
Robinson, D. R., 189–190
Robinson, K. A., 29–30
Robinson, K. M., 291–292
Robson-Doucette, C. A., 258
Rockman, H. A., 116
Rodney, G. G., 281–282
Rodnick, K. J., 273
Rodrigo, M. C., 63
Rodriguez, J., 199–200
Rodriguez, M. C., 131–132
Rodríguez-Fragoso, L., 35–36
Roebuck, K. A., 161–162, 240–241
Roebuck, S. J., 281–282
Rogers, L. C., 6
Rojas, C., 102
Rokach, J., 240–241
Romero, N., 50
Rorth, P., 204–205
Rosemblit, N., 116–117
Rosen, E. M., 223
Ross, R. P., 6
Rothwell, W. F., 207–208
Roveri, A., 11
Rubin, G. M., 198, 200–201, 203, 204–205, 206, 211–212
Rubinsztein, D. C., 221–222, 225
Rucker, J., 112–114
Rudich, A., 272, 273, 275
Rudolph, T. K., 62–63
Rudyk, O., 119–120
Ruffels, J., 29
Rusten, T. E., 224
Rutter, G. A., 258
Rybin, V. O., 82

S

Sabri, A., 82
Sackesen, C., 28
Saegusa, K., 41–42
Saengsirisuwan, V., 271, 273, 274–275
Saha, T., 223
Sahiner, U. M., 28
Said, M., 117
Saint, R., 204–205
Saito, N., 43
Saitoh, M., 15–16, 40–42
Sakai, J., 240
Sakellariou, G. K., 280–282, 290–292, 295–296
Salcini, A. E., 101–102
Salmeen, A., 6, 15
Salsman, S., 29–30

Saltin, B., 285
Sanchez-Gallego, J. I., 131–132
Sanchez-Gomez, F. J., 50
Sandin, A., 130–132
Sandler, L., 200–201
Sano, E., 270
Sano, H., 270
Santaren, J. F., 199–200
Santoriello, C., 238
Santos, F. R., 273, 275
Sarafian, T. A., 225
Sarbassov, D. D., 221–222
Sarkar, D., 231
Sarkar, F. H., 223
Sarkar, S., 218–220, 221
Sarma, G. N., 117–119
Sarraj, B., 240
Sartoretto, J. L., 62–63, 65–66, 69*f*, 76, 77*f*
Sartoretto, S. M., 62–63, 69*f*
Sastre, J., 281, 282
Sathyanarayana, P., 38–39
Sato, K., 117
Savino, C., 101–102
Sawada, Y., 15–16
Schaefer, E., 82
Schäfer, C., 43–44
Schafer, U., 204–205
Scherz-Shouval, R., 218–221, 224
Schiffman, J. E., 82–83, 85, 90, 94–95
Schindelin, J., 245
Schjoerring, J. K., 17–18, 238, 239–240
Schlapbach, R., 30–32
Schlessinger, J., 130–131
Schlossmann, J., 119–120
Schluter, K. D., 63
Schnorrer, F., 205–206
Schoenlein, P., 218–220
Schöffl, H., 43–44
Scholz, R., 273
Schomburg, D., 11
Schrader, M., 11, 17–18
Schramm, V. L., 132–133
Schreiber, D., 63
Schroder, E., 112–114, 117–120
Schröder, K., 43
Schulz, A., 17–18
Schwarzmann, G., 227–228
Schwezer, M., 30–32
Scorrano, L., 94, 105
Scribner, W. M., 88–89
See, W., 272, 273, 275
Seger, R., 28–29, 33–36, 42–43
Seglen, P. O., 229–230
Seiler, A., 131–132
Seiner, D. R., 130–132
Selak, M. A., 223–224
Sen, C. K., 33
Senovilla, L., 222
Seo, G., 220–221
Seo, M. S., 199–200, 221–222
Seo, Y. H., 6
Sessa, A., 242–243
Sessoms, J. S., 86
Shah, A. M., 50, 112–114
Shah, V., 55–56
Shakhbazov, K. S., 19, 68–70, 239, 292
Shandala, T., 204–205
Shattock, M. J., 116–117
Shen, Y. H., 38–39
Shepherd, P. R., 270
Sheveleva, E. V., 159, 167
Shi, G., 281–282
Shi, J., 39–40
Shi, X., 42–43
Shi, Y., 178*f*, 184–185
Shibata, T., 158
Shibata, W., 40–41
Shih, N. L., 42–43
Shimokawa, H., 119–120
Shin, D. H., 131–132
Shingu, M., 240–241
Shiroto, T., 62–63, 69*f*
Shoji, T., 280–281
Siekhaus, D. E., 199–200, 202–203
Sikora, A., 20
Silverman, R. B., 146–147, 151–152
Simon, M. A., 198, 200–201, 204, 206, 211–212
Simpson, P. C., 63
Singer, M. A., 202, 204–205
Singh, A., 223
Singh, B., 282
Singh, C. M., 202, 204–205
Singh, H., 130–132
Singh, N., 55–56
Sivaramakrishnan, S., 130–131

Sjottem, E., 224
Skorey, K. I., 134, 135
Sloniger, J. A., 271, 273, 274–275
Snider, N., 285
Sobotta, M. C., 246
Soetaert, J., 204–205
Soito, L., 6, 11–12
Sokoloski, E. A., 291–292, 295–296
Somma, P., 101–102
Somogyi, K., 204–205
Son, M. J., 225
Sorescu, D., 39–40
Soriano, P., 189–190
Sorrentino, R., 55–56
Sossin, W. S., 82
Spiller, D. G., 280–281, 290–291
Spindel, O. N., 40–41
Spradling, A. C., 204–205
Springate, J. J., 11–12
Springer, M., 200–201
Srivastava, A. K., 42–43
St Johnston, D., 198, 200–201, 204, 205, 206, 211–212
Stadler, W. M., 35–36
Stadtman, E. R., 9
Starke, D. W., 15–16
Starnes, T. W., 239–240
Staroverov, D. B., 19, 68–70, 239, 292
Staudt, N., 204–205
Steegborn, C., 107
Steinberg, S. F., 81, 82
Steinberg, T. H., 119–120
Steipe, B., 231
Steller, H., 207
Stelzer, J. E., 117
Stendardo, M., 107, 108
Stenmark, H., 224
Stern, D., 204–205
Stock, S. N., 42–43
Stocker, R., 19, 39–40, 62–63
Stone, J. R., 6*np*, 131–132, 271
Storer, A. C., 135–136
Storm, T., 242–243, 247
Storz, P., 62–63
St-Pierre, J., 281–282
Strnisková, M., 28–29
Struhl, G., 210*f*
Su, K. C., 101–102, 205–206
Subramanian, K. K., 240
Sueoka, E., 92
Sueoka, N., 92
Suganuma, M., 92
Sugimoto, N., 117
Sugiyama, Y., 17–18
Suh, Y. A., 39–40
Sui, X., 223–224
Sullivan, W., 207–208
Sultan, S., 258
Sun, C., 240
Sun, H., 159, 167
Sun, X., 119–120
Sun, Y., 223–224
Sundaresan, M., 192–193
Sung, H. H., 204–205
Superti-Furga, G., 101–102
Suter, M., 30–32
Suttangkakul, A., 224
Sutton, H. C., 9
Suzuki, S., 223–224
Suzuki, T., 221–222
Sweatt, J. D., 82
Symons, M. C., 280–281
Szabó, C., 282
Szallasi, Z., 81
Szuplewski, S., 205
Szyniarowski, P., 218–220

T

Tabakman, R., 91–92
Tachibana, H., 90
Takahashi, T., 41–42
Takamura, A., 221–222
Takeda, K., 15–16, 40–42
Takemura, Y., 82
Takeuchi, T., 43
Takuwa, N., 117
Tanaka, T., 223–224
Tanaka, Y., 119–120, 227–228
Tang, D., 223
Taniguchi, H., 82
Tanner, J. J., 130–131
Tanner, K. G., 130–132, 141
Tao, X., 222
Tarin, C., 55–56
Tarrant, M. K., 130–131
Tashino, S.-I., 225

Tassa, A., 225
Tavender, T. J., 11–12
Taylor, S. S., 117–119
Teachey, M. K., 271, 273, 274–275
Telford, J. L., 101–102
Telser, J., 30
Templeton, D. J., 158
Tepass, U., 199–200
Terrell, A., 206
Terskikh, A. V., 19, 68–70, 239, 292
Test, S. T., 174–175
Thamsen, M., 198
Therrien, M., 200, 203, 204–205
Thibault, S. T., 202, 204–205
Thomas, L. L., 240–241
Thompson, T. A., 35–36
Thomson, D. M., 270–271
Thomson, L., 11–12
Thresher, J., 117
Thurman, R. G., 273
Tiganis, T., 131–132
Timar, C., 238
Tirosh, A., 272, 275
Tobiume, K., 15–16, 41–42
Toledano, M. B., 13, 15, 50, 62–63, 112–114
Tomilov, A. A., 108
Tong, C. W., 117
Tong, K. I., 158
Tonks, N. K., 6, 15, 130–132
Toppo, S., 11–13, 12*f*
Torii, S., 43
Toro, L., 119–120
Torres, M., 29
Toskulkao, C., 273, 274–275
Tournier, C., 36–37
Tracey, K. J., 223
Trede, N. S., 241
Trewhella, J., 117–119
Tribillo, I., 29
Trindade, D. F., 7
Trinei, M., 101, 102, 107, 108
Trono, D., 70
Trujillo, M., 8, 11–12
Truong, T. H., 6, 15, 130–131
Trushell, D. M., 242–243
Tsang, G., 204–205
Tsuji, Y., 50
Tsuruta, A., 223–224
Tucker, A. L., 116–117
Turell, L., 12–13
Tyurin, V. A., 10
Tyurina, Y. Y., 10
Tzahar, E., 271–272

U

Uchida, K., 158
Udalova, I. A., 184
Ueyama, T., 18–19
Ugochukwu, E., 130–131
Umada-Kajimoto, S., 82
Ursini, F., 11–13, 12*f*, 15
Urso, M. L., 281–282
Ushio-Fukai, M., 50, 130–131

V

Vacher, C., 221–222
Vaishya, R., 282
Valentine, J. S., 225
Valko, M., 30
Van Damme, J., 199–200
van der Meulen, J. H., 280–281, 286–287, 288, 290
van der Wijk, T., 112–114, 130–131
Van Doren, M., 207, 208
Van Eyk, J. E., 112–114
Van Gammeren, D., 283–284
Vandekerckhove, J., 199–200
Vanin, S., 12–13
Vasilaki, A., 280–282, 283–284, 285, 286–287, 288, 289, 290–292, 295–296
Vasquez-Vivar, J., 7, 291–292, 295–296
Veal, E. A., 13
Velez, G., 190, 194
Vepa, S., 88–89
Verbon, E. H., 35–36
Verkman, A. S., 17–18
Verma, P., 205
Verma, R. S., 88–89
Vichaiwong, K., 273, 274–275
Vidal, M. L., 131–132
Vidal, M. N., 223–224
Vierstra, R. D., 224
Vigil, D., 117–119
Villalta, C., 205–206
Villanueva, L., 55–56

W

X

Y

Yano, M., 107
Yao, H., 7
Yarden, Y., 271–272
Yetik-Anacak, G., 119–120
Yick, P. J., 112–114
Yim, S. H., 131–132
Yin, Q., 50
Yong, H. Y., 36
Yong, S., 112–114
Yoo, S. K., 239–240, 246–247
Yoshimori, T., 218–220, 227, 229–230
Yoshioka, K., 117
Yu, D.-Y., 131–132
Yu, K., 199–200
Yu, R., 158
Yu, X., 101–102
Yu, W. H., 221–222
Yu, Z. X., 10, 192–193
Yuan, J., 229–230

Z

Zabouri, N., 82
Zafari, A. M., 50
Zakhary, D. R., 117
Zana, M., 239, 246
Zeh, H. J. 3rd., 223
Zehorai, E., 33–36, 42–43
Zeng, L., 223–224
Zenke, Y., 158
Zhang, C.-Y., 223–224, 258
Zhang, D. D., 158
Zhang, D. X., 119–120
Zhang, G., 231
Zhang, H., 10
Zhang, Q., 258
Zhao, H., 291–292, 295–296
Zheng, B., 225
Zhou, H., 117, 130–132
Zhou, J. Y., 43
Zhou, R., 205–206
Zhou, X., 270
Zhu, J., 38–39
Zhu, L., 131–132
Zibrik, L., 280–282, 290–292, 295–296
Zick, S. K., 117–119
Zick, Y., 130–131, 271–272
Zielonka, J., 20
Zierath, J. R., 270
Zilbering, A., 131–132
Zinkevich, N. S., 119–120
Zong, X., 119–120
Zufferey, R., 70
Zuo, L., 291–292
Zweier, J. L., 174–175

SUBJECT INDEX

Note: Page numbers followed by "*f*" indicate figures, and "*t*" indicate tables.

A

Antioxidant defenses
  catalase, 11
  glutathione peroxidases, 11–12
  peroxiredoxins, 11–12
  rate constants, catalases, 12–13
  redox-regulated pathways, 13
  seleno-glutathione peroxidase, 11–12, 12*f*
Apoptosis signal regulating kinase 1 (ASK1)
  JNK and p38 MAPK pathways, 40–41
  NCC domain, 40–41
  TRAF2, 41–42
Autophagy, ROS
  acidic vesicular organelles, 227–228
  autophagic cell death, 218, 219*f*
  autophagosome formation, electron microscopy, 228
  and Beclin-1, 223
  cell survival, 218
  definition, 218
  detection, *in vivo*, 228–229
  double-membraned autophagosomes, 225
  formation, autolysosomes, 225
  mTOR regulation, 221–222
  oxidation, Atg4, 224
  oxidative stress, 229–231
  and p53, 223–224
  and p62 adaptor, 224
  quantification, GFP-LC3 puncta, 226–227
  regulation, 218–220, 220*f*
  repression, 225
  self-digestion, 218–220
  signaling pathways, 220–221
  stress conditions, 218
  western blotting, Atg8/LC3, 227

B

BAECs. *See* Bovine aortic endothelial cells (BAECs)
Bovine aortic endothelial cells (BAECs), 51, 74

C

Carbon dioxide, 7
Cardiac myocytes
  Cu_2(FL2E), 66–68
  H_2O_2 and NO, 64
  HyPer2 H_2O_2 biosensor, 68–76
  HyPer2 positive cells, 76
  live cell imaging, 65–66
  nifedipine treatment, 66–68, 69*f*
  nitric oxide, 62
  protocols, 63
  redox signaling pathways, 63
  ROS, 62–63
Cell death-related assays, 92
Cell signaling, ROS
  cytosol, 33
  hydrogen peroxide, 32–33
  protein cysteine, 33, 34*f*
Cerebral ischemia
  cadmium chloride, 94–95
  cell death-related assays, 92
  GTPP-induced preconditioning, 92
  H_2O_2, preconditioning, 93
  immunocomplex precipitation assay, 94
  mitochondrial association, 94
  NADPH oxidase, 90–91
  oxidative activation, 93
  oxygen-glucose deprivation/reoxygenation, 91–92
  pan-PKC, preconditioning, 93
Confocal microscopy, H_2O_2 production
  image processing and data analysis, 251–252
  leukocyte recruitment, 248–249
  spinning disk confocal imaging, 251, 252*f*
  traditional laser scanning techniques, 249–250
  wounding and image acquisition, 250–251
Contraction-dependent pathway, 270–271

Cyclic nucleotide dependent protein kinases
description, 112
intermolecular disulfide formation, 112–116
SDS-PAGE, 121–123
tissue and cell lysates preparation, 121

D

Data handling and analysis
bolus addition, 167
cell exposure, 166–167
de novo synthesis, 167, 169*f*
ImageJ software, 166
Nrf2 nuclear translocation, 170
Dichlorodihydrofluorescein (DCFH), 290–291
Differential gene regulation, 174–175
Dihydroethidium (DHE)
and DCFH, 292–294
fibers, 294–295
fluorescence microscopy and image analysis, 295
oxidation, 291–292
superoxide, 291–292
Dominant modifier screens
balancer chromosome, 207
dominant mutations/transgenes, 203
enhancers identification, 208–211
gene copy number, 200–201
germ cell migration, 207–208
jafrac1 germ adhesion defect, 206, 206*f*
lethal mutations, 201
limitations, 211–212
loss-of-function mutation, 203
second-site mutations, 204–206
strains, *D. melanogaster*, 201–202
wild-type levels, 200–201
Drosophila melanogaster (D. melanogaster)
culture and husbandry, 202
dominant modifier screens (*see* Dominant modifier screens)
gene disruption project, 204–205
mutation, 198–199
peroxiredoxin (*see* Peroxiredoxin)
transgenic RNAi, 205–206

E

EGF. *See* Epidermal growth factor (EGF)
Epidermal growth factor (EGF), 35–36
Experimental H_2O_2 exposure
bolus addition, 164
steady-state, 164–166
Extracellular signal-regulated kinases (ERKs)
EGF, 35–36
RTKs, 35–36

G

Gene expression
cell culture preparation, 176–177
cell viability, MCF-7 cells, 177–178, 178*f*
differential gene regulation, 174–175
glucose oxidase, 177–178
H_2O_2 measurement, 178–179
microarrays, 184
proinflammatory cytokines, 174–175
reagents, 175–176
steady-state titration experiments, 181–182
system calibration (*see* System calibration)
Glucose oxidase
activity, 163
method, 272–273
Glucose transport system, H_2O_2
insulin signaling factors, 274–275
measurement, 274
GPCRs. *See* G-protein coupled receptors (GPCRs)
G-protein coupled receptors (GPCRs), 239–240
Green tea polyphenol (GTPP), 82–83
GTPP. *See* Green tea polyphenol (GTPP)

H

Heme peroxidases and metalloproteins, 10
High-mobility group box 1 (HMGB1), 223
HMGB1. *See* High-mobility group box 1 (HMGB1)
Human umbilical vein endothelial cells (HUVECs), 52
HUVECs. *See* Human umbilical vein endothelial cells (HUVECs)
Hydrogen peroxide (H_2O_2)
agarose chemotaxis assay, 238–239
antioxidant defenses, 11–13
biological detection, 19–20
cellular signaling, 271

cellular targets, 20–21
chemotactic signaling, 240–241
chemotaxis, 238
compartmentalization and membrane permeability, 17–19
contraction-dependent pathway, 270–271
cytoplasm, 239
diffusion, 17
enzyme kinetics assays, 141, 141*f*
glucose transmembrane transport, 270
GPCRs, 239–240
gradient, 241–242
HyPer, 239
in vitro generation, 272–273
kinetic constraint, thermodynamically favorable reaction, 4, 5*f*
kinetic heirarchy, cellular targets, 14–15, 14*f*
kinetics and identification, biological targets, 13–15
low-molecular-weight probes, 19–20
LOX, 238–239
mammalian cells, 14–15
metabolic reactions, 4
modeling, mitochondrial environment, 14–15
myocellular mitochondria, 271
PDKs, 270
plasma and skeletal muscle, 273–274
protocol, 142
PTEN, 240
receptor-mediated redox signaling, 4
release and diffusion, superoxide, 18*f*, 20
ROS, 238
signaling mediator
BAECs, 51
endothelial cells, 53
glucose oxidase generation, 54, 55*f*
HUVECs, 52
intracellular detection, 53, 53*f*
LSS, 50
MLECs, 51
ROS, 50
shear stress experiments, 52
siRNA, 57–58
superoxide radical anion detection, 52–53
vascular endothelial cells, 55–56
western blot analysis, 56–57
signals, zebrafish
generating transgenic lines, 247–248
HyPer, 246–247
preparing and injecting HyPer RNA, 247
ratiometric assays, 246
skeletal muscle glucose transport system, 270–271
steady-state concentration, 20
target cells, 241
transition metals and one-electron oxidations, 8–10
transmission, redox signals, 15–17
two-electron oxidations, 5–8
zebrafish, 241
HyPer2 H_2O_2 biosensor
bacterial strains, 70–71
cell culture, 71
lentivirus production, 72–76
plasmid preparation, 71
protocol, 70
virus titration, concentration and infection, 71

I

Immunocomplex precipitation assay, 94
Insulin secretion, 266
Intermolecular disulfide formation
cellular processes, 112–114
PKARIs, 116–119
SDS-PAGE technique, 114–116
thiols, 112–114
thioredoxin, 112–114

J

JNKs. *See* c-Jun N-terminal kinases (JNKs)
c-Jun N-terminal kinases (JNKs), 36–37

K

Keto acids, 7

L

Laminar shear stress (LSS), 50
Lentivirus production
seeding HEK293-T cells, 72
transfection, 72

Lentivirus production (*Continued*)
virus collection, 72–73
virus concentration and injection, 73–76
visual assessment, transfection efficiency, 72
Linear analysis methods
computational abilities, 145
extraction, pseudo-first-order rate constants, 143–145
instrument response, 139
Kitz-Wilson plot, 146–147, 146*f*
time-dependent inactivation, PTP1B, 143, 143*f*
Live cell imaging, cardiac myocytes
HyPer2 H_2O_2 biosensor, 65–66
photosensitivity, 66
LOX. *See* Lysyl oxidase (LOX)
LSS. *See* Laminar shear stress (LSS)
Lysyl oxidase (LOX), 238–239

M

Mammalian target of rapamycin (mTOR)
nutrient-sensing kinase, 218–220
pathway activators, 218–220
regulation, ROS-induced autophagy, 221–222
transcriptional control, 223–224
MAPKs. *See* Mitogen-activated protein kinases (MAPKs)
MBP-C. *See* Myosin-binding protein-C (MBP-C)
Microdialysis techniques
biological systems, 284
cytochrome, 285
experimental procedures, 286–290
hydroxyl radical activity, 288
in vivo microdialysis approaches, 284–285
skeletal muscle, 286
superoxide anion, 289
Migration pathway. *See* Redox-regulated cell adhesion
Mitochondrial function, P66 gene
cell death, 102
MAPK activation, 101–102
oxidative stress, 102
Mitochondrial swelling assay
antioxidant, 105
proapoptotic factors, 104
p66Shc protein, 104–105
PTP, 104
Mitochondrial uncoupling protein-2 (UCP2)
basal respiratory activity, 263–264
glucose-stimulated respiration and coupling efficiency, 261–263
insulin secretion, 266
pancreatic beta cells, 258
protein detection, 259–260
protein knockdown, 260
ROS, 264–266
tissue culture, 258–259
Mitogen-activated protein kinases (MAPKs)
description, 33–35
ERK pathway, 35–36
human alveolar macrophages, 38–39
inactivation, 37–38
JNK pathway, 36–37
p38 MAPK pathway, 36
MLECs. *See* Mouse lung endothelial cells (MLECs)
Mouse lung endothelial cells (MLECs), 52
mTOR. *See* Mammalian target of rapamycin (mTOR)
Myosin-binding protein-C (MBP-C)
phosphorylation, 117
and troponin, 117

N

NADPH. *See* Nicotinamide adenine dinucleotide phosphate (NADPH)
NF-κB family protein levels
cellular transfection and reporter gene assay, 184–186
gene expression microarrays, 184
immunoblot, 182
plasmid constructs, 184
protein extraction, 183
TNF-α, 184
western blot, 183–184
Nicotinamide adenine dinucleotide phosphate (NADPH), 30–32
Nonlinear curve-fitting regression analysis
chi-by-eye approach, 149–150
data fitting, 147–148
graphical analysis, 149–150
inactivation kinetic data, 147–148, 148*f*

integrated rate law, 147
kinetic analysis, 149
Kitz-Wilson plot, 146–147, 148*f*
parameter optimization procol, 147–148

Nrf. *See* Nuclear factor erythroid-2-related factor 2 (Nrf2)

Nuclear factor erythroid-2-related factor 2 (Nrf2)
cell culture, 159
data handling and analysis, 166–170
glucose oxidase activity, 163
HeLa cells, 164
H_2O_2 measurement, 160–161
oxidative stress, 159
proteasomal degradation, 158
protein degradation, 158
protein sample preparation, 161–163
reagents, 160
regulation, 158
steady-state method, 164–166
translocation and synthesis, 159

O

Oxidative stress, ROS
autophagosomes, 230
autophagy detection techniques, 229
cell survival and death, 229–230
chemical inhibitors, 229
inhibitors, SOD, 230
mitochondrial DNA, 231

Oxygen-glucose deprivation/ reoxygenation, 91–92

P

Pancreatic beta cells, 258

PDGFRα. *See* Platelet-derived growth factor receptor α (PDGFRα)

PDKs. *See* Phosphoinositide-dependent kinases (PDKs)

Permeability transition pore (PTP), 104

Peroxiredoxin
cell–cell adhesion, 199–200
Jafrac1, 199
typical 2-cysteine peroxiredoxin II, 199–200

P66 gene
mitochondrial function, 101–102
and protein, 100–101
p52Shc and p46Shc, 101
recombinant preparation, 102–104
ShcA isoforms, 100–101, 101*f*

Phosphoinositide-dependent kinases (PDKs), 270

PKARI. *See* Protein kinase regulatory subunit 1 (PKARI)

PKC. *See* Protein kinase C (PKC)

Platelet-derived growth factor receptor α (PDGFRα)
activation, 190
cell lysates, 192
description, 189–190
Fα and R627 cells, 191–192
H_2O_2 stimulated phosphorylation, 192–193, 193*f*
implication, 194
kinase-inactive mutant, 191
ligand-driven dimerization, 190
plasmid, 191
PVDF, 192
tyrosine phosphorylation, 190–191, 193–194

p-nitrophenylphosphate (pNPP), 138–139

pNPP. *See* p-nitrophenylphosphate (pNPP)

Polyvinylidene difluoride (PVDF), 192

Protein detection, UCP2
sample preparation, 259
TBS, 259–260

Protein extraction, 183

Protein kinase C (PKC)
binding proteins, 80–81
buffers, 84
cerebral ischemia, 90–95
cofactor-independent activity, 86
cytosol-to-membrane translocation, 88–89
dithiothreitol, 85
GTPP, 82–83
hydrogen peroxide, 80
isoenzymes, 83
oxidative activation, 82
oxidative modification, 87–88
phorbol ester binding, 86–87
phosphorylation, 82
reagents and antibodies, 83–84
regulatory and catalytic domains, 82
thiol agents, 85

Protein kinase C (PKC) (*Continued*)
tumor promoters, 81
tyrosine phosphorylation, 89–90
Protein kinase regulatory subunit 1 (PKARI)
cAMP, 116
cyclic nucleotide, 119–120
MBP-C, 117
oxidation state, 117–119
PKG1α, 120
SERCA, 116–117
Protein sample preparation
cytosol/nucleus differential protein extraction, 161–162
detection and protein quantification, western blot, 162–163
total protein extraction, 162
Protein tyrosine kinases (PTKs). *See* Redox regulation, PTKs
Protein tyrosine phosphatase 1B (PTP1B), 143, 143*f*
Protein tyrosine phosphatases (PTPs), 5–6
PTP. *See* Permeability transition pore (PTP)
PTP1B. *See* Protein tyrosine phosphatase 1B (PTP1B)
PTPs. *See* Protein tyrosine phosphatases (PTPs)
PVDF. *See* polyvinylidene difluoride (PVDF)

R

Rapamycin (sirolimus), 221–222
Reactive oxygen species (ROS)
aging, 100
ASK1 activation, 40–42
atmospheric oxygen, 29–30
and autophagy (*see* Autophagy, ROS)
cell signaling, 32–33
cellular antioxidants, 30
cytosolic superoxide activity, 291–292
description, 28
DHE, 294–295
fluorescence microscopy and image analysis, CM-DCF, 295
growth factor receptors, 42–43
HPLC, 295–296
hydrogen peroxide, 292
MAPKs (*see* Mitogen-activated protein kinases (MAPKs))
microdialysis techniques, 284–290
mitochondrial superoxide and hydrogen peroxide activities, 292
mitochondrial swelling assay, 104–105
MKPs, 43–44
Nox proteins, 39–40
P66 gene (*see* P66 gene)
prevention, 29
p66Shc and cytochrome *c* CV, 105–106
single skeletal muscle fibers, 292–294
skeletal muscle tissues, 290–291
superoxide and hydrogen peroxide, 30–32
Receptor tyrosine kinases (RTKs), 35–36
Recombinant p66Shc protein preparation
bacteria, 102–103
cell lysate, 104
E. coli growth, 103, 103*f*
glutathione beads, 104
spectrophotometer, 103
Redox-regulated cell adhesion
dominant modifier screens (*see* Dominant modifier screens)
Drosophila melanogaster, 199–200
germ cell migration, 198–199
oxidants, 198
proteomic methods, 198
Redox regulation, PTKs
assay conditions, 138
cell signaling, 131–132
chemical and biochemical mechanisms, 131–132
covalent enzyme inactivation
first-order process, 133
instrument reading, 134
kinetics assays, 134
noncovalent association, 133
pseudo-first-order conditions, 133–134
enzyme inactivation assays, 134–135
inactivation rate constant
kinetic properties, 150
Michaelis-Menten equation, 151–152
saturation profile, 150–151, 151*f*
instrument response, 139

interference, substrate decomposition, 138
linear analysis methods (*see* Linear analysis methods)
linear response
materials, 135
protocols, 135–137
nonenzymatic hydrolysis, 138–139
nonlinearity, instrument response
enzyme stock solution, 137–138
substrate depletion, 137–138
UV-vis spectrophotometers, 137
oxidative inactivation, 130–131
oxidized enzymes, 130–131
phosphorylation, tyrosine residues, 130–131
pNPP, 138–139
signal transduction, 132–133
time-dependent inactivation, 139–142
xenobiotics, 132–133
Redox signaling pathways, 63
ROS. *See* Reactive oxygen species (ROS)
RTKs. *See* Receptor tyrosine kinases (RTKs)

S

SDS-PAGE. *See* SDS-polyacrylamide gel electrophoresis (SDS-PAGE) technique
SDS-polyacrylamide gel electrophoresis (SDS-PAGE) technique
diagonal gel, 114–116, 115*f*
mercaptoethanol, 122–123
PKARI and PKG1α, 123
proteins detection, 121–122
western blotting techniques, 123–124
Skeletal muscle
contractile activity, mouse muscle, 283
free radicals and ROS, 280–282
lipids, 282–283
microdialysis techniques, 284–290
mitochondria, 281–282
plasma membrane, 281
redox-sensitive transcription factors, 283–284
superoxide, hydrogen peroxide and NO, 280–281
xanthine oxidase, 282
Small interfering RNA (siRNA)
protein sequence, 58
protocol, 57–58
Steady-state titration, 181–182
Superoxide and hydrogen peroxide
chemical effects, 32
mitochondria, 30–32
NADPH, 30–32
ROS, 30–32
Superoxide anion
ascorbate/glutathione, 290
microdialyis probes, 289
molecular weight, 289
System calibration
cellular H_2O_2 consumption, 179–181
glucose oxidase activity, 181

T

TBS. *See* Tris-buffered saline (TBS)
Thiols
high reactivity, thiol peroxidases, 7
pK_a values and reaction rates, H_2O_2, 5–6, 6*t*
PTPs, 5–6
reactivity measurement, 5–6
sulfenic acid, 6, 7*f*
Time-dependent inactivation
activity assay, 139–140
decomposition, agent, 140
dilution, agent, 140
hydrogen peroxide, 141–142
pH perturbation, 140
physical removal, agent, 140
Tissue culture, 258–259
TNF receptor-associated factor 2 (TRAF2), 41–42
TRAF2. *See* TNF receptor-associated factor 2 (TRAF2)
Transition metals and Fenton chemistry
biological damage, 8–9
chelation or sequestration of transition metals, 10
iron/ascorbate system, 9
rate constants, reaction, 9
Transmission, redox signals
direct oxidation, target protein, 15–16, 16*f*
floodgate model, 16–17

Transmission, redox signals (*Continued*)
indirect oxidation, 15
kinetic modeling, 15
Tris-buffered saline (TBS), 259–260

U

UCP2. *See* Mitochondrial uncoupling protein-2 (UCP2)

V

Virus concentration and injection
BAEC, 74
HyPer2 lentivirus, 74–76
mycoplasma infection, 73

W

Western blot analysis
protein electrophoresis, 56
protein staining, 57
whole cell lysates preparation, 56
Wide-field microscopy, H_2O_2 production
fish wounding, 248
image processing and data analysis, 249
microscope settings and fluorescent imaging, 248–249

Z

Zebrafish tail fin wounding assay
experiment setting, 243–245
image analysis and quantification, 245–246
protocol, leukocyte recruitment assay, 242–243
required materials, 243
wound healing and regeneration, 242

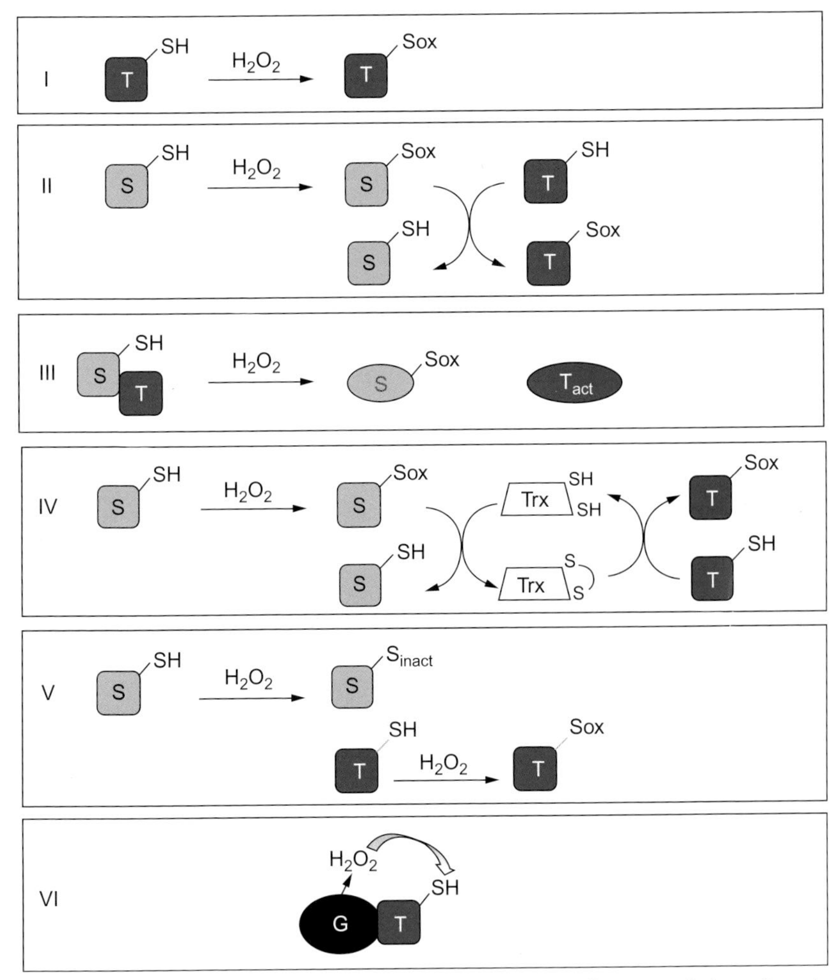

Christine C. Winterbourn, Figure 1.5 Possible mechanisms for transmission of a redox signal initiated by H_2O_2. I, direct oxidation of target protein (T); II, oxidation via a highly reactive sensor protein (S); III, activation of T by dissociation from oxidized S; IV, oxidation of T via a secondary product of S such as thioredoxin (Trx); V, inactivation of scavenging protein S to allow oxidation of T (floodgate model); VI, association of T with H_2O_2-generating protein to allow site-directed oxidation.

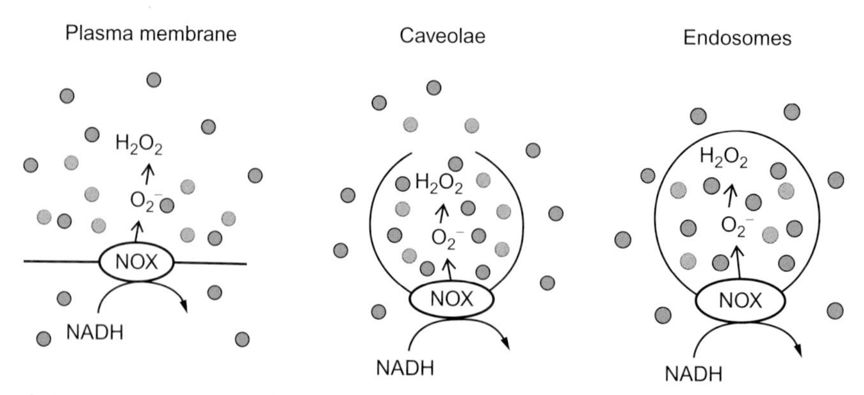

Christine C. Winterbourn, Figure 1.6 Schematic representation of release and diffusion of superoxide (light circles, pink fill) and H_2O_2 (black circles, blue fill) from NADPH oxidase activity at the cell surface, at caveolae, or in endosomes. The membrane is represented as a dark line. NADPH is oxidized on the cytoplasmic surface and superoxide released at the exosurface. Low membrane permeability restricts superoxide to the compartment where it is generated. The membrane retards but does not prevent H_2O_2 diffusion.

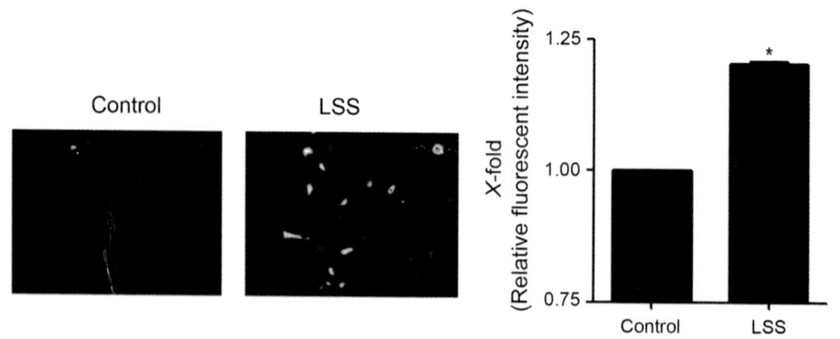

Rosa Bretón-Romero and Santiago Lamas, Figure 3.1 H_2O_2 intracellular detection by HyPer fluorescence. BAECs were transfected with the H_2O_2 biosensor pHyPer-Cyto and exposed to laminar shear stress conditions. Image was obtained 5 min after the exposure and fluorescence intensity was detected using confocal microscopy and quantified by Metamorph and ImageJ software. *, $p<0.05$.

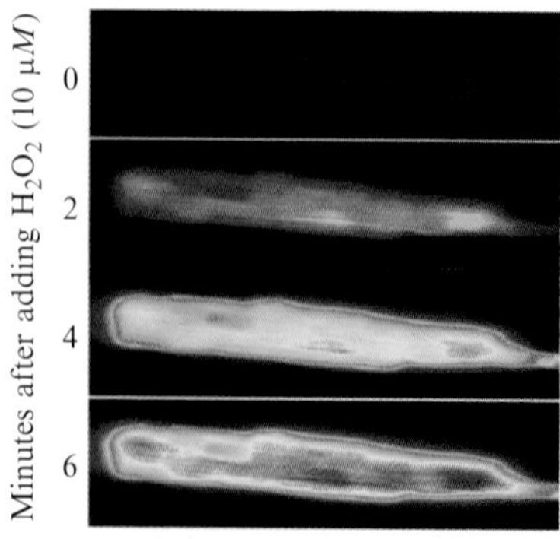

Juliano L. Sartoretto *et al.*, Figure 4.1 Effects of exogenous H_2O_2 on cardiac myocyte NO synthesis. Adult mouse cardiac myocytes were loaded with the NO dye Cu_2(FL2E), and then treated with hydrogen peroxide (H_2O_2, 10 μ*M*) and analyzed by fluorescence microscopy. Fluorescence images obtained at varying times after adding H_2O_2 are shown, as indicated.

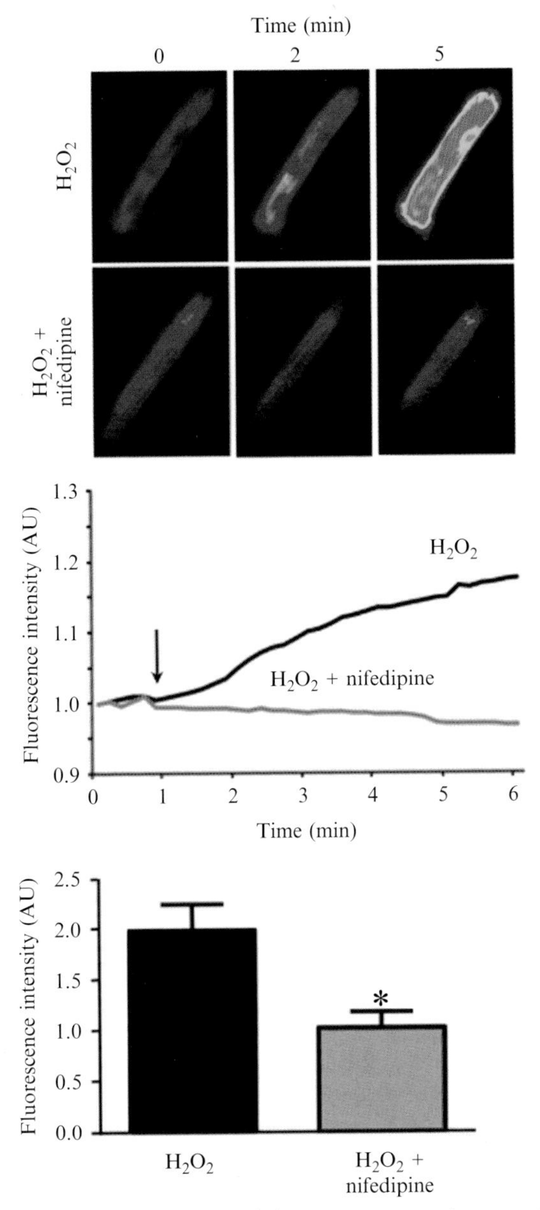

Juliano L. Sartoretto *et al.*, Figure 4.2 Nifedipine treatment abrogates H_2O_2-promoted NO synthesis in cardiac myocytes. Mouse cardiac myocytes were loaded with the NO chemical sensor Cu_2(FL2E), and then treated with nifedipine (100 μM) or vehicle followed by hydrogen peroxide (H_2O_2, 10 μM) treatment. The upper panel shows representative fluorescence images at 0, 2, and 5 min following treatments as indicated. The middle panel shows representative fluorescence tracings of single cells treated with H_2O_2 or H_2O_2 in the presence of nifedipine. The lower panel shows the results of pooled data analyzed from greater than three independent experiments; $^*p < 0.05$. *Adapted with modifications from Sartoretto et al. (2012).*

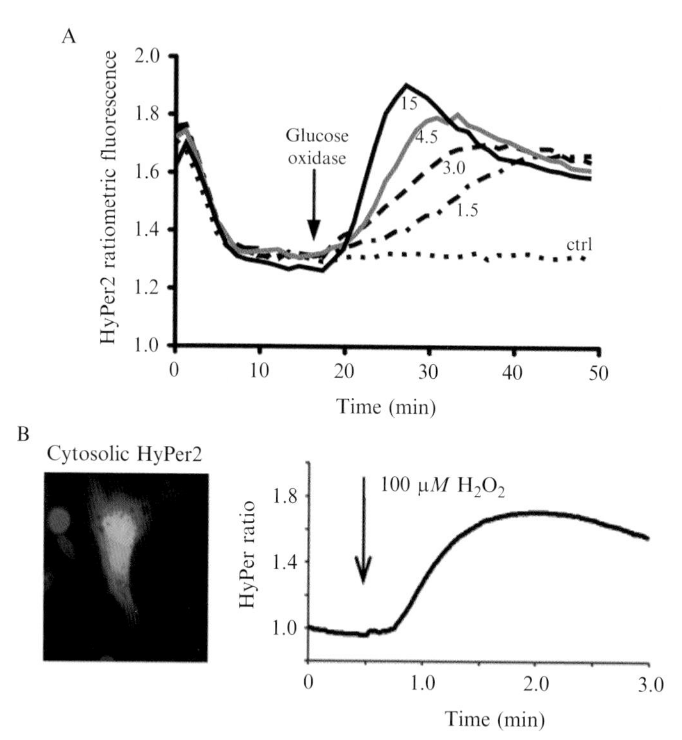

Juliano L. Sartoretto *et al.*, Figure 4.3 Imaging H_2O_2 in cells infected with Hyper2 lentivirus. (A) The results of cellular imaging of HEK293-T cells stably expressing the HyPer2 biosensor. Fluorescence emission at 520 nm was measured using a fluorescence plate reader with 400 and 485 nm excitation filters. After stabilization of the base line, the glucose oxidase was added to the extracellular media to generate H_2O_2 flux (0–15 μ*M*/min), as described in the text. (B) Results obtained in BAEC infected with HyPer2 lentivirus and treated with H_2O_2 (100 μ*M*); images were taken every 5 s. The image shows HyPer2 fluorescence in untreated BAEC that had been infected with the HyPer2 lentivirus; the graph shows a time course of the ratiometric fluorescence change (YFP_{500}/YFP_{420} ratio), revealing a rapid increase in HyPer2 oxidation after adding H_2O_2.

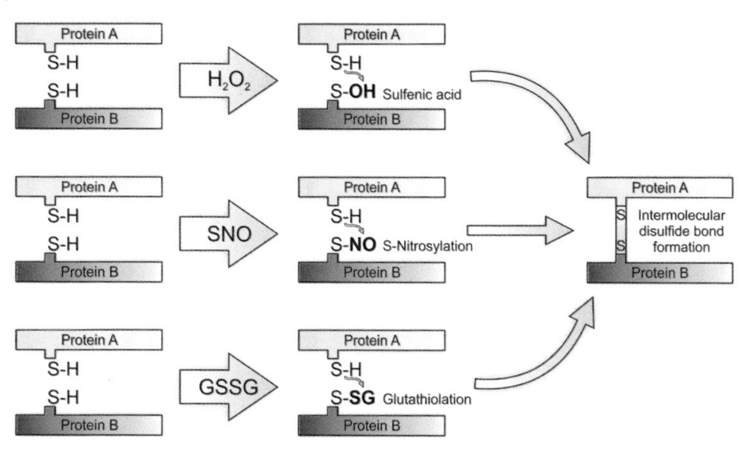

Joseph R. Burgoyne and Philip Eaton, Figure 7.1 Schematic of processes required for intermolecular disulfide formation. For intermolecular disulfide formation, an adjacent cysteine must be present in each protein with close enough proximity for bond formation. In addition, one of the cysteines must contain a reactive thiol (shown here as the S-H in protein B) that can undergo oxidation. The reactive cysteine can be oxidized to a sulfenic acid by H_2O_2, S-nitrosylated by an S-nitrosothiol (SNO) or glutathiolated by oxidized disulfide glutathione (GSSG). Each of these intermediates can then be rapidly resolved by the adjacent cysteine (S-H in protein A) giving rise to an intermolecular disulfide bond.

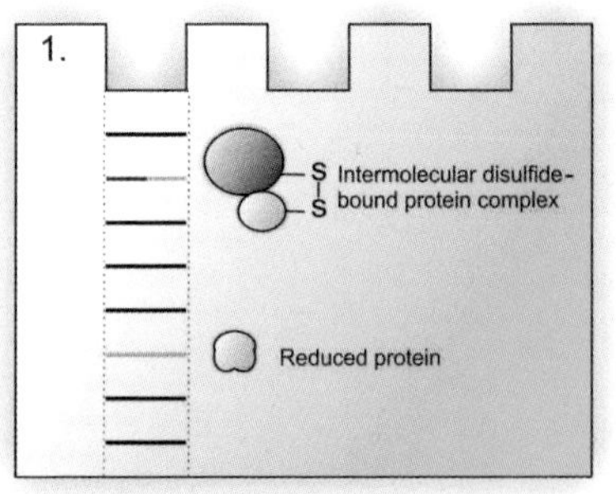

Resolve proteins on SDS-PAGE
under non reducing conditions

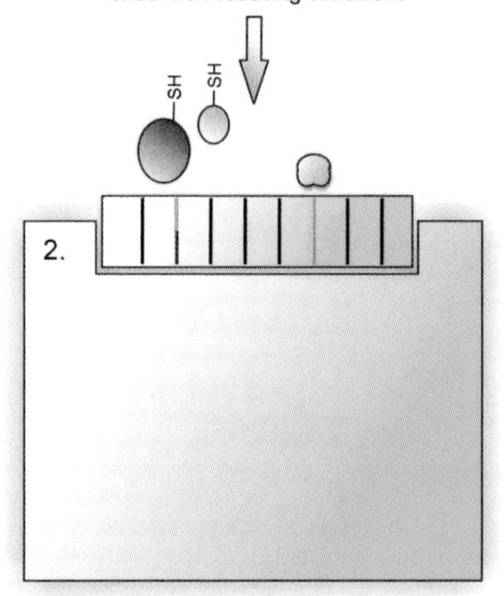

Excise resolved lane, soak in **2-Mercaptoethanol**,
and then place in large well of a fresh SDS-PAGE gel

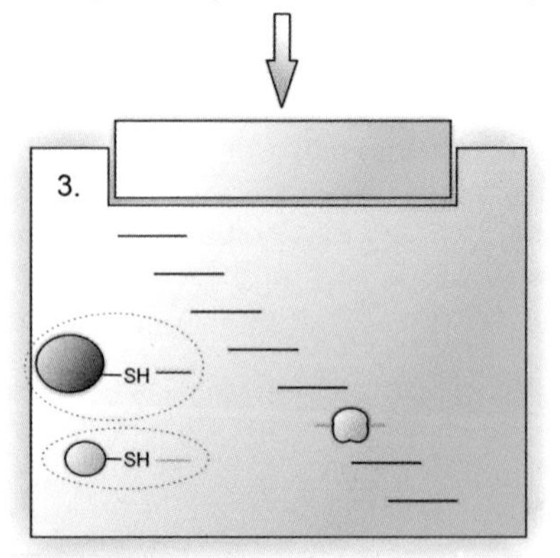

Resolve proteins under reducing conditions and then
stain proteins using colloidal Coomassie Blue

Excise stained proteins resolved below the diagonal
of the gel and identify using mass spectrometry

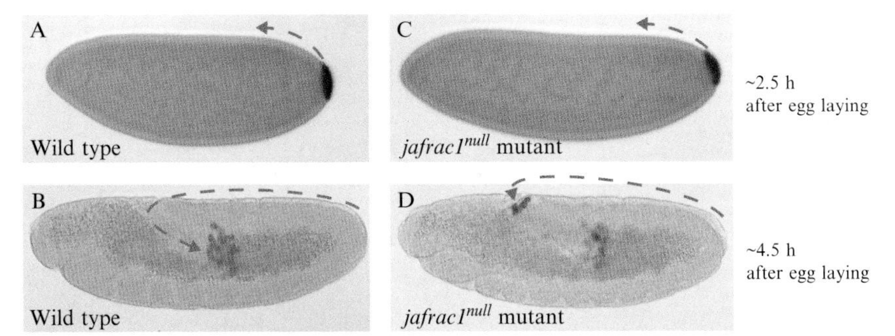

Thomas Ryan Hurd *et al.*, Figure 12.1 The peroxiredoxin, Jafrac1, is required for germ cell internalization during embryogenesis. (A and B) Images of representative wild-type embryos (*w*^{1118}) approximately 2.5 h (A) and 4.5 h (B) after egg laying. (C and D) Images of representative embryos from *jafrac1*^{null} homozygous mutant mothers approximately 2.5 h (C) and 4.5 h (D) AEL. Embryos were stained for Vasa protein to mark germ cells (brown). Embryos are oriented with the posterior to the right, anterior to the left, dorsal up, and ventral down.

Joseph R. Burgoyne and Philip Eaton, Figure 7.2 Diagonal gel SDS-PAGE. Proteins resolved under nonreducing conditions will retain their oxidation status, and therefore, intermolecular disulfide-bound proteins will run at a combined higher molecular weight. Once resolved, protein disulfides are removed from proteins by incubating the excised lane in the reducing agent 2-mercaptoethanol. By resolving a second time under reducing conditions, the proteins then resolve at their individual molecular weights. Once the gel has been stained with colloidal Coomassie Blue, proteins that were once disulfide bound can be identified as those that have migrated below the diagonal of the gel. The identity of these proteins can be determined by excising them from the gel followed by mass spectrometry analysis.

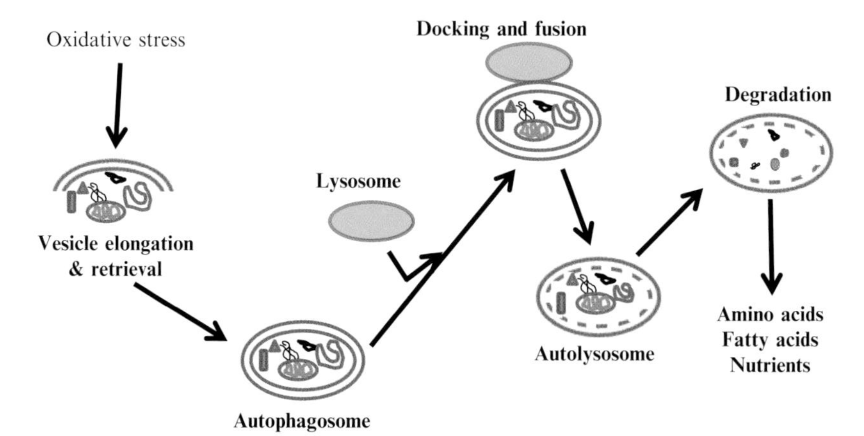

Spencer B. Gibson, Figure 13.1 Model for autophagy and autophagic cell death. Under oxidative stress, autophagy is induced. The autophagy induction process starts with isolation membrane followed by vesicle nucleation, vesicle elongation and retrivial, and the formation of the characterized double-membraned structures, autophagosomes. Then autophagosomes fuse with lysosomes to form autolysosomes. The enclosed cytoplasmic materials and the inner membrane are degraded by the acidic proteases in the autolysosomes. So autolysosomes belong to acidic vascular organelles (AVOs). The degradation process produces amino acids and fatty acids, which can be used for protein synthesis or can be oxidized by the mitochondrial electron transport chain (mETC) to produce ATP.

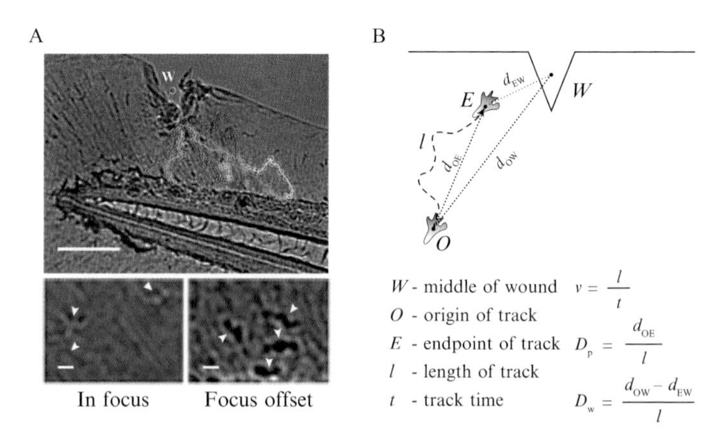

Balázs Enyedi and Philipp Niethammer, Figure 14.1 Quantitative analysis of leukocyte recruitment to incisional tail fin wounds. (A) Transmitted-light images of wounded caudal tail fins of zebrafish. Upper panel: Low magnification image showing tracks of leukocyte migration over the course of 40 min post-wounding (scale bar 100 μm). Lower panel: Zoom into tail fin region. In focus (left) or slightly out of focus (right) images of migrating leukocytes (marked by arrows). Scale bars, 10 μm. (B) Scheme of parameters measured for in-depth analysis of leukocyte trajectories.

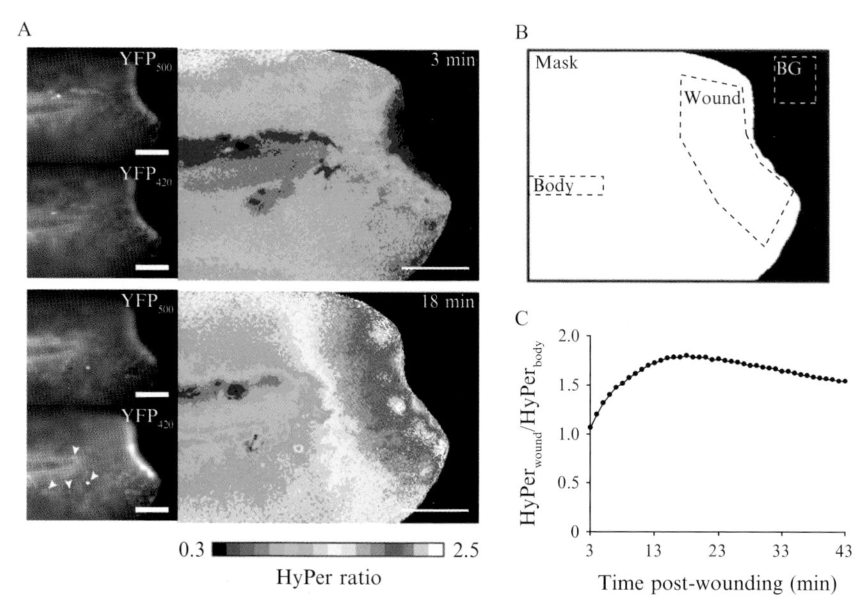

Balázs Enyedi and Philipp Niethammer, Figure 14.2 Wide-field imaging of wound margin H_2O_2 production in TG (bactin2:HyPer) zebrafish. (A and B) HyPer ratio images at indicated time points calculated from YFP_{500} and YFP_{420} images after applying a one-pass median filter, performing background (BG) subtraction and multiplying the ratio images with a mask derived from the thresholded YFP images. (C) Normalized H_2O_2 production is calculated by dividing the ratio values of a ROI along the wound margin by the ratio values of a ROI in the body of the embryo, distant from the wound margin. Scale bars, 100 μm.

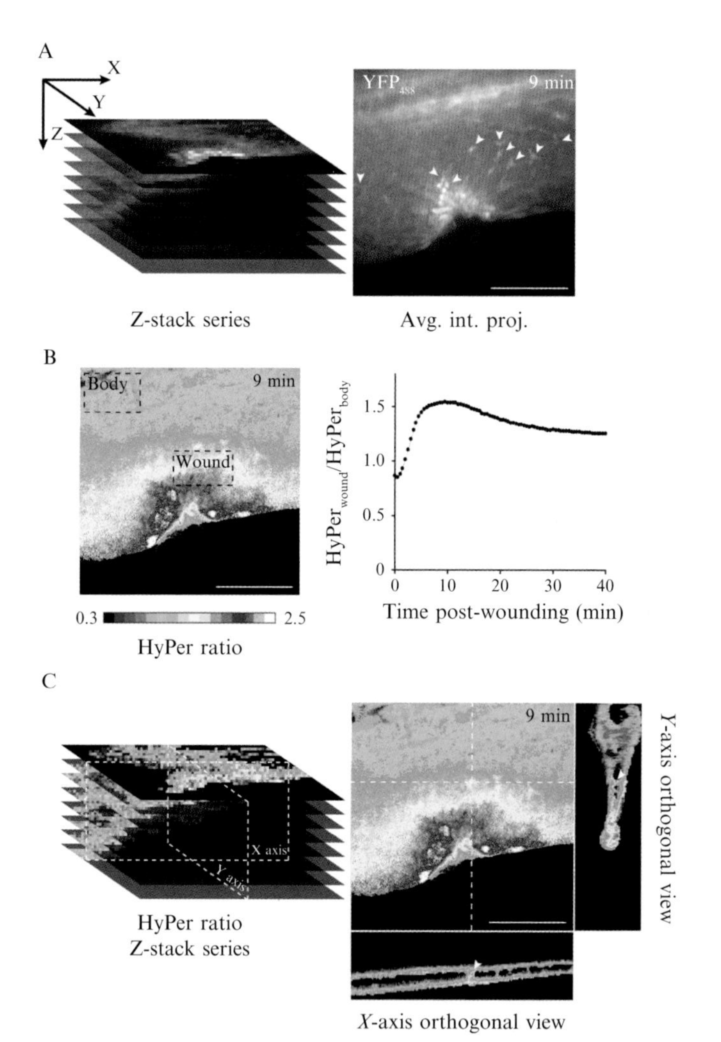

Balázs Enyedi and Philipp Niethammer, Figure 14.3 Spinning disk confocal imaging of H_2O_2 production induced by laser wounding in TG (bactin2:HyPer) zebrafish. (A) Z-stack series (~30 slices/fin, with a resolution of 2 μm/slice) of the YFP_{488} channel of HyPer taken at the indicated time point on a spinning disk confocal microscope. Average intensity projection of the Z-stack series collapses the 3D data into 2D images and allows leukocytes, marked by arrows to be clearly visualized. (B) HyPer ratio image calculated from the average intensity projection of median filtered, background subtracted and thresholded YFP_{488} and YFP_{405} Z-stack series. Normalized H_2O_2 production is calculated by dividing the ratio values of a ROI along the wound margin by the ratio values of a ROI in the body of the embryo. (C) Z-stack series of HyPer ratio images with *X*- and *Y*-axis orthogonal views derived from sections of the 3D image marked by white dashed lines. A leukocyte, marked by an arrow, is visible between the epithelial layers. Scale bars, 100 μm.

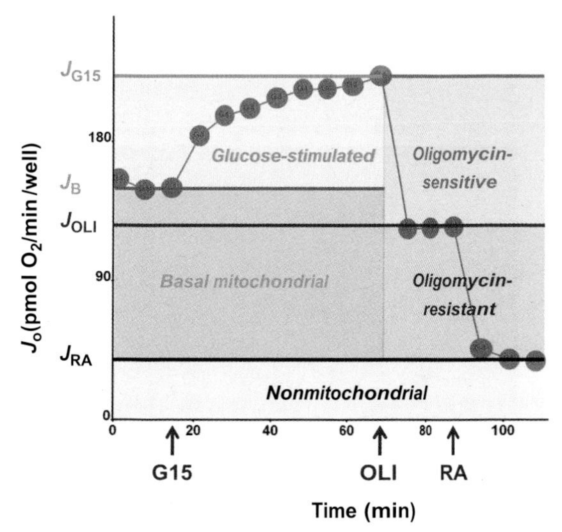

Jonathan Barlow *et al.*, Figure 15.1 Glucose-stimulated mitochondrial respiration and coupling efficiency of oxidative phosphorylation. Cells were seeded in XF24 plates at 6×10^4 cells per well and grown for 48 h, washed into glucose-free KRH, and assayed in a Seahorse XF analyser (see text). J_B, J_{G15}, J_{OLI}, and J_{RA} reflect basal respiration or the respiratory rates observed in the cumulative presence of 15 m*M* glucose, 1 μ*M* oligomycin, and a mixture of 1 μ*M* rotenone and 2 μ*M* antimycin A, respectively. Compounds were added at times indicated by arrows. Data represent oxygen uptake rate (J_o) means of five wells.

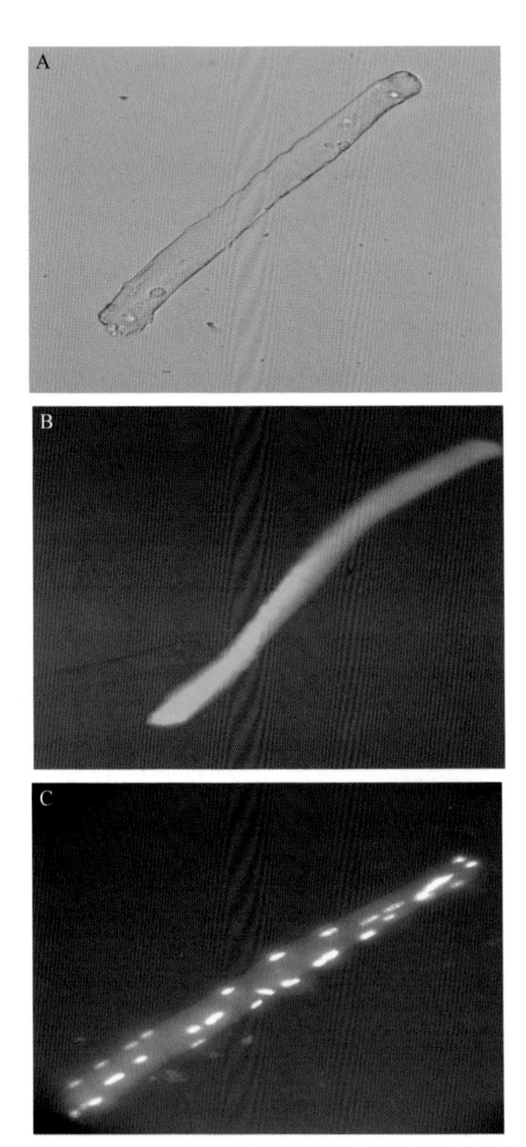

Malcolm J. Jackson, Figure 17.2 Example images of single isolated fibers fro
FDB muscles loaded with ROS sensitive probes. (A) Bright-field image; (B) fibe
with DCFH; (C) Fiber loaded with DHE showing nuclear localization of the
product.

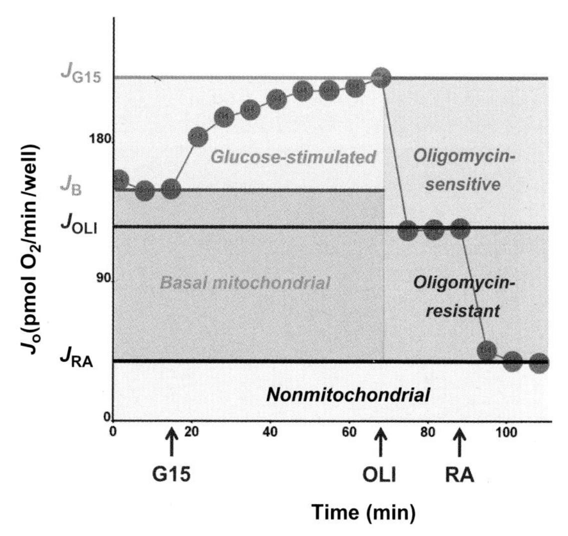

Jonathan Barlow *et al.*, Figure 15.1 Glucose-stimulated mitochondrial respiration and coupling efficiency of oxidative phosphorylation. Cells were seeded in XF24 plates at 6×10^4 cells per well and grown for 48 h, washed into glucose-free KRH, and assayed in a Seahorse XF analyser (see text). J_B, J_{G15}, J_{OLI}, and J_{RA} reflect basal respiration or the respiratory rates observed in the cumulative presence of 15 m*M* glucose, 1 μ*M* oligomycin, and a mixture of 1 μ*M* rotenone and 2 μ*M* antimycin A, respectively. Compounds were added at times indicated by arrows. Data represent oxygen uptake rate (J_o) means of five wells.

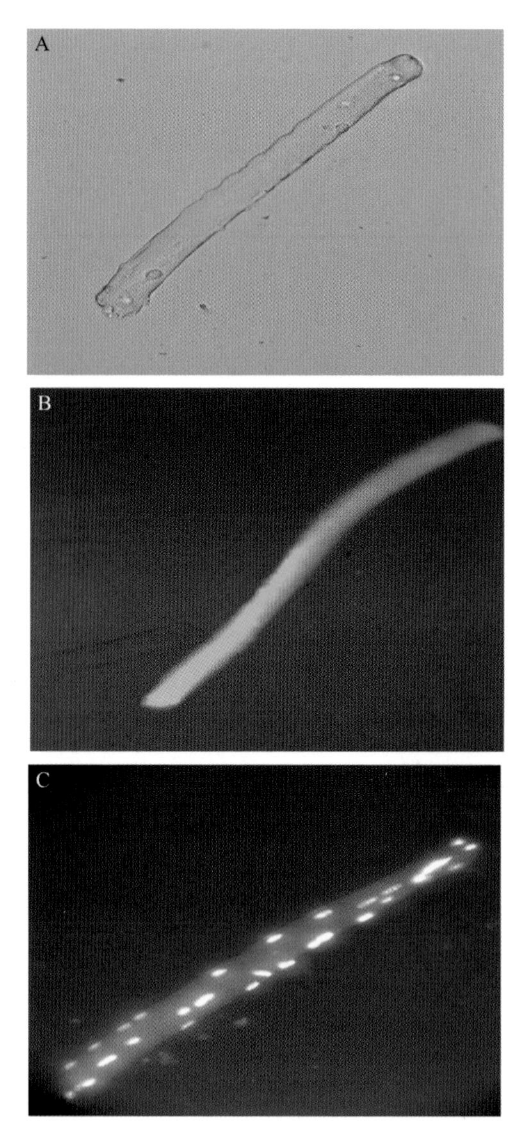

Malcolm J. Jackson, Figure 17.2 Example images of single isolated fibers from mouse FDB muscles loaded with ROS sensitive probes. (A) Bright-field image; (B) fiber loaded with DCFH; (C) Fiber loaded with DHE showing nuclear localization of the ethidium product.